Advances in Intelligent Systems and Computing

Volume 1078

AF806583

The series "Advances in Intelligent Systems and Computing" contains publications on theory, applications, and design methods of Intelligent Systems and Intelligent Computing. Virtually all disciplines such as engineering, natural sciences, computer and information science, ICT, economics, business, e-commerce, environment, healthcare, life science are covered. The list of topics spans all the areas of modern intelligent systems and computing such as: computational intelligence, soft computing including neural networks, fuzzy systems, evolutionary computing and the fusion of these paradigms, social intelligence, ambient intelligence, computational neuroscience, artificial life, virtual worlds and society, cognitive science and systems, Perception and Vision, DNA and immune based systems, self-organizing and adaptive systems, e-Learning and teaching, human-centered and human-centric computing, recommender systems, intelligent control, robotics and mechatronics including human-machine teaming, knowledge-based paradigms, learning paradigms, machine ethics, intelligent data analysis, knowledge management, intelligent agents, intelligent decision making and support, intelligent network security, trust management, interactive entertainment, Web intelligence and multimedia.

The publications within "Advances in Intelligent Systems and Computing" are primarily proceedings of important conferences, symposia and congresses. They cover significant recent developments in the field, both of a foundational and applicable character. An important characteristic feature of the series is the short publication time and world-wide distribution. This permits a rapid and broad dissemination of research results.

**** Indexing: The books of this series are submitted to ISI Proceedings, EI-Compendex, DBLP, SCOPUS, Google Scholar and Springerlink ****

More information about this series at http://www.springer.com/series/11156

Jyrki Nummenmaa · Federico Pérez-González ·
Bruno Domenech-Lega · Jean Vaunat ·
Félix Oscar Fernández-Peña
Editors

Advances and Applications in Computer Science, Electronics and Industrial Engineering

Proceedings of the Conference on Computer Science, Electronics and Industrial Engineering (CSEI 2019)

Springer

Editors
Jyrki Nummenmaa
Faculty of Information Technology and Communication Sciences
Tampere University
Tampere, Finland

Federico Pérez-González
Faculty of Engineering in Bilbao
University of the Basque Country (UPV/EHU)
Bilbao, Spain

Bruno Domenech-Lega
Department of Management,
Escola Tècnica Superior d'Enginyeria Industrial de Barcelona (ETSEIB)
Universitat Politècnica de Catalunya
Barcelona, Spain

Jean Vaunat
Division of Geotechnical Engineering and Geosciences; Department of Civil and Environmental Engineering
Universitat Politècnica de Catalunya
Barcelona, Spain

Félix Oscar Fernández-Peña
Faculty of Software, Electronics and Industrial Engineering
Universidad Técnica de Ambato
Ambato, Ecuador

ISSN 2194-5357 ISSN 2194-5365 (electronic)
Advances in Intelligent Systems and Computing
ISBN 978-3-030-33613-4 ISBN 978-3-030-33614-1 (eBook)
https://doi.org/10.1007/978-3-030-33614-1

This Springer imprint is published by the registered company Springer Nature Switzerland AG
The registered company address is: Gewerbestrasse 11, 6330 Cham, Switzerland

Preface

The impact of the convergence of computer science, electronics and industrial engineering in the efficiency of the industry is undeniable. Nevertheless, the mankind is facing the problem of how to wisely support the technological development in a nature-friendly basement for the sustainability of modern societies. This challenge is even greater for the economies of developing countries.

Within this scenario, the Technical University of Ambato, located in the central, and one of the most important industrial regions of Ecuador, made a call for scholars and researchers from around the world to participate in the First International Conference on Computer Science, Electronics and Industrial Engineering CSEI 2019. With the participation of invited speakers from Chile, Colombia, France, Japan, Spain, Portugal and USA, this academic event was born to become an important forum to discuss recent contributions to the sustainable industrial and social development based on the convergence of computer science, electronics and industrial engineering.

This volume comprises three sections with relevant contributions of special value for scholars and practitioners interested in knowledge representation, the use of metaheuristic for non-deterministic problem solutions, software architectures for supporting e-government initiatives and the use of electronics in e-learning and industrial environments. These contributions are not limited to scientific contributions but take into consideration, as well as, specific technical solutions for issues that impact the sustainable development of ICT-based industry.

In this edition, the CSEI Conference received 159 submissions of authors from 13 different countries. All these papers were peer-reviewed by the CSEI 2019 Program Committee made up of internationally renowned researchers coming from seven different countries. Sixty-six of these submissions were considered for the Springer publication. Based on the results of a double-blind peer review, the program chairs finally accepted the publication of 23 of the submissions as part of this book.

Sponsoring Institutions

Guest Speakers

Artificial Intelligence and Industry 4.0

Néstor Duque

Director of the GAIA Research Group, UNC, Colombia

The development of communications technologies in the twenty-first century, the massification and intensification of using Internet, renewable energy and wide automation of industrial processes give rise to what is known as the *Third Industrial Revolution* or *Third Technological Revolution*. A few years after, with the accelerated scientific and technological changes, a new industrial revolution is happening. Industry 4.0, as this new industrial revolution is known, has its technological bases, among others, on the Internet of things (IoT), additive manufacturing and 3D printing, big data, robotics and augmented reality. Artificial intelligence is pointed out as a central element of this transformation, intimately related to the increasing accumulation of large amounts of data (big data), the use of algorithms to process them and the massive interconnection of digital systems and devices. Advances in AI are not presented as more; they are the agglutinating axis.

Data is central in decision making. They open a new spectrum of possibilities but also responsibilities to the research community; the type of research projects faced nowadays must be adapted to the new conditions that allow the development of the economies of developing countries. It is an opportunity. AI was never before so well known, so named and so necessary.

What are these techniques and how do they manifest in the new industrial revolution? What are the challenges? What are the risks? Are these technologies reducing or widening the gaps among countries? These questions and examples were addressed in this talk.

Building a Temporospatial Software-Defined Network (TS-SDN)

Josua Emele

Senior Software Engineer, Loon LLC, USA

In this talk, an application of temporospatial SDN (TS-SDN) to high-altitude platform station (HAPS) networks was described. Airborne platforms (airplanes, airships, balloons) are used to carry wireless communication systems to provide direct to user as well as backhaul connections. Ground platforms equipped with directional steerable transceivers provide air-to-ground links needed to connect the wireless network to terrestrial networks. Platform movement and the impact of the environment on wireless channels lead to time-dynamic link metrics and availability. As platforms move, the network topology and routing need to adjust to maintain connectivity. Similarly, as the wireless environment changes (due to weather and interference) wireless parameters such as frequency, bandwidth and modulation coding scheme must adjust to maintain connectivity. Physical constraints of the system, such as time required to steer antennas, make reactive repair more costly than in terrestrial applications. Instead, TS-SDN models the physical evolution of the system to proactively adjust the network topology in anticipation of future changes. Using airborne networks under development at Google as an example, the benefits of the TS-SDN approach compared to reactive repair in terms of network availability were discussed. Additional constraints one needs to account for when computing the network topology were also identified, such as non-interference with other stationary and moving sources.

Use of Artificial Intelligence and Data Science in the Future of Online Education

Pablo Moreno

Director of the Chair of Data Science applied to Education, UNIR, Spain

This talk introduced different areas of the educative process in which the artificial intelligence (AI) and data science can involve a significant advance. From the perspective of the educative management, AI can be present from the students recruiting to the optimization of their own learning. From the perspective of the appropriate follow up from the facilitators, AI can be used for detecting students with special needs, anomalies in class performance or for developing automatic assessment techniques (stealth assessment). Finally, from the perspective of the transversal management, AI and data science allow to identify the differences among education districts, education institutions with performance issues, or dispersion patterns of students considering their social–economic profiles. During this talk, different data exploratory techniques were introduced as well as results of their use in recent research projects on educational innovations.

Exploring Millimeter and Terahertz Waves for Communications and Sensing

Tadao Nagatsuma

Professor at Graduate School of Engineering Science, Osaka University, Japan

This talk described how effectively photonics technologies are implemented not only in generation, detection and transmission of millimeter and terahertz (THz) waves, but also in system applications such as communications, measurements and imaging. In addition, some unique approaches, which utilize concepts or physical phenomena established in the light wave region in order to enhance functionality and performance of millimeter-wave and THz applications, were presented. Finally, in order to make millimeter-wave and THz systems more compact and cost-effective, recent challenges in photonic integration technologies were described, which include monolithic and hybrid integration schemes.

Understanding Value Hierarchies and Their Interrelationships—A Glance in the Managerial Multi-objective Processes

Alexis Olmedo

Head of the Engineering School, UNAB, Chile

It is said that the interaction between value hierarchies determines the target involved in decision-making processes. This process is influenced by the type of required strategic decision and how the attraction and commitment effect, described by Simonson, act on the stakeholders. This talk proposed to carry out a fieldwork with manipulation of the factors affecting the goal achievement. This way, the information gathering process is achieved. A focus group research with middle-ranking officials of different companies is proposed for determining the main present factors in the process of managerial decision. Meanwhile, the implementation of a work environment written interview is applied to managers of different organizations and a three-dimensional scale is used for objectives addressing (domain, performance focus and performance avoid).

The use of the analytic network process methodology, proposed by Thomas Saaty, allows us to interpret the gathered information. From the management perspective, based on the consumer behavior theory, it is expected to reliably estimate the relative weight of the meaning and the relationship among the values presented in the choice of managers in the presence of the attraction and commitment effect.

Model for the Quality of Local e-Government Services

Álvaro Rocha

President of AISTI, Portugal

One of the main challenges underlying different electronic government forms is the provision of a quality public service. In the local government context, local authorities allow for an adjustment between the characteristics of public services and the specificities of local communities, letting populations define their own priorities, which vary from community to community based on objective elements but also subjective by nature. The quality of these services in their electronic format should be analyzed and taken into consideration to potentiate and elaborate a strategy capable of improving offered services, increasing the satisfaction of the recipients. This talk presented a new and innovative model for the quality of local e-Government services, based on a literature review, where we analyzed seventeen approaches for e-Government services quality, as well as an empirical study involving a group of experts and users of local government services.

Smart Industry: The 4.0 Data-Centric Revolution

Genoveva Várgas

Senior scientist, CNRS, France

The idea of smart industry is based on the notion of Industry 4.0 that denotes technologies and concepts related to cyber-physical systems and the Internet of things (IoT). In smart industry, there are sensing systems that monitor physical processes, they create a virtual copy of the physical world, a "datified" version of it, and make decentralized decisions. With IoT, monitoring systems communicate and cooperate in real time.

This talk introduced the architecture of a smart industry based on communication layer and software later that integrate physical entities. Each physical entity is seen as an intelligent and autonomous agent that embeds programs for letting it evolve in workshops. Thereby, the physical world composed of connected things and the digital universe consisting of computing, storage and memory resources are combined.

Organization

Program Chair

Félix Oscar Fernández Peña	Technical University of Ambato, Ecuador

Program Committee

Kleber Barcia Higher	Polytechnic School of the Litoral, Ecuador
Julio Barzola-Monteses	University of Guayaquil, Ecuador
Pablo Bengoa	Ikerlan, Spain
Yesenia Cevallos	National University of Chimborazo, Ecuador
Danilo Chavez Garcia	National Polytechnic School, Ecuador
Bruno Domenech	Universitat Politècnica de Catalunya, Spain
Néstor Duque-Méndez	National University of Colombia, Colombia
Gibran Etcheverry	University of the Americas Puebla, Mexico
Félix Fernández-Peña	Technical University of Ambato, Ecuador
Marcelo Garcia	Technical University of Ambato, Ecuador
Carlos Gordon	Technical University of Ambato
Víctor Guachimbosa	Technical University of Ambato, Ecuador
Jorge Hernandez	Polytechnic School of Chimborazo, Ecuador
Jose M. Lavin	Cesine, Spain
Aitziber Mancisidor	University of the Basque Country, Spain
Cesar Martinez	University of the Americas Puebla, Mexico
Fernando Molina-Granja	National University of Chimborazo
Hugo Moreno Aviles	Polytechnic School of Chimborazo, Ecuador
Pablo Moreno Ger	International University of La Rioja, Spain
Roberto Naranjo	Polytechnic School of Chimborazo, Ecuador
Jyrki Nummenmaa	Tampere University, Finland
Alexis Olmedo	University Andrés Bello, Chile
Edelmira Pasarella	Universitat Politècnica de Catalunya, Spain
Victor Peñafiel	Technical University of Ambato, Ecuador

Marcelo Pilamunga Poveda	Technical University of Ambato, Ecuador
Danilo Pástor	Polytechnic School of Chimborazo, Ecuador
Jefferson Ribadeneira	Polytechnic School of Chimborazo, Ecuador
Alberto Ríos	Technical University of Ambato, Ecuador
Asier Salazar-Ramirez	Euskal Herriko Unibertsitatea, Spain
Luis Tello-Oquendo	National University of Chimborazo, Ecuador
Jose Luis Villa	Technological University of Bolivar, Colombia
Corral Vinicio	Armed Forces University, Ecuador

Organizing Committee

Pilar Urrutia	Technical University of Ambato
Julio Cuji	Technical University of Ambato
John Reyes	Technical University of Ambato
Clay Aldás	Technical University of Ambato
Juan Pallo	Technical University of Ambato
Carlos Sánchez	Technical University of Ambato
Franklin Salazar	Technical University of Ambato
Hernán Naranjo	Technical University of Ambato

Technical Program Support Committee

Ronald Guerrero	Technical University of Ambato
David Guevara	Technical University of Ambato
Anita Larrea	Technical University of Ambato
Christian Mariño	Technical University of Ambato
Luis Morales	Technical University of Ambato
Carlos Morales	Technical University of Ambato
Hernán Naranjo	Technical University of Ambato
Cristina Reinoso	Technical University of Ambato

Contents

Computer Science

Method for Edges Detection in Digital Images Through the Use of Cellular Automata 3
Karen Angulo, Danilo Gil, and Helbert Espitia

Semantic Processing Method to Improve a Query-Based Approach for Mining Concept Maps 22
Wenny Hojas-Mazo, Alfredo Simón-Cuevas, Manuel de la Iglesia Campos, and Juan Carlos Ruíz-Carrera

An Ontology-Based Data Management Model Applied to a Real Information System 36
Raúl Comas Rodríguez, Alfredo Simón-Cuevas, Nayi Sánchez Fleitas, and María Matilde García Lorenzo

Image Analysis Based on Heterogeneous Architectures for Precision Agriculture: A Systematic Literature Review 51
Marco R. Pusdá-Chulde, Fausto A. Salazar-Fierro, Lucía Sandoval-Pillajo, Erick P. Herrera-Granda, Iván D. García-Santillán, and Armando De Giusti

Educational Robot Using Lego Mindstorms and Mobile Device 71
José Varela-Aldás, Oswaldo Miranda-Quintana, Cesar Guevara, Franklin Castillo, and Guillermo Palacios-Navarro

Use of E-Learning and Audio-Lingual Method for the Development of Listening Comprehension Skills 83
Ruth Viviana Barona-Oñate, Sonia de los Angeles López-Pérez, Jimmy P. López López, and Julio A. Mocha-Bonilla

Experiments on a Mashup Web-Based Platform for Increasing e-Participation and Improving the Decision-Making Process in the University 99
Víctor Peñafiel and Hernando Buenaño

GLORIA: A Genetic Algorithms Approach to Tetris 111
Diana Patricia Quintero Lorza, Néstor Darío Duque Méndez, and Jacobo Andrés Gómez Soto

Electronics

Analysis and Determination of Minimum Requirements of an Autopilot for the Control of Unmanned Aerial Vehicles (UAV) 129
Hugo Loya, Víctor Enríquez, Franklin W. Salazar, Carlos Sánchez, Fernando Urrutia, and Jorge Buele

Development and Analysis of a PID Controller and a Fuzzy PID 143
Morelva Saeteros, Wilman Paucar, Cristian Molina, and Gustavo Caiza

Real-Time Hand Gesture Recognition: A Long Short-Term Memory Approach with Electromyography 155
Jonathan A. Zea and Marco E. Benalcázar

Development of a Fuzzy Logic-Based Solar Charge Controller for Charging Lead–Acid Batteries 168
Fabricio Paredes Larroca, Erich Saettone Olschewski, Javier Quino Favero, Jimmy Rosales Huamaní, and José Luis Castillo Sequera

Enabling Electronic System for Emergency Alerts 184
Santiago Manzano, César Granizo, Homero Velasteguí, Jaime Guilcapi, Freddy Benalcazar, and Carlos Gordón

Evaluation of Internet of Things Protocols for Shopfloor Communication Integration 199
Carlos S. Leon, David I. Ilvis, Edison G. Remache, Williams R. Villalba, Carlos A. Garcia, and Marcelo V. Garcia

Embedded System for Hand Gesture Recognition Using EMG Signals: Effect of Size in the Analysis Windows 214
Juan Mantilla-Brito, David Pozo-Espín, Santiago Solórzano, and Luis Morales

A Novel Technique for Improving the Robustness to Sensor Rotation in Hand Gesture Recognition Using sEMG 226
Victor H. Vimos, Marco Benalcázar, Alex F. Oña, and Patricio J. Cruz

Industrial Engineering

Optimization of Motorcycle Assembly Processes Based on Lean Manufacturing Tools 247
Jonnathan Quezada, Lorena Siguenza-Guzman, and Juan Llivisaca

A Study on Modeling and Simulation of Automobile Painting Process Based on Flexsim 260
John Reyes, Darwin Aldas, Homer Castelo, Rommel Velasteguí, Nancy Rodríguez, Cristian Suarez, and Kevin Alvarez

Ergonomic Postural Evaluation System Through Non-invasive Sensors 274
Christian Mariño and Javier Vargas

Energy Supply of a Hybrid System of Biomass and Wind Turbines of the Pichacay Landfill Towards an Intelligent Network for the City of Cuenca-Ecuador 287
Daniel Icaza and David Borge-Diez

Fractal Control Design with Anti-windup Effect for Optimal Operation of a Power Flyback Source 308
Jesús Rodríguez-Flores, Víctor Isaac Herrera, Andrés Morocho-Caiza, and Christian Merino

Symptomatology of Musculoskeletal Pain Related to Repetitive Movements. Preliminary Study "Post-harvest in Floriculture Companies" 329
Luis Morales, Diana Silva, Víctor Moreno, and Santiago Collantes

Design of an Ergonomic Prototype for Physical Rehabilitation of People with Paraplegia 341
Franklin W. Salazar, Freddy Núñez, Jorge Buele, Edisson P. Jordán, and Jeneffer Barberán

Author Index 355

Computer Science

Method for Edges Detection in Digital Images Through the Use of Cellular Automata

Karen Angulo[1(✉)], Danilo Gil[1], and Helbert Espitia[2]

[1] Systems Engineering, Universidad Distrital Francisco José de Caldas, Bogotá, Colombia
{kvangulos,dggils}@correo.udistrital.edu.co
[2] Faculty of Engineering, Universidad Distrital Francisco José de Caldas, Bogotá, Colombia
heespitiac@udistrital.edu.co

Abstract. In the present article, an algorithm based on cellular automata for noise elimination and edge detection in grayscale images is proposed. However, the focus of this project will be on the process of identifying contours, since this represents a higher challenge at the research level. Also, the cellular automaton has an adaptive behavior, which allows it to expand when it considers that the information coming from its initial neighbors is insufficient to determine if the pixel in evaluation is a "border pixel" or not, as this is an important feature together with a set of transition rules useful to accentuate relevant details within the image. By integrating these characteristics, we find the results obtained by the Proposed Algorithm presented greater similarity compared to images with ideal borders, since the edge recovery ranges between 92.30% and 97.21%, which indicates that MEDCA is in general terms, more efficient compared to similar algorithms.

Keywords: Cellular automaton · Digital images · Image processing · Edges

1 Introduction

In the field of image processing, edge detection is a relevant and widely used application because it is considered a low-level processing operation [1]. In other words, a contour is located in the first operation of the algorithm on the image, when examining each pixel and determining whether it has the properties of an edge. Thus, the main emphasis of this tool is to focus and identify sharp discontinuities present in the image [2]. These discontinuities are defined by some authors as abrupt changes in the intensity of the pixels and are the main characteristic to find the limits of the objects that are immersed in an image. From this concept arises the need to denote an edge as which limits and segments the different regions or objects within the image under evaluation [2]. Edge detection is usually the first step of many of computer vision algorithms (edge-based facial recognition, edge-based obstacle detection, edge-based objective recognition, and image compression, among others).

Considering that the quality of an identified edge is directly related to the content of the image and the user requirements, the development of a method for edge detection

J. Nummenmaa et al. (Eds.): CSEI 2019, AISC 1078, pp. 3–21, 2020.
https://doi.org/10.1007/978-3-030-33614-1_1

that meets all the desired results is a subjective process [2, 3]. Consequently, it is difficult to compare the performance of two algorithms that have the same purpose. However, it is common for researchers to compare the results with those of Sobel, Prewitt, and especially Canny. The problem of these is that, despite the results obtained by these researchers are remarkable, there is no consistency when comparing the data obtained in algorithms or methods that do not have the same characteristics.

Despite the subjectivity immersed within the algorithms, some authors have chosen to use cellular automata to generate as many possible edges within the image, given its practicality and behavior; an automaton can be seen as a regular grid of cells that contains a finite number of possible states. The state of the cell under evaluation is synchronously updated in discrete time steps and is also determined by the previous states of the surrounding neighborhood [4]. The rules that generate the transition of states of the central cell (in evaluation), are considered as finite state machines, and generally, such rules specify the configuration of the neighborhood states [4, 5].

Thanks to the details provided above and to the fact that cellular automata are discrete dynamic systems used to simulate different behaviors and phenomena in different areas of science, from this arose the idea to Wongthanavasu and Sadananda [6], which use an algorithm to change the pixel in evaluation, if and only if the whole neighborhood has the same value and just one iteration is necessary for the result to be obtained, since the other iterations do not modify the image. In spite of its simplicity and effectiveness, this algorithm shows flaws when the images have noise, since it prevents the proper functioning of the automaton. It should be noted this method implemented a Von Neumann neighborhood.

In another type of approach, they focused on finding optimal transition state rules, for this, the use of genetic algorithms is quite common, which can present considerably satisfactory results. Selvapeter and Hordijk [7] used this method to train cellular automata; in addition, this algorithm has the ability to perform their work on images with different noise levels. At the technical level, Selvapeter and Hordijk used a noise filter based on cellular automata, and they later trained another cellular automaton for edge detection, whose results are comparable with the methods of Sobel, Prewitt and Canny, despite the fact that these methods work directly on the image with noise and it was not previously filtered.

In an analogous way, Batouche, Meshoul and Abbassene [8] generated a more efficient solution, when established a generic algorithm based on a cellular automaton for edge detection, which did not require such a high amount of training patterns (In a Moore neighbor it requires $2^9 = 512$). The operation of this, in general terms, consisted of combining patterns according to a threshold of similarity which reduced to 15 the amount of training patterns, and later the image was rotated every 90° applying the same rules. Thanks to this, processing time was reduced and the results are obtained correctly, although the edge is a little thick.

Within the multiple existing variations for edge detection, Slatnia, Batouche and Melkemi [9] adopted a methodology similar to that used in [8], because they established symmetrical patterns to the same rule; however, the set of rules were trained in different way. Instead, they trained a genetic algorithm with a single generic rule, in which the central pixel changes its state only when its Moore neighborhood matches a specific pattern, and analogously they obtained the same rule as Wongthanavasu and Sadananda. For this case, the results are compared with those of Canny, which visually do not have the same quality.

On the other hand, Yang, Ye, and Wang [10], established an algorithm based on cellular automata taking into account the neighborhoods of both Moore and Von Neumann, which can be defined in two steps: the first step consists of an iteration in which it evaluates if three contiguous neighbors of the central pixel have the value one (1) and the other three neighbors symmetrical respect to the central pixel have the value zero (0), then the pixel in evaluation maintains its state, otherwise it changes it. In the next step, several iterations are performed based on a rule in which the pixel in evaluation is considered edge if exactly two pixels in the Von Neumann neighborhood are pixels, and if the other neighbors in a Moore neighborhood are not edges.

From another point of view, the images are often affected by impulsive noise, which can occur or be caused by different factors such as errors or interference when transmitting the information bits, memory locations defective in the hardware, errors associated to the equipment used for capturing images or pixel malfunction in the camera sensors [11]. An image, as well as any type of data can reveal valuable information in any field and the presence of noise can significantly affect the results that are expected from it, thus, it is important to efficiently suppress these interferences [12]. Within the classification of impulsive noise there is a special and well known case named "salt and pepper noise". This can be perceived in a contaminated image when the pixels take maximum or minimum values of the scale; that is, they usually appear as black and white dots on the image [11, 13].

One of the solutions to this type of situation can be represented by means of cellular automata through the coupling of different methods that allow not only the identification of noise, but an efficient restoration of the image quality. This work, in different cases, can be carried out by means of a series of iterations on the image by the cellular automaton. In the first instance, for this particular case, the cellular automaton advances through each pixel, identifying the maximum and minimum values to modify the value depending on the configuration of their neighbors and, subsequently, different iterations are performed (depending on the noise level) where the quality of the image is improved thanks to the use of the average.

Like the previous ones, there are countless of other works in which the subject of contour detection is covered, which encompass this situation from different perspectives, keeping in mind that today there are different types of images with scales of varied intensities. Then, given the diversity and challenges in this field, in addition to the strong relationship between cellular automata and image processing, this paper aims to generate a proposal for an algorithm based on cellular automata that eliminates noise and detects edges of digital images.

2 Background

2.1 Cellular Automata

The concept of cellular automaton was introduced and put into practice in the computational field of physics by the Hungarian-American mathematician Von Neumann in the 1950s [14]. This is defined as a simple mathematical model that is based on the concepts of self-organization and self-reproduction [14]; moreover, conceptually it is seen as a dynamic system that is not only discrete in time, but also in space [15].

Regarding the operation of a cellular automaton it can be said that the state of the cell under evaluation, at time t + 1, will be determined by the current state of the surrounding neighbors. This will be updated synchronously in discrete time intervals [15, 16]. That is, from a more formal and mathematical perspective, an automaton is defined as 3-tuple like this:

$$A = (S, N, \delta) \tag{1}$$

Where, S it is a set of non-empty states, N is the neighborhood, and $\delta : S^N \rightarrow S$ is the local transition function (or rule) that defines how a cell changes its state.

As mentioned above, the automata can be seen as a set of $D - dimension$ cells, where the cells that will be part of it are located. Figure 1 shows cellular automata of dimension, $D = 1$, $D = 2$ and $D = 3$. Currently, the case that is most frequently used in the various applications of the investigative sciences, particularly in the processing of images, is simulated in a two-dimensional space.

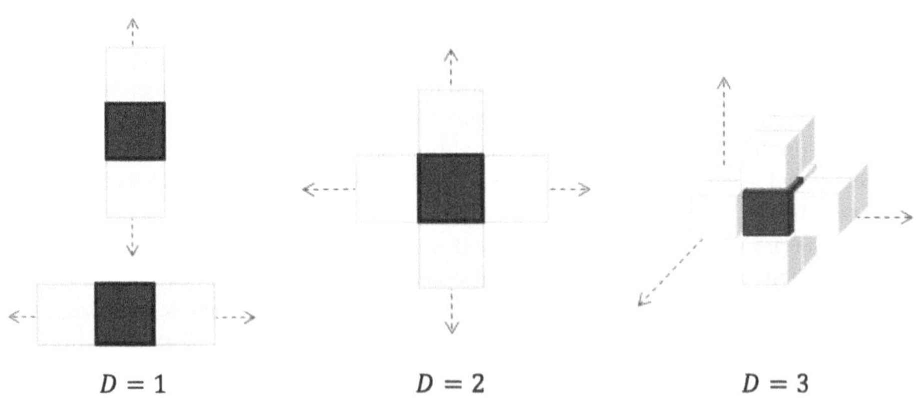

Fig. 1. Dimensions of cellular automata.

Regardless of the chosen dimensions, every cellular automaton has a certain basic structure, which is composed of three parts: (a) set of cells or lattice, (b) a set of adjacent neighbors or neighborhood, and (c) a set of rules for local transitions [17].

a. Set of cells or lattice, which has a finite number of possible states or values for each cell, for binary images 1 or 0, and for grayscale images from 0 to 255.

b. Set of adjacent neighbors or neighborhood, such as the Von Neumann neighborhood (4 neighbors) or the Moore neighborhood (8 neighbors).
c. A group of rules for local transitions (local functions of transition, transition of the state of the cells or table of transitions of the states of the cells). In synthesis, within the groups of rules there are two particularities to be highlighted: (i) the communication between the central cell and its neighbors is local, uniform and synchronous, and (ii) the global evolution of the system through a discrete time step is deterministic [17].

Over the last 60 years, a number of researchers (including Stanislaw Ulam and John Von Neumann, John Holland, Stephen Wolfram, and John Conway) have investigated issues related to the properties and variations that a cellular automaton can have, but in particular there are large studies on neighborhoods, of which the neighborhoods of Von Neumann and Moore stand out, since these have been the most used applications over the years.

- Von Neumann Neighborhood: A cross-shaped neighborhood on a grid of two-dimensional with $l \times l$ size is used to define the set of cells that surrounds the cell under evaluation (x_0, y_0). The Von Neumann neighborhood of rank r is defined by Eq. (2).

$$N_{x_0,y_0} = \{(x, y) : |x - x_0| + |y - y_0| \leq r\} \quad (2)$$

This means that at its most elementary level it has 4 neighbors from the central pixel as shown in Fig. 2. However, in many cases this neighborhood can be extended n times, searching so that by means of the previously established rules a greater amount of information can be handled as shown in Fig. 3.

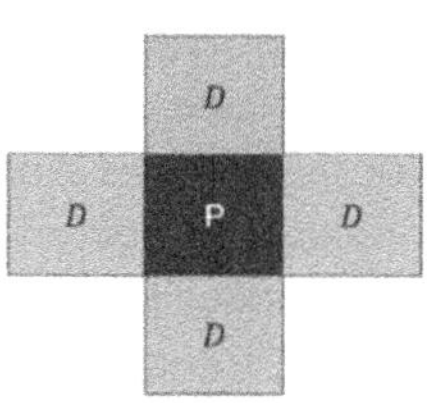

Fig. 2. Von Neumann neighborhood.

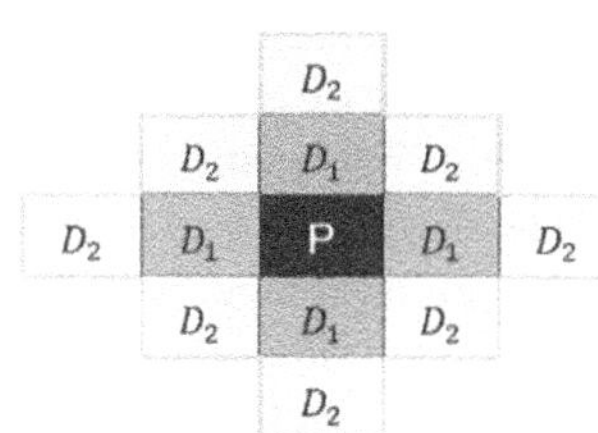

Fig. 3. Von Neumann extended neighborhood ($n = 2$).

- Moore Neighborhood: A square neighborhood on a two-dimensional $l \times l$ grid is used to define the set of cells that surrounds the cell under evaluation (x_0, y_0). The Moore neighborhood of rank r is defined by Eq. (3).

$$N_{x_0,y_0} = \{(x, y) : |x - x_0| \leq r, |y - y_0| \leq r\} \quad (3)$$

In the case of the Moore neighborhood, its essential form considers 8 neighbors from the base cell, which facilitates the handling of edges and corners; in addition, the information that can be managed with this neighborhood is greater than the one of Von Neumann due to the consideration of four additional cells, as shown in Fig. 4. In the same way as it happens with the configuration of Neumann, this neighborhood also presents extended versions in n cells, which obey to the formula $(2n+1)^2-1$. In the following example, the expanded neighborhood can be appreciated in $n = 2$ (see Fig. 5).

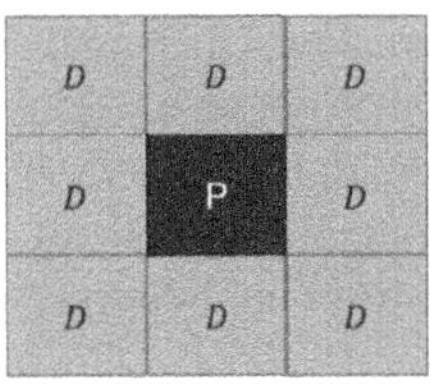

Fig. 4. Moore neighborhood.

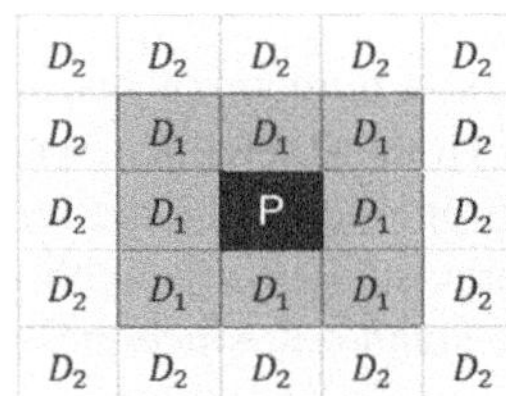

Fig. 5. Moore extended neighborhood ($n = 2$).

Having clear the neighborhood types and their concepts, it was determined that in accordance with the needs of the algorithm, the Moore neighborhood is more useful and efficient, since it offers greater advantages over the others, in terms of the amount of information that you can get from your neighbors; thus giving the pixel under evaluation the option of being detected as a possible "edge pixel" and of being corrected efficiently.

2.2 Edge Detection

Within the sensorial capacities that human beings possess, the sight is probably the most important among all of them, thanks to the fact that through this we acquire a considerable quantity of information of our environment, such as colors, shapes, sizes, distances, etc. Of the above, the variation between intensities is one of the features that stands out, even though an image lacks three-dimensional information such as texture or shadows, it is possible to identify objects by means of borders or silhouettes [1, 18]. This is why the borders detection has a key role in image processing as through the existing methodologies it is possible the identification of objects, pattern definition or segmentation of information inside the images [2].

Within the scope of this research work, a pixel can be considered as an edge, when there is a noteworthy difference between the levels of gray intensities within the image [1, 2, 19], said the change is recognized as a boundary between two different regions in an image.

In gray scale images edge detection can be done by generating an identification of changes in light intensity over the number of pixels under evaluation [20]. That is, if there is a linear series of pixels with intensities of: 255, 248, 252, 76, 73, 79 an edge or discontinuity would be expected between pixels with intensities of 252 and 76. Within some texts, these pixels are referred to as "edge points" and these may be useful at the time of analyzing the image.

Given the importance of this process of digital analysis in the images, over the years several basic methods for edge detection have emerged, among the most outstanding, the following can be mentioned:

- Method based on the first derivative (gradient): In this method the intensity has greater magnitude than the predefined threshold of the image. The largest peaks in the image are searched. The gradient of an image at a point indicates the maximum variation of the function at that point [21], which is defined by:

$$\nabla f(x,y) = \left[G_x G_y\right] = \left[\frac{\delta f}{\delta x}\frac{\delta f}{\delta y}\right] \tag{4}$$

  Where, $G_x = \frac{\Delta f}{\Delta x}$ y $G_y = \frac{\Delta f}{\Delta y}$.

- Method based on the second derivative (Laplacian): The intensity has crossing by zero. Unlike the first derivative method, it does not require a threshold value to have a more effective approach. The signs changes where an edge is found [21]. Mathematically it is defined by Eq. (5):

$$\nabla^2 f(x,y) = \left[\frac{\delta^2 f}{\delta x^2}\frac{\delta^2 f}{\delta y^2}\right] \tag{5}$$

Figure 6 shows a relation of the behavior of these methods when variations in intensity are detected.

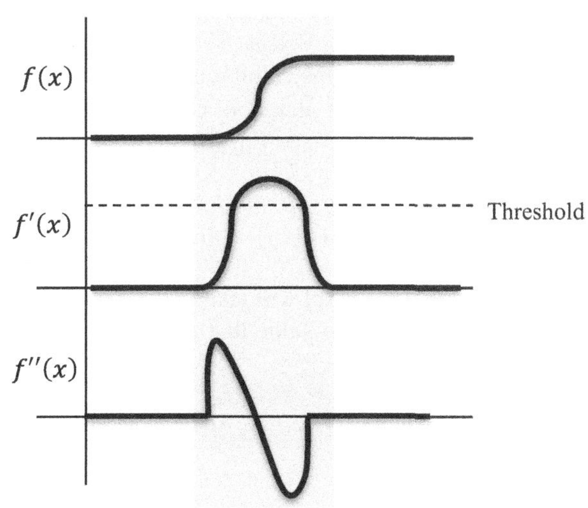

Fig. 6. Methods of the first and second derivatives for edge detection.

- Method of Canny: This algorithm for the identification of contours was developed in 1986 by John F. Canny, it presents better results than those previously explained methods, although the computational complexity is higher [21]. This algorithm is based on three basic criteria:
  - Detection: It avoids the elimination of relevant edges, as well as the generation of false edges that can harm the final result.
  - Location: It stipulates that the distance between the real position of the edge and the generated one, must be reduced.
  - Unique answer: Express that the algorithm must return an edge pixel for each true pixel, that is, there must not be groups of pixels where only one must exist.

3 Cellular Automata Design

In this work, the experiments were done with grayscale input images and an initially determined neighborhood with a 3×3 size, that is, Moore neighborhood, which travels the whole image pixel by pixel and determines if the pixel in evaluation, according to the configuration that its neighbors have (transition rules) is defined as edge. When this edge is identified, it changes its current value (by 1 or 0) and the automaton advances to the next position. The input values (initials), are provided by the image and having that data, the automaton can start to iterate all over the image. For fixed value limits, the conditions apply in which transition rules can only be designated to non-limit cells.

When talking about a Moore neighborhood and gray scale images, it is said that: (i) there is a neighborhood of fixed size of 3×3, described by Eq. (3), and, (ii) a pixel within the image you can take discrete values between 0 and 255, that is, 256 intensity values. Conceptually, the salt and pepper noise is scattered randomly over the whole image, affecting it with the maximum (white) and minimum (black) values, which, for this case, would be represented by $min = 0$ y $max = 255$. Then, when describing in a mathematical way the proposed rule for the cellular automaton, we have that the value of the pixel in evaluation in the next iteration $(C^{t+1}(i,j))$ can take two values depending of the current status $(C^{t}(i,j))$ (see Eq. (6)).

$$C^{t+1}(i,j) = \begin{cases} C^{t}(i,j), \; if \; min < C^{t}(i,j) < max \\ Cn^{t}(i,j), \; if \; C^{t}(i,j) = min \; or \; C^{t}(i,j) = max \end{cases} \tag{6}$$

Where, $C^{t+1}(i,j)$ is the value that the pixel will take in the next iteration, $C^{t}(i,j)$ is the current value of the pixel, $Cn^{t}(i,j)$ is the value that the noisy pixel will take.

$$Cn^{t}(i,j) = \begin{cases} C(2n+1)2{-}1, \; if \; \forall C^{t}(i,j) \in N, C^{t}(i,j) \neq min \; or \; C^{t}(i,j) \neq max \; and \; N = \emptyset \\ avg(C^{t}(i,j)), \; if \; \forall C^{t}(i,j) \in N, \exists C^{t}(i,j) \neq min \; or \; \exists C^{t}(i,j) \neq max \end{cases} \tag{7}$$

Where, N is the neighborhood in the current iteration y $C(2n+1)2{-}1$ indicates that the neighborhood will be expanded by $n = 2$.

Applying Eq. (6), the primary focus of this methodology is divided into two broad strands. The first one avoids acting on those pixels that are not considered noisy, which decreases the operations carried out by the algorithm and, consequently, the computational cost. The second one examines whether the affected pixel has the value of 255 (maximum or white) or 0 (minimum or black), in the affirmative case, the sub-rule described in Eq. (7) is applied, which specifies that all the minimum and maximum values will be eliminated to calculate an average of the rest and thus obtain the new value of the cell in case the Moore neighborhood lacks information, that is, that it only has maximum or minimum values in its configuration, it will adapt and it will be extended in $n = 2$ (see Fig. 5 for Moore extended neighborhood) and the cell state will be the average with the additional information excluding the maximum and minimum values. To conclude, the cellular automaton is evaluated a finite number of times depending on the value of the noise immersed in the image, in this way, if the noise level is equal to n, the cellular automaton will iterate $(n/10) + 1$ times. Upon completion of this process, the image will be ready for the edge detection process.

As explained above, when working with a 3×3 neighborhood with cells (pixels) that can take intensity values from 0 to 255, there are 256^8 possible patterns or rules that the neighborhood can have (this without considering the value of the central pixel). However, they can be reduced considerably by applying the concept of rotational symmetry. It is said that a flat figure has rotational symmetry when a center (called the center of rotation) can be found so that if the whole figure is rotated at a certain angle (greater than or equal to 0° and less than 360°), the rotated figure matches the original figure. The number of times that the rotated image can be matched with an original figure is called the order of the rotation [22].

A	*B*	*A*
B		*B*
A	*B*	*A*

A	*B*	*C*
D		*D*
C	*B*	*A*

A	*B*	*A*
C		*C*
D	*B*	*D*

Fig. 7. Patterns of a neighborhood of 3×3 that remains invariant with a rotation of ±90°, 180° and mirror-like symmetry [4].

Considering the above and applying the concept of a Moore neighborhood, it is possible to affirm that the same rule can be discarded if the image is rotated and the pattern remains the same, reducing the number of rules by approximately 5 times its size [4, 5]; to identify the number of different resulting patterns after the elimination of equivalent symmetric pairs, it is possible to apply the enumeration method described by Pólya-Burnside [23], for all G that is a set of permutations of a set A, so that the number of equivalence classes is:

$$N = \frac{1}{|G|}\sum_{g \in G}|Fix(g)| \tag{8}$$

Where, $Fix(g)$ is the number of elements of A that are invariant (they do not modify their configuration in spite of the different rotations) in g (see Fig. 7). By obtaining the value of the number of different rules N, it is possible to define what would be the number of N in terms of the number of possible intensities n, as shown in the Eq. (9) [4].

$$N = \frac{n^8 + 2n^2 + n^4 + 4n^5}{8} \tag{9}$$

From the above, the numbered is expressed as the sum of the identical rotation (e.g. 0°), two rotations of ±90°, a simple rotation of 180° and four mirror-like rotations (through a horizontal reflection line, vertical and diagonal) [4]. Despite the fact that when carrying out this procedure, the possible rules amount to $N > 2 \times 10^{18}$, in that investigation, a two-pass processing was carried out to simplify the number of operations, rules and computational expenditure.

In the first iteration, the average value of the neighborhood is calculated excluding the central pixel. This value serves as a basis to obtain the existing variation between the intensity values of the pixel under evaluation $(C(i,j))$ against the average of the neighborhood. Having these values it is possible to determine if a pixel is considered as an edge or not, by means of the rule presented below:

$$C_{(i,j)}^{t+1} = \begin{cases} background, \ if \ z = C(i,j) > average \\ background, \ if \ z = variation < 4 \\ edge, \ for \ other \ z \end{cases} \tag{10}$$

As shown in the Eq. (10), $C_{(i,j)}^{t+1}$ is the value that will be assigned to the cell under evaluation at time $t+1$, this can take three values depending on its current state $C(i,j)$, at the end of this process the first pass is finished.

The values of background or edge are directly related to 0 (for this case background) or 1 (for this case edge), which leads to think of a binary image. That is to say, that the set of possible rules was restricted to $2^8 = 256$; however, when applying the concept of similarity and eliminating the symmetries, these are reduced to $N = 51$.

Taking into account the reduced number of rules and that the borders on the image should be continuous, eleven (11) rules were proposed that thanks to the extension of the neighborhood, allow to determine, when having more information, if these really are edges or not. It should be noted that for the neighborhoods that lacked information because they were in the corners or in the borders of the image the first and last row and column were duplicated. In general, the pseudocode of the Proposed Algorithm is presented below:

```
Proposed Algorithm or MEDCA

Step 1: Read original image I(x,y)
Step 2: If I(x,y) It is not grayscale, convert it to this kind of
image.
Step 3: Normalize I(x,y)
Step 4: From I(x,y) create a 3x3 matrix, which simulates Moore's
neighborhood.
Step 5: The central pixel of the matrix will be represented by C(i,j)
Step 6: If C(i,j) is greater than or equal to the average of neigh-
boring pixels excluding C(i,j), then:
        It is background.
else,
        If the variation between pixels is less than 4, then:
                It is background.
        else,   It is edge.
        Endif.
Endif.
Step 7: Apply the transition rules until the isolated points are
removed and the edges delineated.
```

4 Discussions and Results

Taking into account that the proposed model is composed by two different techniques but at the same time complementary as they are, noise elimination and edge detection, this section addresses two areas from which a detailed observation of the virtues of each algorithm can be made, these are qualitative and quantitative analyzes; this allows to create an adequate base when confronting the results, since by establishing a comparative framework that shows the greatest possible impartiality, the characteristics used by the different algorithms to achieve their objective become more relevant.

The above is achieved because this comparison covers different parameters such as, the one-to-one comparison of the pixels (resulting and original), continuity in the strokes, non-existence of false borders, defined edges, etc., where each algorithm can have its best performance in both the graphic and numerical part. In addition, typical metrics for edge detection are used, which implies that the data thrown are similar to those expressed by other authors.

4.1 Qualitative Results

The results of this section are aimed to graphically show the visual faculties of the proposed development. In this sense, a scenario is initially presented that allows to show the functionalities of the cellular automaton and of the whole algorithm, for this reason the images presented below were subjected to an initial transformation that includes simulating the salt and pepper impulsive noise in a level of 10% on them.

It should be noted that the algorithms with which the Proposed Algorithm is compared (Canny, Sobel and Prewitt reported in [24], do not have a process that allows them to purify the noise; however, this approach serves as a starting point to establish similar conditions (images with and without noise) and likewise, observe their behavior.

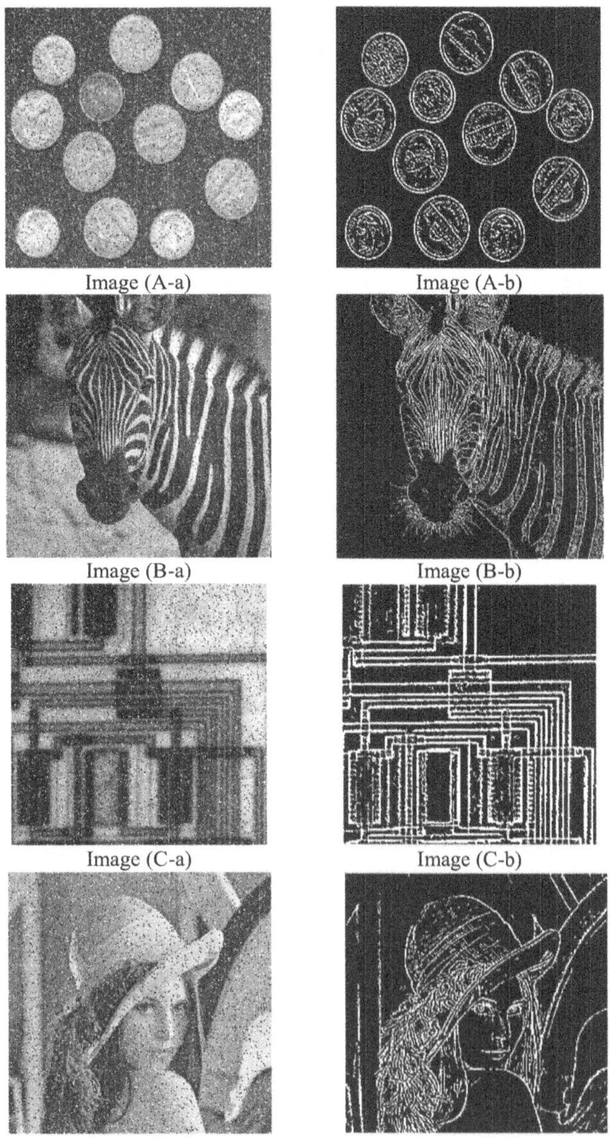

Fig. 8. Images with noise at 10%. (A) Coins. (B) Zebra. (C) Circuits. (D) Lena. Columns (a) and (b) show the original image and the obtained using the Proposed Algorithm with CA.

As can be seen in Fig. 8, the Proposed Algorithm has the ability to obtain visually acceptable results despite the noise that the image has, this is due to the different qualities immersed in the operation of the cellular automaton, where the adaptive behavior allowed two aspects of image processing to be executed satisfactorily, detecting the edges in a suitable way and without having major affectations in the resulting image due to the salt and pepper noise present in the image.

In parallel, in Fig. 9 a comparison was made of the commonly used methods for edge detection versus the Proposed Algorithm (Method for Edge Detection with Cellular Automata MEDCA). The images used show the application of this method in the field of medicine and specifically, the field of radiology.

Furthermore, in Fig. 9, where only images without noise were considered, it can be highlighted that the resulting images of the Proposed Algorithm identify more details than those presented by the other algorithms, showing relevant details in the objects present in the image and, in turn, this discards those that in the other methodologies are considered as necessary within the image (e.g. Canny), considering this as a remarkable advantage of the methodologies that make one of the cellular automata.

Finally, it is noticeable that the edges generated by the Proposed Algorithm are in some points thicker than the others. However, this is the result of the different operations carried out by the cellular automaton to always try to identify the relevant information of the image and thus perform the relevant tasks, such as not leaving loose pixels, identifying edges regardless of their shape, avoid the creation of both the incorrect and duplicated edges.

4.2 Quantitative Results

For the calculation of the quantitative results, three images were used that can be considered as a laboratory case or ideals, since they have lines (horizontal, vertical, and diagonal) and clearly defined curves, which facilitates the comparison of the algorithm against the usual methods (Canny, Sobel and Prewitt described in [24] and generates some coherence compared to the exposed conditions. Figure 10 shows the pictures utilized for this process (ideal cases).

Although the formulation of test cases is a suitable way to measure the effectiveness of the Proposed Algorithm, it is also convenient to establish a series of metrics that serve as a solid basis for an approximate comparison. Three comparison metrics were used to determine the contrast: the Peak Signal to Noise Ratio (PSNR), the Maximum Error (in this context, MAXERR) and the Ratio of Squared Norms (in this context, L2RAT) [25]. For PSNR calculation is used the Mean Square Error (MSE).

The Mean Square Error is a statistical estimator that is in charge of measuring the average squared errors, that is the squared norm of the difference between the expected value and the real value. The calculation of this value is thus given as follows:

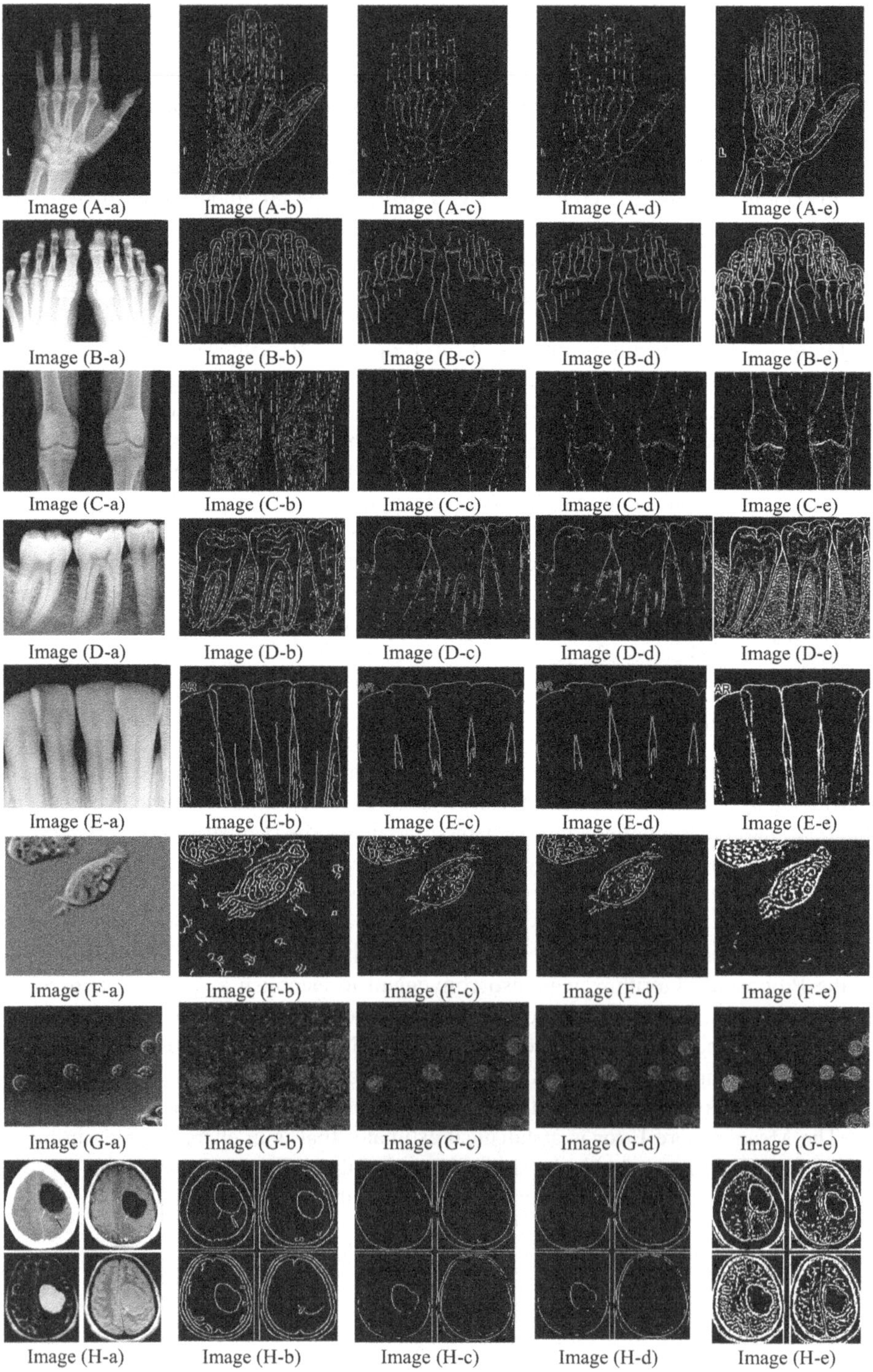

Fig. 9. (*continued*)

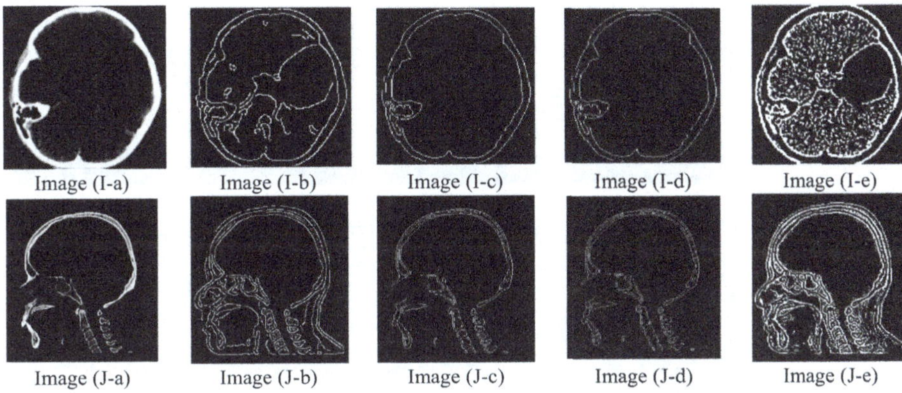

Fig. 9. X-ray. (A) Left hand with the presence of a cancerous tumor. (B) Right and left foot. (C) Knees. (D) Lower molars. (E) Lower incisors. (F) Cell. (G) Microorganisms. (H) Tomographies of the brain with tumor presence. (I) Tomography of the skull base tumor presence. (J) MRI of the head from the lateral perspective. Columns (a) to (e) show the different algorithms: Canny, Sobel, Prewitt and Proposed Algorithm (MEDCA).

(a) Image 1 (b) Image 2 (c) Image 3

Fig. 10. Test cases for edge detection.

$$MSE = \frac{\|X - Y\|^2}{N} \tag{11}$$

Where, X is the base image (ideal), Y is the image to compare, and N is the number of elements of the image (pixels). Once the MSE is calculated, the PSNR (relative measurement of image quality) can be obtained using (11).

$$PSNR = 20\log_{10}\left(\frac{2^B - 1}{\sqrt{MSE}}\right) \tag{12}$$

Where B represents the bits per sample, in this case $B = 8$. The Maximum Error or MAXERR corresponds to the maximum deviation to the absolute square of the image to be compared, and finally, the Ratio of Squared Norms or L2RAT, is defined as the relation of the squared norm of the image to compare regarding the base image. Figure 11 shows the obtained images.

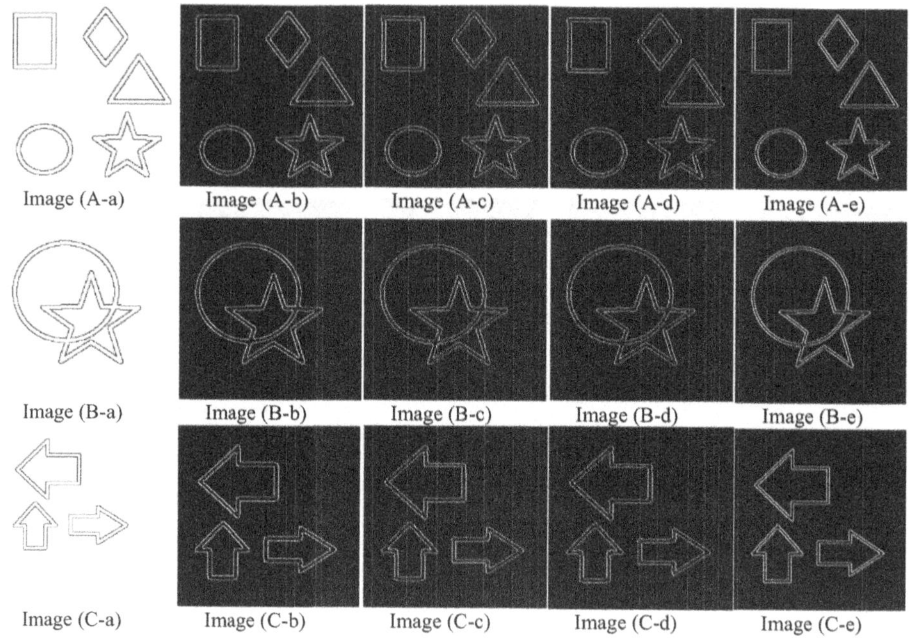

Fig. 11. (A) Image 1. (B) Image 2. (C) Image 3. Column (a) image with ideal edges. Columns (b) to (e) show the algorithms results: Canny, Sobel, Prewitt and the Proposed Algorithm (MEDCA).

In accordance with the objective of each of these metrics and taking into account their application in this process, is made the comparison of the images resulting from the methods of Canny, Sobel, Prewitt and the Proposed Algorithm (MEDCA). The data presented in Table 1 shows the results of the considered metrics.

Table 1. Results of the metrics applied for images 1, 2 and 3.

		Canny	Prewitt	Sobel	MEDCA
Image 1	PSNR	0.4332	0.4332	0.4332	0.4332
	MAXERR	255	255	255	255
	L2RAT	1.4949×10^6	1.8454×10^6	1.8393×10^6	1.3155×10^6
Image 2	PSNR	0.2921	0.2919	0.2919	0.2923
	MAXERR	255	255	255	255
	L2RAT	2.0426×10^6	2.5480×10^6	2.5399×10^6	1.7602×10^6
Image 3	PSNR	0.3323	0.3321	0.3321	0.3323
	MAXERR	255	255	255	255
	L2RAT	2.1148×10^6	2.5739×10^6	2.5490×10^6	1.9663×10^6

Another variable that was taken into account when evaluating the performance of the methods used was CPU time spent and the performance of the algorithm, in seconds, when executing the task of edge detection within the image. This measure is of great importance, since it specifies the execution time of the program in the simulated environment. That is, this interval time starts with the execution of the program in the operating system and concludes with the printing of the resulting images with the edges detected. In Table 2 and Fig. 12, the time used by each of the methods to detect the contours in each of the proposed laboratory cases can be observed.

Table 2. Execution time when applying the methods in images 1, 2 and 3.

(in sec.)	Canny	Prewitt	Sobel	MEDCA
Image 1	0.205723	0.170806	0.168754	0.209583
Image 2	0.171584	0.159336	0.162779	0.179615
Image 3	0.160017	0.148048	0.151929	0.175637

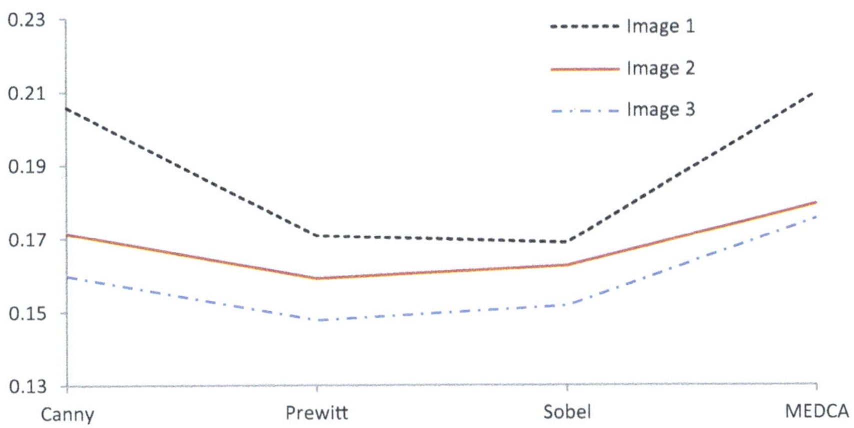

Fig. 12. Graph of the execution times of each method.

In addition, the percentage difference of the proposed method compared to Canny ranges between 0.39% and, 1.59%, which means that despite making a pixel to pixel revision with cellular automata, the execution time of the Proposed Algorithm is quite acceptable and generates a large computational load when executed.

5 Conclusions

Through the research, it was clearly noted that an algorithm has not yet been developed to obtain perfect results, that is, to serve as a fundamental and solid basis for establishing a completely objective comparison between the methods developed. However, there are algorithms such as Canny, Sobel or Prewitt, which are used frequently to establish a comparative base, without forgetting that the conditions of each of these algorithms are different.

A series of "ideal cases" was considered to perform a quantitative comparison of results with different estimators using elementary shapes. Using these cases the results show that the Proposed Algorithm is able to detect edges between 92,30% and 97,21%. The results of PSNR indicate a suitable degree of coincidence between the image with ideal edges and the obtained using MEDCA.

Regarding the results of L2RAT, these fluctuated between 3,1% and 4,6%, which represents that the results of Proposed Algorithm are quite approximate to the edges of reference images. On the other hand, the Maximum Error (MAXERR) in all cases is showing that the process is not perfect for all algorithms.

The Proposed Algorithm was able to eliminate the amount of corrupted pixels in the image and at the same time to identify the edges.

References

1. Bhardwaj, S., Mittal, A.: A survey on various edge detector techniques. Procedia Technol. **4**, 220–226 (2012)
2. Diwakar, M., Kumar, P., Gupta, K.: Cellular automata based edge-detection for brain tumor. In: 2013 International Conference on Advances in Computing, Communications and Informatics (ICACCI), India, pp. 53–59. IEEE (2013)
3. Rosin, P., Sun, X.: Edge detection using cellular automata. In: Cellular Automata in Image Processing and Geometry, pp. 85–103. Springer, Cham (2014)
4. Rosin, P., Sun, X.: Cellular automata as a tool for image processing. Emerg. Top. Comput. Vis. Appl. **1**, 233–251 (2011)
5. Rosin, P.L.: Training cellular automata for image processing. IEEE Trans. Image Process. **15**(7), 2076–2087 (2006)
6. Wongthanavasu, S., Sadananda, R.: A CA-based edge operator and its performance evaluation. J. Vis. Commun. Image Represent. **14**(2), 83–96 (2003)
7. Selvapeter, J., Hordijk, W.: Genetically evolved cellular automata for image edge detection. In: International Conference on Signal, Image Processing and Pattern Recognition, India (2013)
8. Batouche, M., Meshoul, S., Abbassene, A.: On solving edge detection by emergence. Adv. Appl. Artif. Intell. **4031**, 800–808 (2006)
9. Slatnia, S., Batouche, M., Melkemi, K.E.: Evolutionary cellular automata based-approach for edge detection. Appl. Fuzzy Sets Theory **4578**, 404–411 (2007)
10. Yang, C., Ye, H., Wang, G.: Cellular automata modeling in edge recognition. In: Seventh International Symposium on Artificial Life and Robotics, pp. 128–132 (2002)
11. Tourtounis, D., Mitianaudis, N., Sirakaulis, G.: Salt-n-pepper noise filtering using cellular automata. J. Cell. Automata **13**(1–2), 81–101 (2018)
12. Chickerur, S., Kumar, A.: Color image restoration using neural network model. J. Univ. Comput. Sci. **17**(1), 107–125 (2011)
13. Betancourt, A., Mujica, A., Tapias, H.: Procesamiento difuso de imágenes: filtro difuso para eliminar el ruido impulsivo. Rev. Ing. **8**(2), 40–46 (2003)

14. Dhillon, P.K.: A novel framework to image edge detection using cellular automata. In: IJCA Special Issue on Confluence 2012 - The Next Generation Information Technology Summit, Confluence, vol. 1, pp. 1–5 (2012)
15. Neumann, J.V.: Theory of self-reproducing automata. In: Burks, A.W. (ed.) Machine Theory, pp. 91–131. University of Illinois Press, Urbana (1966)
16. Shukla, A.P.: Training cellular automata for image edge detection. Rom. J. Inf. Sci. Technol. **19**(4), 338–359 (2016)
17. Pan, P.Z., Feng, X.T., Zhou, H.: Solid cellular automaton method for the solution of physical field. In: 2009 WRI World Congress on Computer Science and Information Engineering, Los Angeles, CA, USA, pp. 765–768. IEEE (2009)
18. Kumar, T., Sahoo, G.: A novel method of edge detection using cellular automata. Int. J. Comput. Appl. **9**(4), 38–44 (2010)
19. Hasanzadeh, M., Sadeghi, S., Rezvanian, A., Reza, M.: Cellular edge detection: combining cellular automata and cellular learning automata. AEU – Int. J. Electron. Commun. **69**(9), 1282–1290 (2015)
20. Wongthanavasu, S.: Cellular automata for medical image processing. In: Salcido, A. (ed.) Cellular Automata - Innovative Modelling for Science and Engineering, pp. 395–410 (2011)
21. Chacón, M., Sandoval, R., Vega, J.: Percepción visual aplicada a la robótica, 1st edn. Alfaomega Grupo Editor, Mexico (2015)
22. Costa, A.F.: Una Introducción a la Simetría. UNED - Universidad Nacional de Educación a Distancia, Bogotá (2009)
23. Roberts, F., Tesman, B.: Applied Combinatorics, 2nd edn. Chapman & Hall/CRC, Boca Raton (2009)
24. Öztürk, Ş., Akdemir, B.: Comparison of edge detection algorithms for texture analysis on glass production. Procedia Soc. Behav. Sci. **195**, 2675–2682 (2015)
25. Patil, M., Dhopeshwarkar, M., Sathe, P.: Calculate the quality measures on classification of continuous EEG without trial structure EEG dataset. Int. J. Comput. Appl. **147**(10), 32–35 (2016)

Semantic Processing Method to Improve a Query-Based Approach for Mining Concept Maps

Wenny Hojas-Mazo[1], Alfredo Simón-Cuevas[1](✉), Manuel de la Iglesia Campos[1], and Juan Carlos Ruíz-Carrera[2]

[1] Universidad Tecnológica de La Habana José Antonio Echeverría, Ave. 114, No. 11901, 19390 La Habana, Cuba
{whojas,asimon,miglesia}@ceis.cujae.edu.cu
[2] Universidad Técnica de Ambato, Ambato, Ecuador
jc.ruiz@uta.edu.ec

Abstract. Concept maps are powerful visual tools for organizing and representing knowledge. The development of computational systems for building and managing concept maps has facilitated the creation of concept maps repositories, which can be considered valuable knowledge models. In this paper, a semantic processing method to improve a query-based approach for mining concept maps repositories is proposed. The method is based on a mechanism of semantic extension of the represented concepts in the concept maps and a set of rules for guiding the semantic integration of the conceptual structures and retrieval information processes. WordNet is used for the semantic extension of concepts, which is supported in a disambiguation algorithm. Results of experimental evaluations of the disambiguation algorithm applied to several concept maps are presented and compared with the state of the art. The application of the proposed method is illustrated through a study case with promising results.

Keywords: Concept maps · Semantic processing · Mining concept maps · Word sense disambiguation

1 Introduction

Concept maps (CM) are powerful graph-based knowledge representation scheme, for organizing and representing conceptual meanings through a proposition structure [17]. CM is composed by concepts and labeled relationship between them form propositions. Propositions contain two or more connected concepts using linking words or phrases to form a meaningful statement, and sometimes these are called semantic units, or units of meaning [6]. Concept mapping [17] has been widely used for knowledge elicitation, for encouraging knowledge construction by students and others, and for making internal conceptualizations explicit to facilitate knowledge sharing, comparison, and assessment [11]. The development of editors and management systems for CM, such as CmapTools and CmapServer [4, 6], as well as, the development of several proposal for their automatic or semi-automatic construction from texts [1, 10, 15, 21], not only facilitates

J. Nummenmaa et al. (Eds.): CSEI 2019, AISC 1078, pp. 22–35, 2020.
https://doi.org/10.1007/978-3-030-33614-1_2

the concept mapping process, but also the constructs of CM Repositories (CMR). CMR constitute knowledge models where the CM represents knowledge of specific topic or domain [4]. The CMR can be considered an explicit, structured and not formal knowledge base. The relevant information retrieval and the useful knowledge discovery from CMR constitutes great challenges, not only due to the inherent complexity of the graph-based representation models, but also the underlying ambiguity the knowledge represented. Several approaches for carried out mining processes over CMR or knowledge models have been reported, among them: query-based [10] and data mining techniques-based [9, 22]. In this sense, the effectiveness improvement of concept maps querying process arises as an important necessity for take advantage from this valuable knowledge source and to increase the exploitation, analysis and discovery capacities over the represented knowledge.

Most of the reported proposals for information retrieval in CMR (IR-CMR) have fundamentally focused on the retrieval of concepts, propositions or CM [3, 4, 7, 8, 13]. Nevertheless, there is another reported alternative which proposes several query operations for obtaining different views of the CMR, concretely the Concept Map Query Language (CMQL) [20] (a refined version is reported in [10]). CMQL formalizes a set of query operations that facilitate to obtain different views of the stored knowledge in a CMR, including an information integration mechanism as part of the information retrieval process. Through CMQL the system can retrieve information about concepts and propositions according to different types of query operations, and the results of each query are shown by means of a new automatically generated CM. However, in all these proposals [3, 4, 7, 8, 13, 20], the information retrieval process is supported in the syntactic matching between concepts in the query and those included in the search space (CMR), which is a weakness due to the concepts and propositions in CM are expressed in natural language and they are usually subjected to ambiguity problem. Therefore, the effectiveness of the concept map querying process can be reduced if a semantic analysis is not considered as part of these querying process. The application of the word-sense disambiguation task [16] constitutes an alternative to deal with this problem, and several disambiguation algorithm for CM has been reported [5, 19].

In this paper, a semantic analysis method to improve a query-based approach for mining concept maps repositories is proposed. The method is based on a mechanism of semantic extension of concepts represented in the CMR using WordNet [14] and the definition of a set of rules for guiding in the processes of integration and retrieval information included in the queries processing. The semantic processing is fundamentally supported in the use of the disambiguation algorithm reported in [10], which allows reduce the ambiguity associated to the concepts represented in the CM. This algorithm uses WordNet as source of meanings and improves the disambiguation results of other reported proposals [5, 19]. The new disambiguation algorithm was evaluated using a dataset of 50 CM in English language, which were obtained from literature. Results of experimental evaluations are presented and compared with the state of the art. Besides, several example cases were included for illustrating the application of the proposed method.

The rest of the paper is organized as follows: Sect. 2 discusses related work; Sect. 3 describes the proposed semantic processing method; Sect. 4 presents the results of the experiments carried out and the developed study cases; and conclusions arrived are given in Sect. 5.

2 Related Works

The IR-CMR has been focused in the retrieval of concepts [13], propositions (using the Knowledge Soup of CampTools) [4] or CM [3, 7, 8], in which the CM are retrieval in the same form that they are stored. In [20], and most recently in [10], the mining CM for retrieving relevant information and knowledge from the CMR is carried out through several types of queries defined in CMQL and several mechanisms to filter and integrate concepts and propositional structures. CMQL offers the formalization of a set of query operations, such as: *union*, *intersection*, *projection* (or sub-map) and *extension*, which facilitates the IR-CMR from different perspectives [10, 20]. In each query processing, the concepts and propositions including in the search source (set of selected CM from the CMR) are processed as independent units and, at the same time, they can be integrated in the answer generation; aspect not considered in other proposals.

A common factor in these proposals is that the information retrieval and integration processes (also in CMQL [20]), are carried out considering the syntactic similarity between the concepts included in the query and the stored concepts in the CMR, without considering the existence of a possible ambiguity problems and semantic similarity among them. This situation can imply that it is not possible to retrieval of potentially relevant information associated to concepts that are not syntactically equivalent or similar to the included ones in the query, although they can be semantically similar. The information integration process can also produce not appropriate results, for example: the integration of concepts written in the same way, but with different senses, and the not integration of concepts semantically similar, although writings in a different way. The knowledge in CM is expressed in natural language, so, in contexts in which the computational processing of CM is required, as in the case of the IR-CMR, the semantic processing of the knowledge represented constitutes an important aspect that should be considered due the inherent ambiguity to the natural language. The word-sense disambiguation constitutes a successful alternative to deal with this problem.

The word sense disambiguation in unstructured texts has been broadly studied [16], but there are few reported works that approaching this problem in the CM context [5, 19]. Cañas et al. [5] exploits the topology and the semantic of the CM, by trying to determine which senses in WordNet [14] best matches the context of the CM, which is done constructing hypernym sequences, using hypernymy relations from WordNet [14]. In this proposal, only words from key concepts are included as part of the disambiguation process. In the algorithm reported in [19], the disambiguation process in CM was improved increasing the information being considered in the process, in addition to the contextual analysis (reported in [5]). Specifically, the use of domain information, according to Bentivogli et al. [2], and the gloss were included in the disambiguation process, as well as, others relations from WordNet, such as hyponymy, and meronymy/holonymy were considered. Therefore, the disambiguation process is carried out through heuristics based on *domain*, *context* and *gloss*, which are applied sequentially [19]. Nevertheless, the sequential use of these heuristics constitutes a limitation because the sense of the concept was determined according one of them and not taking advantage of the combination of the results obtained from each one [12, 16].

3 Semantic Processing Method Proposed

The proposed semantic analysis method is based on the semantic extension of concepts, and a set of rules for guiding the integration, search and retrieval processes. In this method, the semantic information associated to the concepts is captured from WordNet and a concept sense disambiguation algorithm is used for reducing the ambiguity that may emerged. The semantic extension of concepts is defined as the process of associating to one concept other synonym terms identified in WordNet. The rules utilize the information of the semantic extension of concepts to expand the number of search elements in the retrieval process (the identification of relevant concepts to retrieval combines syntax and semantic information) and to return a semantically integrated result. In CMQL, the search query Q can be defined through four components (Eq. 1):

$$Q = (Cq, Pq, SCM, T) \tag{1}$$

where Cq is a set of concepts of interest, Pq is a set of proposition of interest, SCM is a set of CM $\{cm_1, cm_2, \ldots, cm_n\}$ which constitute the search space for querying, and T specifies the type of query, according to CMQL (union, intersection, projection, or extension). The SCM can be defined as the tuple (C, P), where C is the set of concepts $\{c_{11}, c_{12}, c_{13}, \ldots, c_{ij}\}$ and P the set of propositions $\{p_{11}, p_{12}, p_{13}, \ldots, p_{ij}\}$ included in these cm_i. The Fig. 1 shows the workflow of the proposed method.

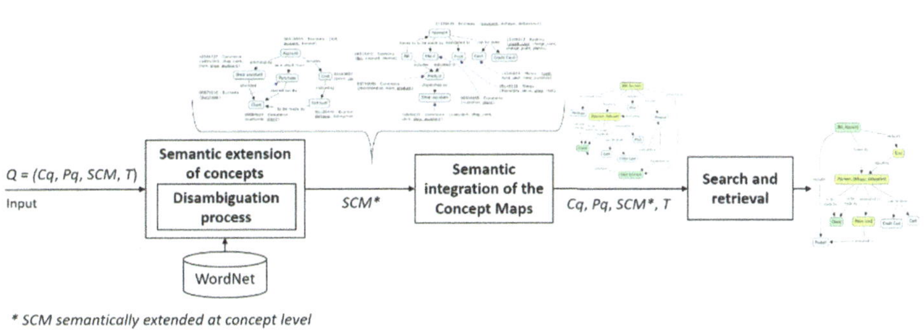

Fig. 1. Workflow of the semantic processing method

In this analysis the semantic information associated to the concepts is retrieved from WordNet, according to the senses that are been used in the representation. This information is used later in the integration and search process carried out in the queries processing. This process includes a method for the semantic extension or enrichment of concepts using WordNet, and a set of rules for identifying concepts of interest (Cq) in the search space (SCM), and determine when two or more concepts (included in SCM) can be unified, considering their semantic equivalence. The use of WordNet is supported with a disambiguation algorithm for CM, and the defined rules are integrated in each one of the queries in CMQL.

3.1 Semantic Extension of Concepts

The semantic extension is defined as the process of associating to one concept other synonym terms identified in WordNet, and it is applied to all represented concepts in the CMR. Initially, the synsets in which each concept appears in WordNet are captured, and then the concepts are classified in: ambiguous - *AC* - (those having more than one associated synsets), not ambiguous - *NA* - (only one associated synset) or unknown - *UC* - (not associated synset). Next, the disambiguation algorithm (describe below) is applied to identify the most appropriated sense (or senses) for the ambiguous concepts. This algorithm improves the disambiguation results, respect to [5, 19], fundamentally through combining the results obtained for each heuristics for determining the sense of the concept; inspired in [12, 16]. The disambiguation algorithm is defined in seven steps as follows:

Input: the CM; the ambiguous concept (c_a); the set of *synset* associated to c_a ($S(c_a)$).
Output: set of *synset* more appropriated to c_a.

Step 1. Capturing preliminary information from WordNet: For each concept $c_i | c_i \neq c_a$ and linking phrase ($l - p$) of CM, the sets of *synset* in which the concepts and linking phrases appear are obtained from WordNet and the sets of synsets $S(c_i)$ and $S(l - p)$ are created.

Step 2. Selecting the most representative domains: From the domains associated to the *synsets* of each c_i and $l - p$ are selected the most frequent ones (according to the presence in their *synsets*) and these domains are stored in D_{cm}.

Step 3. Domain analysis: For each *synset* $s_i | s_i \in S(c_a)$, the influence grade that each domain $d \in D_{cm}$ exercise on the sense s_i ($h_d(s_i)$) is calculate, through the sum of occurrence frequencies of those domains associated to s_i; the resulting values are normalized.

Step 4. Context analysis: For each *synset* $s_i | s_i \in S(c_a)$, the influence grade ($h_c(s_i)$) that the context (propositional structure in which c_a appear) exercises on the sense s_i is calculated as follows:

$$\mathrm{h_c(s)} = w_c * \frac{\sum_{\mathrm{c_i \in C_r}} \mathrm{rel(s, c_i)}}{|\mathrm{C_r}|} + w_r * \frac{\sum_{\mathrm{r_i \in R_r}} \mathrm{rel(s, r_i)}}{|\mathrm{R_r}|} \tag{2}$$

where C_r and R_r are the set of concepts and linking phrases, respectively, being included in the propositional structure of a vicinity of radius r (starting with $r = 3$) having c_a as it centroid. Initially, $rel(s, e)$ is the value obtained from the application of a metric for measuring the semantic relatedness between the *synset* s and each *synset* associated to concepts and linking-phrases included in the context.

In the Eq. 2, w_c and w_r are weights whose values can be assigned according to the user's interests in order to define the relevance degree that the information from concepts and linking phrases, respectively, will have for the heuristic. The sum of w_c and w_r should always be 1 to guarantee that the value of $h_c(s)$ be between 0 and 1. These weights are associated to the problematic use of linking phrases for the automatic processing of CM [18].

Step 5. Gloss analysis: For each $s_i | s_i \in S(c_a)$, the influence grade ($h_g(s_i)$) of the information contained in the definition (gloss) of the synset s_i in WordNet, considering its relation with the elements of context of c_a, is calculated as follows:

$$h_g(s) = w_c * \frac{|C_r \cap G(s)|}{|C_r|} + w_r * \frac{|R_r \cap G(s)|}{|R_r|} \quad (3)$$

being *G(s)* the set of words composing the gloss of *synset s*. The use of weights w_c and w_r in the formula have the same purpose described in the previous step.

Step *6. Heuristics combination:* For each $s_i | s_i \in S(c_a)$, the global influence of the different heuristics ($h_{dcg}(s_i)$) is calculated as follows:

$$h_{dcg}(s_i) = w_d h_d(s_i) + w_c h_c(s_i) + w_g h_g(s_i) \quad (4)$$

where w_d, w_c and w_g are weights with assigned values of 0.75, 0.5 and 0.25, in that order. These values represent the influence degree that each one of the combined heuristics has in the precision of the complete algorithm and were defined according the precision results obtained by domain, context and gloss heuristics reported in [19].

Step 7. Selection of the resulting synset. The more appropriated *synset* for c_a is the *synset* having a greater $h_{dcg}(s_i)$. In case of more than one *synset* in this condition, all of them are considered, and the other ones are discarded.

At the end of the disambiguation process, the lists of *ambiguous* and *not ambiguous* concepts are updated. Finally, each one of those concepts are extended with the terms included in their associated *synset*.

3.2 Semantic Integration

The semantic integration task is aimed at explicitly integrating propositional structures (initially disconnected) through the unification of concepts represented in different CM and it is applied when the query is performed on more than one CM. The concepts unification process is carried out through the identification of synonymous concepts in the selected CM as the search source of the query, and considering several defined rules (R) to determinate when two concepts can be integrated in only one concept (in a unique node). Considering that $S(c_i)$ is the set of synsets *s* associated to a concept c_i, and c_1 and c_2 two concepts included in different CM, then c_1 and c_2 are integrated if:

- *R1:* $(c_1, c_2 \in NAC) \wedge (S(c_1) = S(c_2))$; *or*
- *R2:* $(c_1, c_2 \in AC) \wedge (\exists s' | s' \in S(c_1) \wedge s' \in S(c_2))$; *or*
- *R3:* $((c_1 \in NAC \wedge c_2 \in AC) \vee (c_1 \in AC \wedge c_2 \in NAC)) \wedge (\exists s' | s' \in S(c_1) \wedge s' \in S(c_2))$; *or*
- *R4:* $(c_1, c_2 \in UC) \wedge (c_1 = c_2)$.

The R4 rule does not consider the semantic, however, it is fundamentally included because could be useful the integration of concepts not included in WordNet, for example, named entities and other concepts. As result, if *R4* is triggered or the labels of c_1 and c_2 (in the case of other triggered rules) are the same, then this label is used for representing the unified concept in the query results. In other cases, the label used for representing the unified concept is constructed with the labels of both concepts, separating by a comma and enclosing in [] (e.g. *[c1, c2]*, as shown below). Finally, the synset associated to the unified concept is determined according the following rules:

- *R5: if R1 is triggered, then the synset is the same to the c_1 or c_2*;
- *R6: if R2 is triggered, then the synsets are the common ones between the associated to c_1 and c_2*;
- *R7: if R3 is triggered, then the synset is the one associated to the $c_i \in NAC$.*

3.3 Search and Retrieval

The defined retrieval model is fundamentally applied in the in the projection queries. In the process of query *Q* specification, the definition of one or more interesting concepts $CQ = \{c_1, c_2, \ldots, c_n\}$ by the user is required, as well as defining the search source *(SS)* selecting a set of CM $\{cm_1, cm_2, \ldots, cm_n\}$ from the CMR. The CM included in *SS* are semantically integrated through a union query as internal task in the query processing. Therefore, *SS* can be formally defined by the tuple (C^{ss}, P^{ss}), where C^{ss} is the set of concepts $\{c_1, c_2, \ldots, c_n\}$ and P^{ss} is the set of propositions $\{p_{11}, p_{12}, p_{13}, \ldots, p_{ij}\}$, included in the selected CM. For the query evaluation process, three rules were defined for identifying if a concept $c_j/c_j \in C^{ss}$ is retrieved or not, according to a concept $c_i \in CQ$, where syntactic and semantic analysis are combined.

These rules are described below and are executed following the same order in which they appear. However, it is possible to parameterize the combination of the analysis type considered, according to: using the syntactic analysis, using the semantic analysis, or combining both analyses. The rules are described below:

Being a concept $A/A \in CQ$, a concept $B/B \in C^{ss}$, $W(c_i)$ the set of words included in the label of c_i, and $S_w(c_i)$ the set of synonym terms included in $S(c_i)$. The concept *B* is retrieved from *SS* if:

- *R1:* $A \equiv B$ *(syntactically equivalent)*; or
- *R2:* $A \in W(B)$; or
- *R3:* $A \in S_w(B)$.

4 Evaluation and Results

The evaluation of the proposed method turns out complex because a methodology to carry out the evaluation of this type of method has not been reported. The application of metric such as precision and coverage commonly used in information retrieval area could be applied but neither reference corpus has been identified. According to this situation, the experimental evaluation was focused in the proposed disambiguation algorithm, considering it a key task inside the semantic processing method. The practical application of the method is exemplified through example cases.

4.1 Evaluation of the Disambiguation Algorithm

In a different way as occurs in word sense disambiguation for texts, there is no corpus suitable for the evaluation of disambiguation algorithms in CM. Therefore, a dataset constituted by 50 CM written in English language were constructed to carry out the evaluation task. Those CM were selected from various sources, mainly scientific papers published in the CMC[1] proceedings, where each CM having at least one ambiguous concept. From all concepts, 65% of them had at least one synset in WordNet, with 24 different domains tagging those synsets. A characterization of this set of CM is shown in Table 1.

Table 1. Characterization of the 50 CM used in the experiments.

	C	*CS*	*SCWN*	*AC*	*SAC*	*DWN*
Average	22.98	15.32	4.88	9.72	7.05	24.20
Std. Dev.	10.75	9.56	2.20	7.63	3.00	13.70
Total	1149	766	–	486	–	–

Notation: C: concepts; CS: concepts with at least one synsets in WordNet; SCWN: synsets of concepts in WordNet; AC: ambiguous concepts; SAC: synsets of ambiguous concepts; DWN: domains in WordNet.

The results of the algorithm were evaluated using precision (*P*), recall (*R*), coverage (*C*) and *F1* measures [16], and compared with those obtained by the other reported proposals [5, 19]. The results are shows in Table 2. In this experiment, the new disambiguation algorithm was evaluated considering $w_c = 0.75$ and $w_r = 0.25$.

Table 2. Comparative evaluation of the disambiguation algorithms

Algorithms	P	R	C	F_1
Cañas et al. [5]	0.538	0.538	**1.000**	0.538
Simón et al. [19]	0.745	0.656	0.880	0.698
New proposal	**0.864**	**0.833**	0.965	**0.848**

In Table 2, the best results shown in the disambiguation process were obtained with the new proposal, although the coverage is slightly lower than the obtained with [5]. The obtained results of precisions and coverage indicates a significant reduction of the ambiguity associated to the concepts represented in the CM, contributing to improve the quality of the semantic processing carried out in the information retrieval and integration processes. If the senses that are identified as most appropriated for concepts are correct, then the terms included in the *synset* resultant of the disambiguation are effectively their synonyms. Moreover, as the semantic integration is based on the

[1] International Conference on Concept Mapping. URL: http://cmc.ihmc.us/cmc-proceedings/.

semantic similarity, if the quality of the disambiguation is high, then the effectiveness of the integration is increased. The effectiveness of the disambiguation process largely ensures the quality of results to be obtained applying the semantic processing method proposed.

4.2 Application of the Proposed Method: A Case Study

As example cases for illustrating the application of the proposed method two CM were selected (see Figs. 2 and 3), which constitutes fragments from others CM that represent conceptualizations of some services in the hotel domain. These conceptualizations were the result of a knowledge elicitation process from hotel domain experts, as part of a terminological ontology construction process. The proposed method was implemented in Java language and executed in this case using the English WordNet 3.0. In this method, WordNet versions of other language can be used too, according to the languages used the CM.

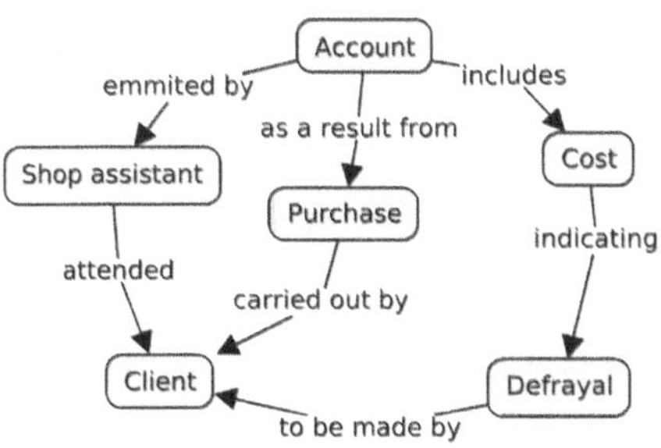

Fig. 2. Concept map of 'Receipt'

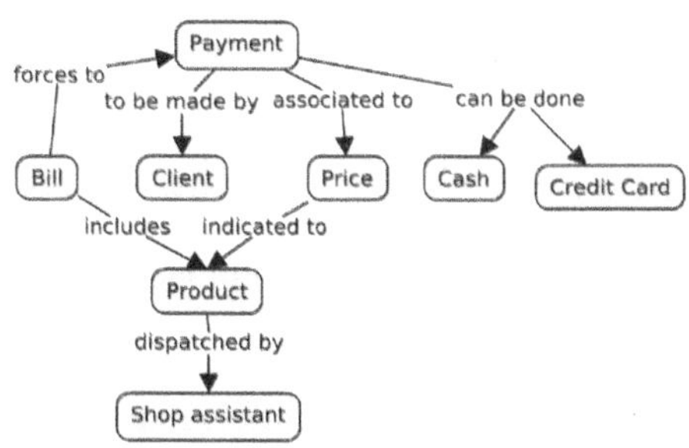

Fig. 3. Concept map of 'Paid'

Two tasks that the knowledge engineer can performs as part of the conceptualization analysis process required for obtaining the ontology were selected to exemplify the proposed method. These tasks are defined as follows:

1. to recover a global view of the conceptualization, and
2. to recover relevant information about the concepts *'defrayment'* and *'cost'*.

To achieve the defined tasks the following CMQL queries were used [10]: *Union/CMUnion({Receipt, Paid})* and *Projection/CMProj*$^{+1}$*({Receipt, Paid}, {'defrayment', 'cost'}),* where 'Receipt' and 'Paid' are the names of the CM shown in Figs. 4 and 5, respectively.

The first step of the execution of any queries is the semantic extension of the concept included in the CM that conform the search space ('Receipt' and 'Paid'). The Figs. 4 and 5 show the different synset identified in WordNet for each concept represented in these CM, and the results of the disambiguation algorithm for the 11 ambiguous concepts (78.5% of the concepts) is also shown.

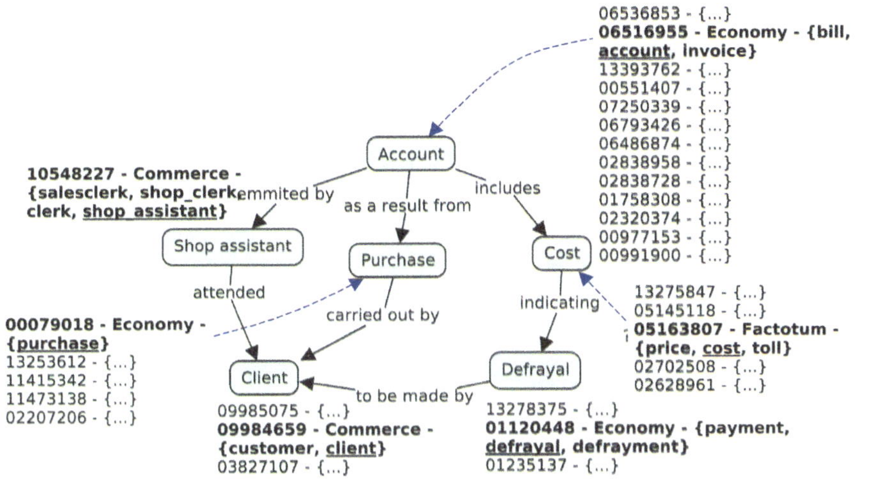

Fig. 4. Semantic extension of the CM 'Receipt'

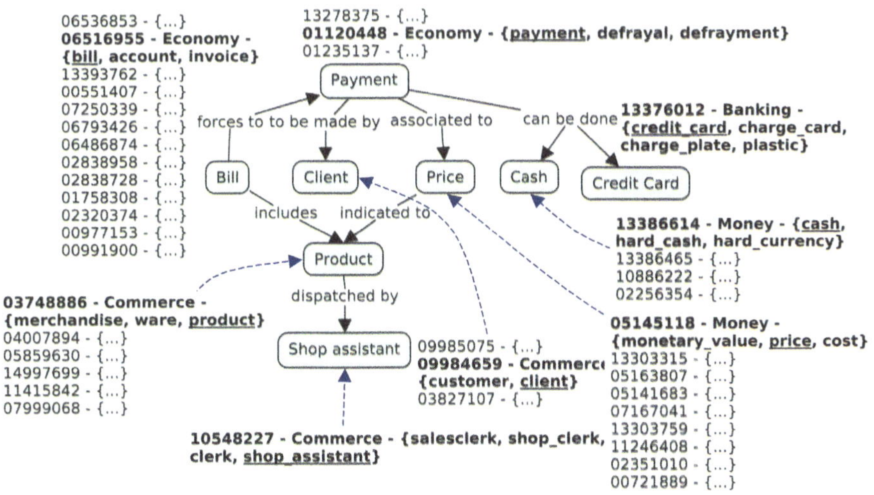

Fig. 5. Semantic extension of the CM 'Paid'

In the cases shown, the proposed disambiguation algorithm produced a 100% of precision and coverage in both CM. This result is very relevant, taking into account the high average of synset for ambiguous concepts that is shown in the Figs. 4 and 5. As result of the semantic extension, the following concepts are extended: *'account (synonyms: 'bill'* and *'invoice'), 'cost' (synonyms: 'price'* and *'toil'), 'defrayal' (synonyms: 'payment'* and *'defrayment'), 'shop assistant' (synonyms: 'salesclerk', 'shop clerk'* and *'clerk'), 'client' (synonym: 'customer'), 'payment' (synonyms: 'defrayal'* and *'defrayment'), 'bill' (synonyms: 'account'* and *'invoice'), 'product' (synonyms: 'merchandise'* and *'ware'), 'cash' (synonyms: 'hard cash'* and *'hard currency')* and *'price' (synonyms: 'monetary value'* and *'cost').*

Two variants of the queries were executed: without applying the semantic processing method and applying the method. The results of *Union* queries are shown in Figs. 6 and 7 and the results of *Projection* queries are shown in Figs. 8 and 9.

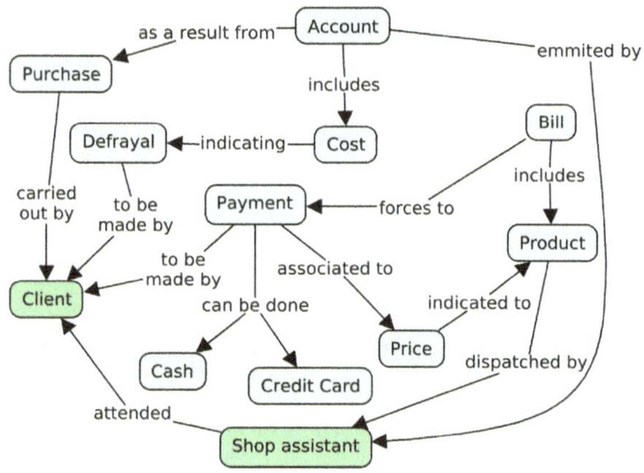

Fig. 6. Result of the *CMUnion({Receipt, Paid})* without applying the semantic processing

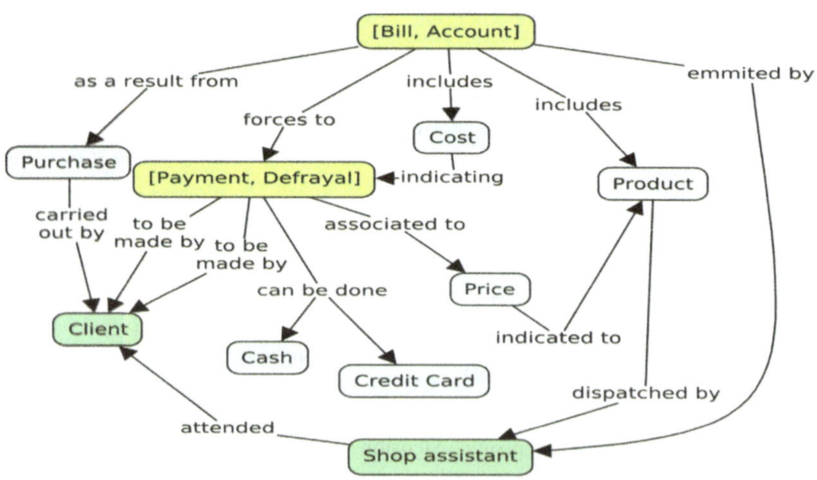

Fig. 7. Result of the *CMUnion({Receipt, Paid})* applying the semantic analysis

In the Fig. 6 it can be appreciated as the CM are only integrated by means of '*client*' and '*shop assistant*' concepts. However, in the Fig. 7, the CM are furthermore integrated by means of '*bill*'-'*account*', and '*payment*'-'*defrayal*' concepts, which have the same sense in both CM, but without a semantic processing they would not be integrated. In the Fig. 9, it is appreciating that applying the semantic processing method the amount of retrieved information was increased, respect the result shown in Fig. 8, which facilitates to increase the identification possibilities of relevant information by the end user.

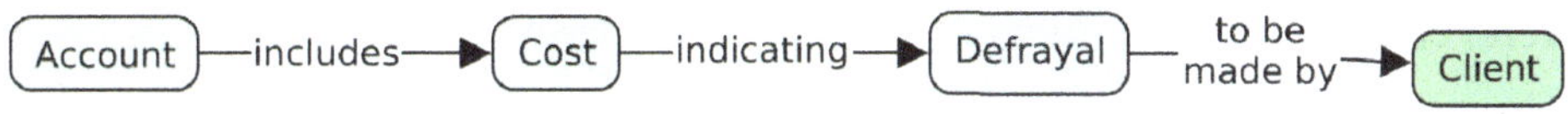

Fig. 8. Result of the *CMProj*$^{+1}$*({Receipt, Paid}, {'defrayment', 'cost'})* without applying the semantic processing

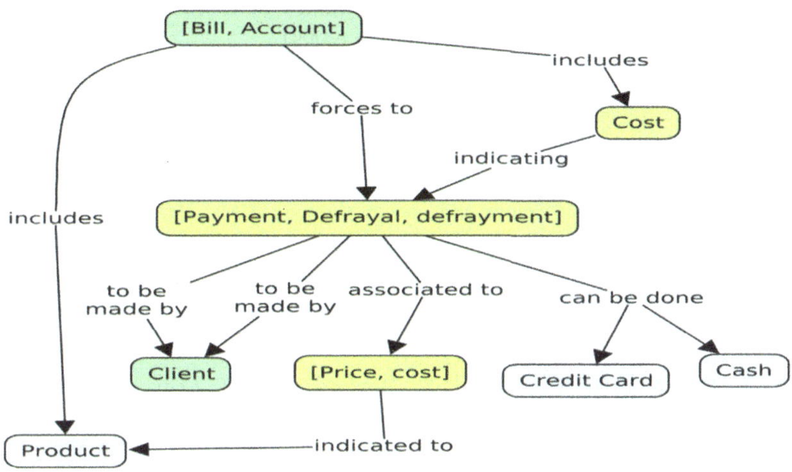

Fig. 9. Result of the *CMProj*$^{+1}$*({Receipt, Paid}, {'defrayment', 'cost'})* applying the semantic processing

5 Conclusions

This paper presented a semantic processing method applied to a query-based approach for mining CMR or CM-based knowledge models, which is based on the semantic extension of concepts, and a set of rules for guiding the integration, search and retrieval processes. In this method, the semantic information associated to the concepts is captured from WordNet and a concept sense disambiguation algorithm is applied. The combination of syntactic and semantic information processing, associated to the concepts represented in the CM stored in CMR and integrated to the query processing defined in CMQL, allowed improve the results of information integration and retrieval processes carried out for mining this valuable knowledge source. The application of a disambiguation algorithm contributes to reduce the inherent ambiguity in CM and increase the effectiveness of the semantic processing method. The proposed disambiguation algorithm improves the results of those obtained by other proposal, through combining the results obtained from domain, context and gloss heuristics for determining the sense of the concept. The reached accuracy by this disambiguation algorithm largely ensures the quality of results to be obtained applying the proposed method. Through the example cases developed, the usefulness of the semantic processing method was demonstrated.

References

1. Aguiar, C.Z., Cury, D., Zouaq, A.: Automatic construction of concept maps from texts. In: Proceedings of the 7th International Conference on Concept Mapping (CMC 2016), pp. 20–30 (2016)
2. Bentivogli, L., Forner, P., Magnini, B., Pianta, E.: Revising WordNet domains hierarchy: semantics, coverage, and balancing. In: Proceedings of COLING 2004 Workshop on Multilingual Linguistic Resources, pp. 101–108 (2004)
3. Cañas, A.J., Leake, D.B., Maguitman, A.G.: Combining concept mapping with CBR: towards experience-based support for knowledge modeling. In: Proceedings of the FLAIRS Conference, pp. 286–290. AAAI Press (2001)
4. Cañas, A.J., Hill, G., Carff, R., Niranjan, S., Lott, J., Eskridge, T.C., Gómez, G., Arroyo, M., Carvajal, R.: Cmaptools: a knowledge modeling and sharing environment. In: Proceedings of the 1st International Conference on Concept Mapping (CMC 2004), pp. 125–133 (2004)
5. Cañas, A., Valerio, A., Lalinde, J., Carvalho, M., Arguedas, M.: Using WordNet for word sense disambiguation to support concept map construction. LNCS, vol. 2857, pp. 350–359. Springer, Heidelberg (2003)
6. Cañas, A.J., Novak, J.D., Reiska, P.: How good is my concept map? Am I a good Cmapper? Knowl. Manag. E-Learn. **7**(1), 6–19 (2015)
7. Carmona, E.J., Marrero, S., Nelson, J.C., Rubio, E.: Uso de Mapas Conceptuales como Soporte en Actividades de Gestión del Conocimiento en una Comunidad Virtual: Centro I + D+I. In: Proceedings of the 2nd International Conference on Concept Mapping (CMC 2006), pp. 307–310 (2006)
8. Eskridge, T.C., Granados, A., Cañas, A.J.: Ranking concept map retrieval in the CmapTools network. In: Proceedings of the 2nd International Conference on Concept Mapping (CMC 2006), pp. 477–484 (2006)
9. Hojas Mazo, W., Simón Cuevas, A., Rodríguez Blanco, A.: Aplicación de Técnicas Minería de Grafos para el Análisis de Textos. III Congreso Internacional de Ingeniería Informática y Sistemas de Información (CIIISI 2016), Cuba (2016)
10. Hojas-Mazo, W., Simón-Cuevas, A., de la Iglesia Campos, M., Romero, F.P., Olivas, J.A.: A concept-based text analysis approach using knowledge graph. In: Medina, J. et al. (eds.) IPMU 2018. CCIS, vol. 854, pp. 696–708 (2018)
11. Katagall, R., Dadde, R., Goudar, R.H., Rao, S.: Concept mapping in education and semantic knowledge representation: an illustrative survey. Procedia Comput. Sci. **48**, 638–643 (2015)
12. Klein, D., Toutanova, K., Ilhan, H.T., Kamvar, S.D., Manning, C.D.: Combining heterogeneous classifiers for word-sense disambiguation. In: Proceedings of the ACL 2002, pp. 74–80 (2002)
13. Leake, D., Maguitman, A., Reichherzer, T., Cañas, A.J., Carvalho, M., Arguedas, M., Eskridge, T.: Googling from a concept map: towards automatic concept map-based query formation. In: Proceedings of the 1st International Conference on Concept Mapping (CMC 2004), pp. 409–416 (2004)
14. Miller, G., Fellbaum, C.: WordNet: An Electronic Lexical Database. The MIT Press, Cambridge (1998)
15. Ng'asia Wafula, B.: Automatic construction of concept maps. Master's thesis, University of Eastern Finland, Faculty of Science and Forestry (2016)
16. Navigli, R.: Word sense disambiguation: a survey. ACM Comput. Surv. **41**(2), 1–69 (2009)
17. Novak, J.D., Gowin, D.B.: Learning How to Learn. Cambridge University Press, Cambridge (1984)

18. Reichherzer, T., Leake, D.: Towards automatic support for augmenting concept maps with documents. In: Proceedings of the 2nd International Conference on Concept Mapping (CMC 2006), pp. 566–573 (2006)
19. Simón, A., Ceccaroni, L., Rosete, A., Suárez, A., de la Iglesia, M.: A concept sense disambiguation algorithm for concept maps. In: Proceedings of the 3rd International Conference on Concept Mapping (CMC 2008), pp. 14–21 (2008)
20. Simón, A., Ceccaroni, L., Rosete, A., Suárez, A., Victoria, R.: A support to formalize a conceptualization from a concept maps repository. In: Proceedings of the 3rd International Conference on Concept Mapping (CMC 2008), pp. 68–75 (2008)
21. Valerio, A., Leake, D., Cañas, A.J.: Using automatically generated concept maps for document understanding: a human subjects experiment. In: Proceedings of 5th International Conference on Concept Mapping (CMC 2012), pp. 438–445 (2012)
22. Yoo, J.S., Cho, M.H.: Mining concept maps to understand university students' learning. In: Proceedings of the 5th International Conference on Educational Data Mining (EDM), Greece, pp. 184–187 (2012)

An Ontology-Based Data Management Model Applied to a Real Information System

Raúl Comas Rodríguez[1], Alfredo Simón-Cuevas[2](✉), Nayi Sánchez Fleitas[3], and María Matilde García Lorenzo[4]

[1] Research Directorate, Universidad Regional Autónoma de los Andes, Vía a Baños km 5 ½, Ambato 180150, Tungurahua, Ecuador
raulcomasrodriguez@gmail.com

[2] Universidad Tecnológica de La Habana José Antonio Echeverría, Ave. 114, No. 11901, 19390 La Habana, Cuba
asimon@ceis.cujae.edu.cu

[3] Empresa de Tecnología de la Información y la Automática, Ave. de los Mártires esq. Circunvalación, 60100 Sancti Spíritus, Cuba
nayi78@gmail.com

[4] Universidad Central "Marta Abreu" de Las Villas, Carretera de Camajuaní km 5,5/Santa Clara, 54830 Villa Clara, Cuba
mmgarcia@uclv.edu.cu

Abstract. In scenarios where capturing and efficiently managing data from heterogeneous data sources can be an advantage, the development and use of ontologies is increasingly common. In this paper, an ontology-based data management model applied to the Geographic Information System denominated SIGOBE as part of the Business Management System of the Cuban Electric Union (SIGE), is presented. The proposed data management model combines the use of a developed domain ontology (OntoSIGOBE), with an intelligent query answering process based on the case-based reasoning technic. OntoSIGOBE represents the conceptualization associated to the distribution and transmission processes of electrical energy, and the captured knowledge from the heterogeneous data sources of SIGE. OntoSIGOBE allows achieve the semantic interoperability between the data sources and the query answering process. The obtained results evidence that the proposed model provides a flexible and integrated data access in SIGOBE reaching high satisfaction to the end users.

Keywords: Ontology-based data management · Ontology engineering · Geographic information systems · Case-based reasoning

1 Introduction

The Electrical Union (UNE, Spanish acronym) in Cuba develops the Business Management System of the Electric Union (SIGE) for the automation of electrical processes [3]. SIGE integrates of two main subsystems: Integral System of Network Management (SIGERE) and Integral Management System of the Electric Industry Construction Enterprise (SIGECIE). The functions of SIGERE and SIGECIE are to collect technical,

J. Nummenmaa et al. (Eds.): CSEI 2019, AISC 1078, pp. 36–50, 2020.
https://doi.org/10.1007/978-3-030-33614-1_3

economic and management data to convert them into key information for supporting the decision making.

The collected data facilitate and improve the efficiency in the analysis, planning, operation, and control of the distribution and transmission electricity networks. SIGERE and SIGECIE are considered complex systems because they have 36 modules and a database compose of: 716 tables, 1303 stored procedures and 74 functions. Both systems establish the databases of a Geographic Information System (GIS) of the SIGE, which is denominated SIGOBE. An average query in the SIGOBE involves approximately nine tables with different attributes and the queries on a specific topic requires knowledge of the database organization. Despite the number of stored queries in SIGOBE, they still do not cover the needs of the customer due to the operational dynamics of the national electro-energy system and several weaknesses identified, such as: absence of semantic correspondence between the databases of SIGE and the digital cartography of the electric system; low integration and heterogeneity of the key concepts for the electric system in the database, as well as between the geographical objects in cartography; and the develop of a static query for each problem which is very inefficient.

In order to address these challenges, the Ontology-Based Data Management (OBDM) paradigms [1, 7] were considered to define a new data management model approach for SIGOBE. OBDM has recently emerged as a general paradigm based on the assumption that a domain ontology capturing complex knowledge can be used for data management by linking it to data sources using declarative mappings [1, 12]. In these systems, the ontology is a key resource and constitutes a conceptual, formal description of the domain of interest to a given organization (for instance, the transmission and distribution processes for the Cuban Electric Union), expressed in terms of relevant concepts, attributes of concepts, relationships between concepts, and logical assertions characterizing the domain knowledge [2].

In this paper, an ontology-based data management model applied to SIGOBE is presented. This model combines the use of a manually developed domain ontology (denominated OntoSIGOBE) with an Intelligent Query Answering Process. OntoSIGOBE formalize the semantic meaning of the conceptualization associated to the processes of distribution and transmission of electrical energy, as well as represents the implicit captured knowledge from SIGERE and SIGECIE Databases, and SIGE Cartography, in order to achieve semantic interoperability between these data sources in the query answering process. The Intelligent Query Answering Process exploits conceptual schema in OntoSIGOBE as an intermediate layer for accessing and querying the data source, through the query preprocessing applying several basic natural language processing (NLP) tasks and a case-based reasoning process, for increasing the efficiency and flexibility of the querying process by end users. The proposed model and the constructed ontology were evaluated from different perspective.

OntoSIGOBE was evaluated considering a task-based approaches to measure how this ontology helps improving the querying process associated to customers' information needs, and using the OOPS! (OntOlogy Pitfall Scanner!) system[1] to diagnose

[1] http://oops.linkeddata.es/.

aspects such as: structural, functional, usability-Profiling, consistency, completeness, and conciseness [13]. Additionally, a software-quality evaluation process was carried out to evaluate the quality of the implemented system for applying the proposed model. The obtained results in the experimental and quality evaluations processes carried out evidence that the proposed model provides a flexible and integrated data access to the end users of SIGOBE, improving the performance of this system in the querying process and reaching high satisfaction of the end users.

The rest of the paper is organized as follows: Sect. 2 summarizes the theoretical and scientific foundation associated to the fundamental interest topics for the research carried out, Sect. 3 describes the proposed data management model for SIGOBE; Sect. 4 presents the experimental results and the corresponding analysis; and conclusions arrived and future works are given in Sect. 5.

2 Background

In the geographic data domain, ontologies are used to formalizing the semantic meaning of geospatial concepts and data, categories and spatial relations in a machine-understandable manner. Geospatial semantics can also facilitate the design of GIS by enhancing the interoperability of distributed systems and developing more intelligent interfaces for the user interactions [6]. Ontology-Based Data Management (OBDM) has recently emerged as a general paradigm.

The OBDM is based on the assumption that a domain ontology capturing complex knowledge can be used for data management by linking it to data sources using declarative mappings [1, 12]. The key idea of OBDM is to resort to a three-level architecture: the ontology, the sources, and the mapping between them, where the ontology is a formal description of the domain of interest, and is the heart of the whole system [7]. The ontology is a conceptual, formal description of the domain of interest to a given organization, expressed in terms of relevant concepts, attributes of concepts, relationships between concepts, and logical assertions characterizing the domain knowledge [2].

In OBDM, Ontology-Based Data Access and Integration (we refer to OBDI) is one of the distinctive dimensions, which deals with capturing implicit knowledge from heterogeneous data sources in order to achieve semantic interoperability between them at access and integration levels. Ontologies are the key resource in these systems, since they allow to management the implicit semantic meaning of the data source, at the same time that facilitate the retrieval, integration and maintenance of the information [17]. In recent years, research on OBDI applications from heterogeneous data sources in different real-world scenarios has been intensified. One of the most prominent scenarios concerns the geospatial data domain, in which ontologies deals with the totality of geospatial concepts, categories, relations and processes and with their interrelations at different resolutions [10]. According to the reported in [17], the semantic data integration can be defined in three variants, which are based on what kind of ontologies are used and how these ontologies relate to each other: (i) the single-ontology, (ii) the multiple-ontology, and (iii) the hybrid OBDI. Our proposed data management model would classify to the first variant because a single domain ontology is developed and used to integrate all data sources.

The ontology offers the conceptualization and the required semantic description to reach a high level of data integration. However, we still have a problem associated to the static behavior of the actual querying process in the SIGOBE system. If a static query is developed for each problem that arises, the database begins to store a group of scarcely-used queries. In order solve the problem, the system must be able to generate intelligent queries in real time, in which the knowledge obtained from previous ones is used and the Case-Based Reasoning (CBR) constitute a promising alternative for this propose. CBR is based on the premise that similar previous problems will have similar solutions. CBR have three main components: a user interface, a knowledge base (cases database or cases repository) and an inference engine and contributes to progressive learning, so that the domain does not need to be fully represented. The CBR allow quickly solutions to the problems, because: the answers are not derived from scratch, but from previously solved cases; propose solutions in domains not fully understood; they offer a means of evaluating solutions when an algorithmic method is not available; focus on the most important characteristics or parts of the problem; and lighten the knowledge engineering required to build the knowledge base, because it works directly with cases or examples of the problem to be solved, without mediating a particular form of knowledge representation. An analysis of the queries carried out, including those for SIGERE, allows establishing the structure of a case to solve the problem, which is divided into predictive traits and objective traits; as it will be shown ahead.

3 Data Management Model Proposed for SIGOBE

In the proposed model, the information needs by end users are formulated in terms of natural language sentences and using a technical vocabulary, instead of the data sources. The sentences are automatically translated into operations (queries) over the data sources, which results are graphically visualized in the geographic interface of SIGOBE. A domain ontology, named OntoSIGOBE, was manually developed to represents the conceptualization associated to the distribution and transmission processes of electrical energy, as well as the captured knowledge from the heterogeneous data sources of SIGE, in order to achieve semantic interoperability between them in the query answering process. An Intelligent Query Answering Process exploits the conceptual schema (OntoSIGOBE) as an intermediate layer for accessing and querying the data source that given information to SIGOBE. This process is carried out combining the query preprocessing with the case-based reasoning, for increasing the efficiency and flexibility of the querying process by end users. An overview of the proposed model is shown in Fig. 1.

3.1 OntoSIGOBE Construction

OntoSIGOBE constitute a domain ontology, and formalize the semantic meaning of the conceptualization associated to the processes of distribution and transmission of electrical energy in the SIGE context. The conceptualization represents the implicit captured knowledge from the SIGERE and SIGECIE Databases, and the SIGE Cartography, in order to achieve semantic interoperability between these data sources in the query answering process in SIGOBE. This conceptualization was enriched and refined considering the captured knowledge from several domain experts in the energy-electro sector which were interviewees. The construction process of this ontology was carried out using Methontology [4] and Protégé, and OWL was used for coding the formalized knowledge.

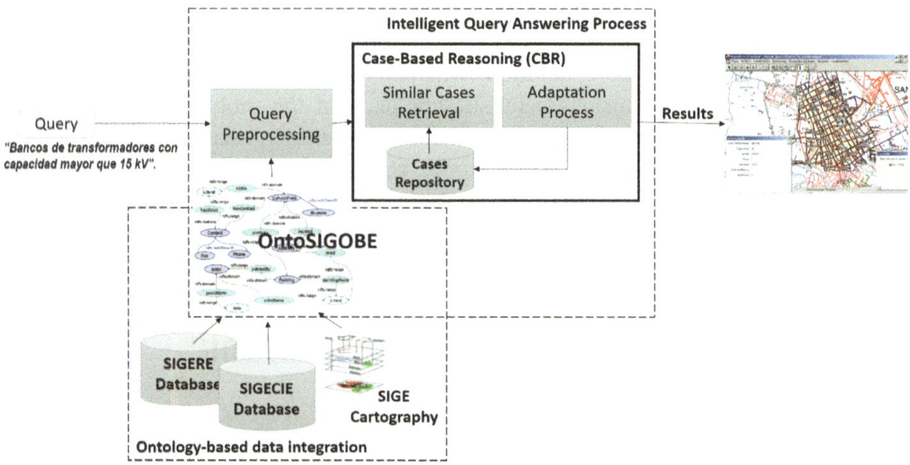

Fig. 1. Overview of the data management model proposed

In OntoSIGOBE, different concepts types are represented as *owl:Class* and integrated in the taxonomy: geographic and electrical domain concepts. Specifically, three concepts groups were represented and integrated: electrical concepts defined as nomenclatures in the databases, such as: administrative and political structures (according to the administrative organization of the Cuban Electrical Union), capacities, voltages, manufacturer, among others; electrical elements geographically represented in the cartography, which have spatial and alphanumeric information, such as: circuits, transformer installations, lamps, disconnectors, among others; and geographic concepts, such as: geographical areas, regions, natural geographical accidents or built by the man with impact in the electric system operation, among others. The basic elements of the electric network are represented in correspondence with their geographical characteristics (e.g. position in the geographic space). From the perspective of the knowledge representation architecture, OntoSIGOBE has been modelling according to the schema shown in Fig. 2.

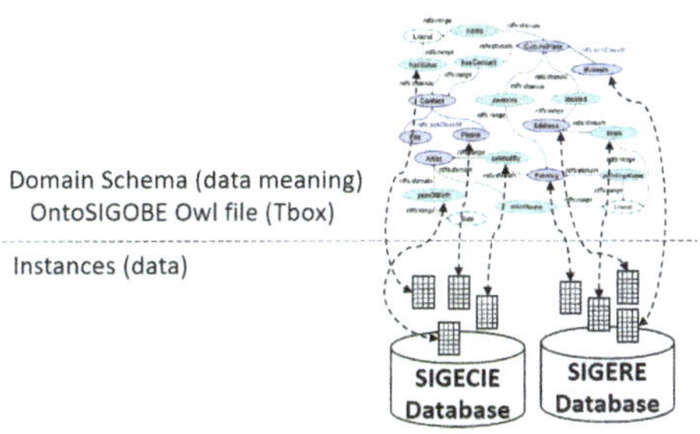

Fig. 2. Overview of the knowledge representation architecture

The TBox defines concepts and relationships between them (taxonomy and semantic relationship) and is stored in the OWL file, whereas the instances (data) are not included in the OWL file, these are stored in the data sources. The instances retrieval process in the proposed model is carried out through SQL queries, which are defined in the specification of each case stored in the knowledge base of the CBR Module (descriptions are given in above section). Some classes of the taxonomy of OntoSIGOBE are shown in Fig. 3.

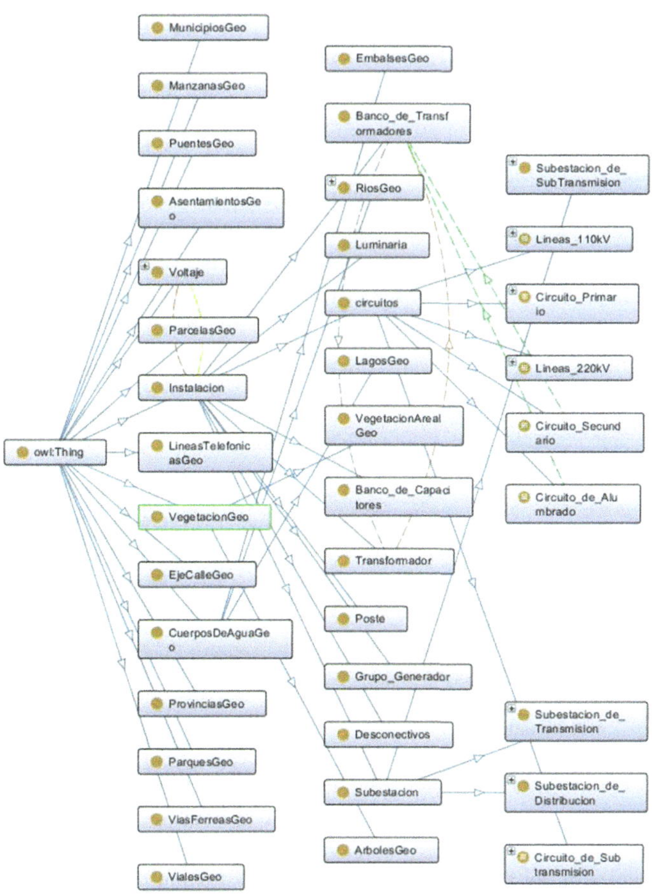

Fig. 3. Portion of the taxonomy of the OntoSIGOBE

In this taxonomy some examples of electrical elements are represented as *owl: Class*, such as: "Instalación" (*Instalation*, in English) and represented as *rdfs:-subClassOf* of it are "Circuitos" (*Circuits*, in English) and "Banco_de_Capacitores" (*Capacitors Bank*, in English), and from "Circuitos" others *rdfs:subClassOf* are derived: "Circuito_Primario" (*Primary Circuit*, in English) and "Líneas_220 kV" (*Line 220* kV, in English). Several types of non-taxonomy or semantic relationship between concepts are represented in the conceptualization of OntoSIGOBE, which are formally defined as Object Properties and coding using *owl:ObjectProperty* specifications. These relationships have been defined from electric and geographic perspective. One example of electric domain relationship is "alimenta" (*feeds*, in English), useful to modelling the case of exists some sub-stations and transformers bank that feeds different circuits, and an example of geographic domain relationship is "intersecta" (*intersect*, in English), useful to modelling the case of the "RiosGeo", "VialesGeo" or "ViasFerreasGeo" (captured concepts from the SIGE Cartography) that intersect circuits. The Object Properties in OntoSIGOBE are shown in Table 1. Several property restrictions, such as: *hasValue* and *someValuesFrom*, have been specified in the formalization of the defined Object Property (some of them are shown in Fig. 4),

Table 1. Examples of semantic relationship

Properties	Inverse (*owl:inverseOf*)	Examples of real situations
alimenta	alimentada_por	A circuit feeds to a transformers bank
contiene	esta_contenido_en	A Substation contains a control hut
dista	dista_de	A transformers bank is at a distance of a road
intersecta	intersectada_por	The line phone intersect the electric line
pertenece_a	tiene	A transformer belongs to a specific transformers bank

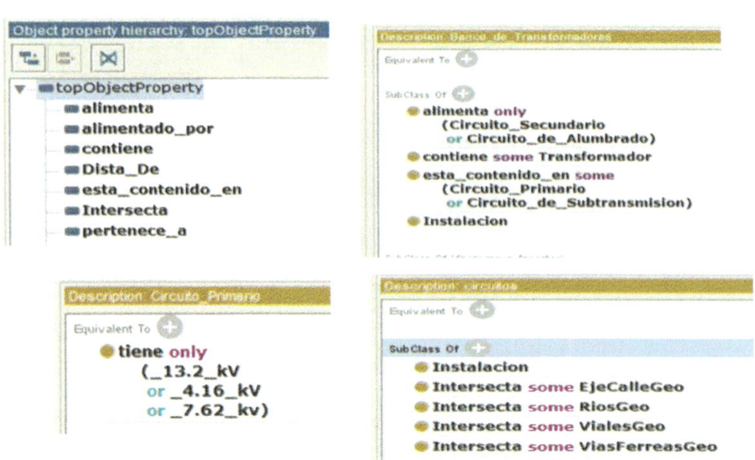

Fig. 4. Examples of object properties and property restrictions included in OntoSIGOBE

for increasing the semantic expressiveness of the OntoSIGOBE and the capabilities to answer more complex queries. Finally, in the Table 2 some composition elements of the OntoSIGOBE are shown.

Table 2. Composition of the OntoSIGOBE

Classes	SubClassOf relationship	Equivalent classes	Object properties	Axiom	Semantic relationship
53	63	6	10	155	929

3.2 Query Preprocessing

The users have the possibility of querying the SIGOBE system by free sentences in natural language using a technical vocabulary; for example: *"Bancos de transformadores con una capacidad mayor que 15 kV"*. In order to increase the efficiency in this querying process, the new proposed data management model for SIGOBE includes an intelligent query answering process, which is based on the use of case-based reasoning technic. In this sense, the query processing task is carried out for extracting the relevant information and predictive features from queries for understanding the information necessities of users based on the application of the case-based reasoning technic.

In this process, basic NLP tasks, like lexical-syntactic and semantic analysis, are applied to the query. The tokens without meaning are removed from the query sentence, e.g.: stop words, articles, and punctuation signs. Next, the relevant concepts and the relationship between them are identified through the syntactic and semantic analysis, which is supported by the OntoSIGOBE developed. These analyses include the identification of entities, the type of relationship between entities, and the search or filter requirements. This identification process is carried out by using the represented knowledge in the OntoSIGOBE as reference vocabulary to interpret the information available in the data sources. As a result of this process, the user's query is transformed in a set of predictive features that considers the information elements captured from queries as values of those features. In Table 3, the specifications and meaning of the predictive and objective features are described.

The ON and OG ontological features are represented using a description logics approach. A possible value of the ON would be: $T \sqcap TPot \sqcap TMonophasic \sqcap \neg SSecondary$. This range expresses that the element is a monophasic primary transformer without secondary output. OG works similarly but considering spatial relationship. An example that refers to the location of an element would be: $P \sqcap Prov \sqcap Muncp$, which expresses that an element belongs to the country (*P*), to a province (*Prov*) and to a municipality (*Muncp*). The structure of the cases in this proposal is represented in Fig. 5, and it is based on the cases repository structure reported in [15].

Table 3. Universe of discourse of predictive and objective features

Features	Possible values	Type
Predictive features		
NV	Secondary, Primary, Subtransmission, Transmission	Symbolic and single-valued
EB	Posts, Transformer banks, Capacitor banks, Generator groups, Disconnectors, Structures, Lamps, Transmission Circuit, Subtransmission Circuit, Primary Circuit, Secondary Circuit, Lighting Circuit, Distribution Substation, Transmission Substation	Symbolic and single-valued
AT	Attributes to be returned by the inquiry (code, voltage, name, etc.)	Set
Tables	Tables of the SIGERE involved in the inquiry (Accessories, Actions, Connection, Interruptions, Line, CurrentSupplyPrimary)	Set
CA	Element to compare (Attribute being compared)	Symbolic and single-valued
OP	Operator ($\cup$, $\cap$, $\leq$ $\geq$ =, like , ...)	Symbolic and single-valued
ON	Ontology (description logics)	OntoSIGOBE
OG	Spatial constraint (description logics)	OntoSIGOBE
Objective features		
From	Returns the From of the inquiry to the SIGERE	String
Where	Returns the Where of the inquiry to the SIGERE	String
CE	Returns the GIS inquiry	String

Specifications: NV: voltage level; EB: Base element; AT: Attributes in analysis; T1... T10: Tables involved; CA: Analysis Conditions; OP: Operator analysis; ON: Ontology Base; OG: Ontology Space.

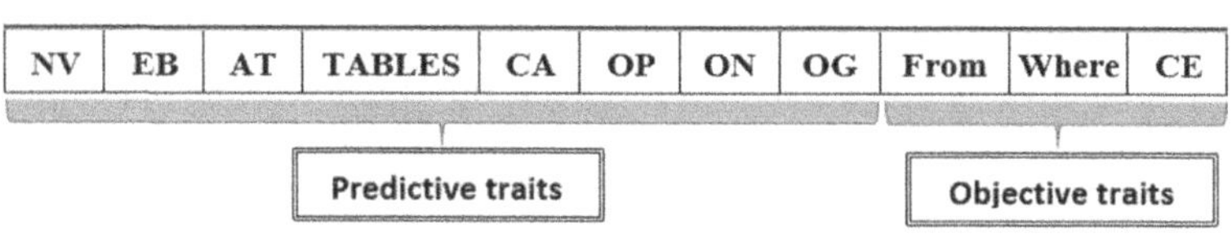

Fig. 5. Structure of the cases.

3.3 Case-Based Reasoning

Taking as a premise that similar problems will have similar solutions, Case Based Reasoning (CBR) is used as a tool in order to define an intelligent querying mechanism. The CBR is a method of artificial intelligence to solve unstructured problems, where the reasoning is made from an associative memory, which uses an algorithm to determine a measure of similarity between two objects. In cases-based systems, the domain does not have to be completely represented [8] and learning is progressive [16].

The CBR is a cycle so-called 4R that has the following stages [11]: retrieves, reuse, revise and retain.

Global dissimilarity is determined by Eq. 1. The distances are weighted considering the expert criterion, with a weight w_i, the greater w_i, the greater the importance of the trait. The most important traits are the ontological ones.

$$\mathrm{DisSimGlobal}(X, Y) = \sum_{i=0}^{m} w_i * d_i\left(x_j, y_j\right)/n \quad (1)$$

where $\sum w_i = 1$.

Local distance $d_i = (x_i, y_i)$ is determined by the type of data x_i, y_i. In the case presented here there are three types of data for which different distance measures are used.

The defined recover process in Algorithm 1 obtains the k cases closest to the inquiry requested by the user, using Eq. 1 to calculate the distance. The result of the evaluation of the selected k system is 3 by default. The set of cases obtained by Algorithm 1 (Table 4) constitutes the input of the adapter module. Based on the transformational analogy, to propose an initial solution is developed [15]. The new case is evaluated and adapted to the conditions on the recovered cases. The pre-set consultations are not necessarily identical to those stored in previous cases. To develop an initial solution, all recovered cases are considered and a combination of the recovered solutions is taken as a starting point.

The input of the adaptation module is an initial solution of the three objective traits. This module allows to reuse and adapt based on transformational analogy, which implies structural changes in the solution. Transformational adaptation is guided by common sense where the rules were defined and used in the adaptation process. This process is considered a T-space, where the known solution (KS) is going to be transformed with the use of T-operators, until it becomes the solution of a new problem.

The retention is induced from the cases, so it will be necessary to redefine it periodically. The efficiency of the system is affected when the number of cases grows excessively, therefore, it is important to avoid including cases that do not contribute new information to the system. To carry out the retention of cases, the following steps are followed:

- The degree of information provided by the case to the system is calculated. This degree of information is estimated by the number of T-operators applied between the set of T-operators in the T-space.
- It is considered feasible to retain the case whose degree of information provided is greater than α (α represents an information threshold).
- If the case is feasible, it is retained in the corresponding sub-base according to the value of the *NV*, *EB* and *OP* predictive traits, given the calculation of the degree of information provided by an objective trait.

The degree of information provided to the system by the value of an objective trait is calculated as the minimum number of T-operators applied.

Table 4. Algorithm 1: Recover algorithm of *k* most similar cases

Input: P; problem to solve and $BC' = \{bc_1, bc_2, \ldots, bc_n\}$; sub-set of cases obtained from the sub-base that corresponds to the P case
Output: Sets of K more similar cases
$\forall\, bc_j \in BC'$ is calculated the $DisSimGlobal(bc_i, P)$. **Where:**

$$DisSimGlobal(bc_j, P) = \sum_{i=0}^{8} w_i * d_i(bc_{ji}, P_i)/n$$

$$d_i(bc_{ji}, P_i) = \begin{cases} d_i(bc_{ji}, P_i) = \begin{cases} 0 Si\ bc_{ji} == P_i \\ 1 Si bc_{ji} \neq P_i \end{cases} i = 1, 2, 5, 6 \\ d_i(bc_{ji}, P_i) = \frac{|bc_{ji} \cap P_i|}{|bc_{ji} \cup P_i|} i = 3, 4 \\ d_i(bc_{ji}, P_i) = \frac{1}{3}\left(\frac{c}{|bc_{ji}|} + \frac{c}{|P_i|} + \frac{c - t/2}{c}\right) i = 7, 8 \end{cases}$$

The most similar K cases are returned with a K = 3

4 Experimental Results and Quality Evaluation

SIGOBE is considered a support decision system with national scope, therefore applicable to different areas inside the Cuban National Electric Union. In order to apply the proposed data management model in SIGOBE, the module SICUNE (Intelligent System of Consultation for the UNE) was implemented. SICUNE implement the proposed model according to a structure of six packages: Set, String and Position, for grouping the functionalities and procedures associated to the sets, strings, and position similarity processing, respectively; Structure and Useful, for grouping the functionalities and procedures associated to the access and management of the cases base, respectively; and Visual CBR for establishing the link between the interface and the application. The system was designed with the possibility of adding new measures of similarity by attributes in the CBR process, although the Jaro-Winkler distance [18] was used in the experiment carried out (97% accuracy was obtained with this function).

Application or task-based evaluation constitutes one of those classification referred to mostly by recent research in ontology evaluations [5, 14]. Task-based approaches try to measure how far an ontology helps improving the results of a certain task [14]. In the proposed model, OntoSIGOBE was designed for improving the performance of the SIGOBE system in the querying process associated to customers' information needs about the distribution and transmission processes in the electrical sector, through a CBR approach and the SICUNE module.

Inspired in the task-based approach, we evaluate the proposed ontology-based data management model using SICUNE in the context of the daily operational work of three departments of the Electric Union: Office, Engineering Department, and Customer Service Area, and measuring the precision of the proposed model in the question answering process. These departments use information from different areas of the databases and achieve greater coverage in the information contained. A period of one month was considered as time interval for the evaluation tests; an average of 175 queries were carried out on SIGOBE from those departments. Additionally, a knowledge base composed of 265 cases was developed and used in this evaluation task.

Table 5 shows the obtained results. The engineering area was the one with the lowest representation in the knowledge base, because version one of SIGOBE focused on the Office, attention to complaints from the population and the investment area. However, 94.18% of precision was obtained.

Table 5. Results of the proposed ontology-based data management model using SICUNE

Areas	Queries	Correct predictions	Incorrect predictions	Learned	Precision (%)
Office	230	221	9	23	96.08
Engineering department	189	178	11	40	94.18
Customer service area	147	140	7	18	95.23
Total	566	539	27	81	
Precision Ave.					95.16%

OntoSIGOBE was also evaluated using the OOPS! (OntOlogy Pitfall Scanner!) system. OOPS! is actually the most complete available system for the (semi)automatically diagnose of ontologies; this system implements the quality model for ontology diagnose proposed in [13]. The quality model aligns the pitfall catalogue to the existing quality models for semantic technologies, which describe 41 evaluation pitfalls classified in the following dimensions: structural, functional, usability-profiling, consistency, completeness, and conciseness. The pitfalls report generated by the OOPS! system is shown in Table 6.

Table 6. Evaluation report of OntoSIGOBE from OOPS! system

Pitfalls	Definition	Category of evaluation	Important level	Argumentation
P08	Missing annotations	Usability-profiling	Minor	–
P10	Missing disjointness	Completeness	Important	OntoSIGOBE not include disjoint axioms because no classes fulfilled this characteristic
P22	Using different naming conventions in the ontology	Usability-profiling	Minor	–

Considering that SIGOBE constitutes a real system whose results have a direct impact in the decisions making of the Cuban Electric Union, a software-quality evaluation process was carried out. Specifically, the model reported in [9], which is based on ISO-9126:2002 was used, and the results are shown in Table 7.

Table 7. Results of the software-quality evaluation

External and internal evaluation		Usability evaluation	
Attributes	Values	Attributes	Values
Functionality	10,0	Efficacy	9,0
Reliability	10,0	Productivity	10,0
Usability	10,0	Satisfaction	10,0
Efficiency	10,0	Security	9,5
Maintenance capacity	10,0		
Portability	8,0		
Subtotal	**58,0**	**Subtotal**	**38,5**
Total			**96,5**

The results were computed from the captured valuation of five main specialists of technical departments of the Cuban Electric Union. The specialists gave a weight to each attribute in a value range of 1 to 10, being 10 the maximum score. ISO-9126: 2002 is an official, approved and validated standard that aims to establish an international standard for the evaluation of the quality of computer systems using metric indicators. The obtained results are satisfactory because the system complies with 96.5% of the indicators.

5 Conclusions and Future Works

This paper presented an ontology-based data management model applied to the Geographic Information System denominated SIGOBE, which combines the use of a developed domain ontology with the application of the case-based reasoning technic for defining an intelligent query answering process and improving the performance of SIGOBE system. The application of the proposed model to SIGOBE allowed to increase the spectrum of successfully answered requests of the specialists and to facilitate several functions and operations such as: (1) to locate objections from the population associated to failed installation or abnormal parameters, (2) to organize the routes of the technical vehicles (displaying the customer voltages on the map, (3) to develop studies of equipment faults in rural areas, and (4) to optimize the electrical networks and their use.

Through the OntoSIGOBE and NLP tasks, applied to the query processing, comprehensive access to the database and dynamic queries are achieved, which are fundamental in high demand stages of services due to the speed of the link between alphanumeric and geographic information, increasing the ease and convenience with which requests are made. The obtained results in the evaluation process are promising because high accuracy rates and high satisfaction of the software quality indicators were achieved. On the other hand, satisfactory results were obtained in the automatic evaluation of OntoSIGOBE using OOPS! system.

As future work, several techniques that allow the ontology to be further exploited will be evaluated, as well as the increase of the represented knowledge, with the objective that the ontology itself becomes a query interface to satisfy more complex user requests. Furthermore, others alternative for obtaining the weights of the features will be analyzed in order to improves the global dissimilarity function and the results of the similar-cases retrieval process.

References

1. Abiteboul, S., Arenas, M., Barceló, P., Bienvenu, M., Calvanese, D., David, C., Hull, R., Hüllermeier, E., Kimelfeld, B., Libkin, L., Martens, W., Milo, T., Murlak, F., Neven, F., Ortiz, M., Schwentick, T., Stoyanovich, J., Su, J., Suciu, D., Vianu, V., Yi, K.: Research Directions for Principles of Data Management (Dagstuhl Perspectives Workshop 16151). arXiv:1701.09007v1 [cs.DB], Dagstuhl Manifestos, pp. 1–28 (2017)
2. Calvanese, D., De Giacomo, G., Lembo, D., Lenzerini, M., Rosati, R.: Ontology-based data access and integration. In: Liu, L., Tamer Özsu, M. (eds.) Encyclopedia of Database Systems, pp. 1–7. Springer, New York (2017)
3. Fernández, R.: Informatización de la Gestión de las Redes Eléctricas. Ph.D. thesis, Facultad de Ingeniería Eléctrica, Universidad Central "Marta Abreu" de Las Villas, Santa Clara, Cuba (2011)
4. Fernandez, M., Gómez, A.: Overview and analysis of methodologies for building ontologies. Knowl. Eng. Rev. **17**(2), 129–156 (2002)
5. Hlomani, H., Stacey, D.: Approaches, methods, metrics, measures, and subjectivity in ontology evaluation: a survey. Semant. Web J. **1**(5), 1–11 (2014)
6. Hu, Y.: Geospatial semantics. In: Huang, B., Cova, T.J., Tsou, M.-H., et al. (eds.) Comprehensive Geographic Information Systems. Elsevier, UK (2017)
7. Lenzerini, L.: Ontology-based data management. In: Proceedings of the 6th Alberto Mendelzon International Workshop on Foundations of Data Management (AMW 2012), CEUR Workshop Proceedings, vol. 866, pp. 12–15 (2012)
8. Liu, X., Wang, X., Wright, G., Cheng, J.C., Li, X., Liu, R.: A state-of-the-art review on the integration of building information modeling (BIM) and geographic information system (GIS). ISPRS Int. J. Geo-Inf. **6**(2), 53–74 (2017)
9. Macías Rivero, Y., Sánchez, G., Victoria, M., Martínez Suárez, Y.: Modelo de evaluación para software que emplean indicadores métricos en la vigilancia científico-tecnológica. ACIMED **20**(6), 125–140 (2009)
10. Mostafavi, M., Edwards, G., Jeansoulin, R.: An ontology-based method for quality assessment of spatial data bases. In: Proceedings of the 3rd International Symposium on Spatial Data Quality (Geoinfo series), vol. 1/28a, pp. 49–66 (2004)
11. Ortiz, Y.K., Bañuelos, P., Rodas, J.: Razonamiento basado en casos. CULCyT **14**(63), 48–56 (2017)
12. Poggi, A., Lembo, D., Calvanese, D., De Giacomo, G., Lenzerini, M., Rosati, R.: Linking data to ontologies. J. Data Semant. 133–173 (2008)
13. Villalón, M.P.: Ontology evaluation: a pitfall-based approach to ontology diagnosis. Ph.D. thesis, Departamento de Inteligencia Artificial, Escuela Técnica Superior de Ingenieros Informáticos, Universidad Politécnica de Madrid, España (2016)
14. Raad, J., Cruz, Ch.: A survey on ontology evaluation methods. In: Proceedings of the International Conference on Knowledge Engineering and Ontology Development (IC3K 2015). ACM DL, pp. 179–186 (2015)

15. Sanchez, N., Comas, R., García, M., Carrera, F.: Geographical information system based on artificial intelligence techniques. In: Proceedings of International Conference on Technology Trends (CITT 2018), CCIS, vol. 895, pp. 446–461. Springer, Cham (2018)
16. Tah, J., Oti, A., Abanda, F.: A state-of-the-art review of built environment information modelling (BeIM). Organ. Technol. Manag. Constr. **9**(1), 1638–1654 (2017)
17. Wache, H., Voegele, T., Visser, T., Stuckenschmidt, H., Schuster, H., Neumann, G., Huebner, S.: Ontology-based integration of information - a survey of existing approaches. In: IJCAI 2001 Workshop: Ontologies and Information, pp. 108–117 (2001)
18. Winkler, W.E.: Using the EM algorithm for weight computation in the Fellegi-Sunter model of record linkage. In: Proceedings of Section on Survey Research Methods. American Statistical Association (2000)

Image Analysis Based on Heterogeneous Architectures for Precision Agriculture: A Systematic Literature Review

Marco R. Pusdá-Chulde[1,2(✉)], Fausto A. Salazar-Fierro[1], Lucía Sandoval-Pillajo[3], Erick P. Herrera-Granda[1], Iván D. García-Santillán[1], and Armando De Giusti[2]

[1] Faculty of Engineering in Applied Sciences, Universidad Técnica del Norte, Ibarra, Ecuador
{mrpusda, fasalazar, epherrera, idgarcia}@utn.edu.ec
[2] Universidad Nacional de La Plata, La Plata, Argentina
marco.pusdac@info.unlp.edu.ar,
degiusti@lidi.info.unlp.edu.ar
[3] Carrera de Software, Universidad Regional Autónoma de Los Andes, Ibarra, Ecuador
ui.anasandoval@uniandes.edu.ec

Abstract. Precision agriculture (AP) is a management strategy that uses ICT (Information and Communication Technologies) to obtain information from different sources in order to support decision-making, considering environmental and economic aspects to optimize the Farmer's tasks and provide quality products to the costumer. The application of AP in agriculture can reduce time spent in manual activities, avoid the indiscriminate use of chemicals, increase production costs, soil deterioration and environmental pollution. Nowadays, AP is a booming area that, taking advantage of technological advances, in computer vision, heterogeneous architectures (Multicore, GPU, FGPA) and artificial intelligence techniques (Machine learning, Deep learning), has allowed to systematize a variety of agricultural activities, such as disease detection, plant counting, and identification of weed, pests and insects in different crops. This paper presents a systematic review of literature (SRL) of image analysis and processing techniques applied in precision agriculture using heterogeneous technologies. Therefore, 32 scientific articles of the last five years from four relevant bibliographic databases (Scopus, ScienceDirect, IEEE Xplore, SpringerLink) were analyzed and synthesized. The selected publications answer to four research questions proposed in this study. From the obtained results, great opportunities for image analysis (segmentation), machine learning and the use of graphic accelerators (GPU) were identified, which stand out as promising techniques and tools for the development of efficient and precise automatic systems, with the perspective to its application in real time for many agricultural tasks.

Keywords: Image analysis · Heterogeneous architectures · GPU · Multicore · FPGA · Precision agriculture · SLR

J. Nummenmaa et al. (Eds.): CSEI 2019, AISC 1078, pp. 51–70, 2020.
https://doi.org/10.1007/978-3-030-33614-1_4

1 Introduction

Currently, precision agriculture (PA) plays an important role in the agricultural context through the use of technological and electronic resources, which allow improving production by adjusting fertilizers, chemical products and other agricultural inputs, avoiding water and environmental pollution, and soil quality deterioration. The image analysis and computer vision techniques are used in different areas, including many agricultural applications that require an effort in computing, in order to automate different activities [1]. These techniques have increased their application in precision agriculture due to different factors, such as: equipment and accessories cost reduction, increase in computational calculation efficiency, camera quality improvement and a growing interest in different agricultural tasks, such as automatic crop lines detection [2], weed detection [3, 4], pests and diseases detection [5], people, animals and obstacles detection in different crop fields [6]. However, there still are many challenges to overcome related to accuracy, execution time and real time application [7].

The technological change, since multicore processors appeared, has imposed the need to use new paradigms of hardware and software (programming techniques), in which shared memory schemes with messages coexist [8], with the utilization of graphic accelerators (GPU, FPGA, Xeon Phi), which represent an alternative to achieve high performance in certain computer applications [9, 10].

Graphics processing units (GPU) is a type of hardware born to improve the performance of computer games. Currently, they have evolved to be used in other areas such as precision agriculture [11], health, industry, commerce, computer modeling, among others. Due to their computational structure, GPUs can offer superior performance compared to the classic central processing units (CPU) [9]. Additionally, programming languages such as CUDA (Computed Unified Device Architecture) [12], can be adapted to parallel algorithm programming, which can be used in agricultural image analysis applications, achieving shorter execution times. The integration of computer vision and high-performance computing (HPC), represent an encouraging answer to solve specific problems in precision agriculture [13].

In precision agriculture, different jobs have been carried out using agricultural image analysis in many types of crops. Some of these works are researches and literary reviews that deal with agricultural tasks, such as: rice quality identification [14], grain classification [15], plant disease detection [7], citrus disease detection [16], obstacle detection for agricultural videos [6], and weed identification in corn and potato crops [2, 4, 17, 18]. In addition, the advances achieved in techniques and artificial intelligence tools, were applied in precision agriculture for plant identification using machine learning [19], and food production employing Deep learning [20].

The objective of this work is to carry out a systematic literature review (SLR) to obtain updated information of the work done, on image analysis based on heterogeneous architectures applied in precision agriculture, including the detection of weeds in different crops. Existing works mentioned above, as far as is known, use limited heterogeneous architectures to systematize agricultural processes. Therefore, it is necessary to identify new challenges, opportunities and trends in the application of heterogeneous architectures for precision agriculture with the perspective to the development of more robust, efficient and accurate autonomous systems. The research questions that were addressed in this study are the following:

- Which are the agricultural tasks that have been automated using analytical techniques?
- Which are the image analysis techniques that have been used to detect crops and weeds?
- Which heterogeneous architectures have been used in PA?
- Which are the challenges and trends in PA?

The manuscript is organized as follows: In Sect. 2, the methodology applied in this study is presented, including research questions, document search, paper selection and data extraction. Section 3 indicates the results obtained, including the answers found to research questions. In Sect. 4, the discussion of the own results is carried out indicating some limitations in this work. Finally, Sect. 5 presents the main conclusions and future work.

2 Methodology

The methodology selected for the present systematic literature review (SLR) was the proposed by [21–23]. The review protocol is presented in Fig. 1 and consisted of four steps: (i) Research Questions, (ii) document search, (iii) paper selection, and (iv) relevant data extraction. Each of the phases is explained below:

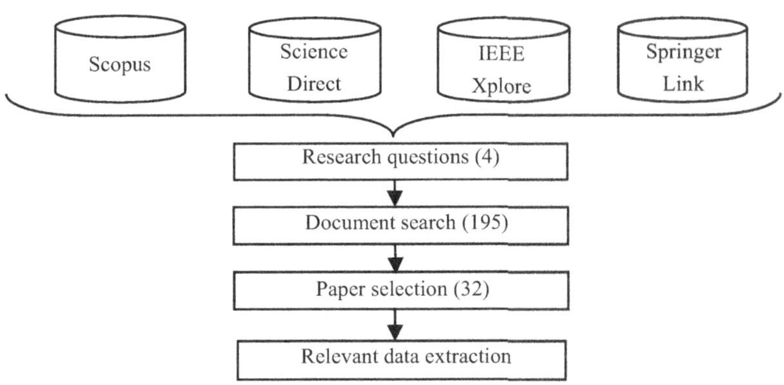

Fig. 1. Protocol diagram used on SLR

2.1 Research Questions

We established four research questions (RQ), Table 1, which are the guidelines in the review process on the subject study. These questions address the analysis of images and heterogeneous architectures in PA, specifically for crop identification and bad herbs in different crop fields. The investigation team consisted of six researchers from Argentina and Ecuador. Four scientific databases were considered: IEEE Xplore

Digital Library [24], ScienceDirect [25], Scopus [26] and SpringerLink [27], which the Técnica del Norte University (Ecuador) has access.

Table 1. Research Questions (RQ)

Number	Research Question	Motivation
RQ1	Which image processing and analysis techniques were used for crop and weed detection?	Identify image analysis techniques that were used
RQ2	Which agricultural tasks have been automated using image processing and analysis techniques?	Identify the agricultural processes that have been addressed through image analysis
RQ3	Which heterogeneous architectures are used in precision agriculture?	Identify the types of heterogeneous architectures that are used in the context of image analysis for PA applications
RQ4	Which are the challenges and trends in precision agriculture?	Know challenges, trends and future work to be applied in precision agriculture

2.2 Document Search

For document search stage, a basic search string was used: ((“automatic detection” OR “Image processing” OR “Image segmentation”) AND (“Graphics Processing Unit” OR “Multiprocessor”) AND (“precision agriculture”)). Additionally, some variants were used in the search chains, in order to obtain at least 10 documents in each bibliographic database. Table 2 presents the search chains in each database and the combinations used, considering that the databases have different search criteria and capabilities. A total of 195 documents were found, from which 89 belongs to Scopus, 18 to ScienceDirect, 62 to IEEE Xplore and 26 to SpringerLink.

2.3 Paper Selection

For paper selection, three phases were considered. In the first phase, inclusion and exclusion criteria were applied. The inclusion criteria considered by the authors were: (i) scientific articles, (ii) reviews, (iii) conferences evaluated by peers, (iv) studies focused on the use of heterogeneous architectures in image processing for agricultural tasks. All the works are related to the disciplines of computer science and engineering (Computer Science and Engineering) published during the last 5 years (2014–2019) in English. The exclusion criteria considered by the authors were: (i) duplicate work, (ii) technical reports, (iii) book chapters, (iv) thesis, (v) studies published in other areas of knowledge.

In the second phase, criteria related to the search strings were applied to give greater relevance to the literary review and to answer to the four proposed research questions. The selected documents were sorted by year of publication and title, summary and keywords were initially reviewed.

Table 2. Search strings used in scientific databases

Criterion	Scopus	ScienceDirect	IEEE Xplore	SpringerLink
Search string	((“Images processing” OR “automatic detection” OR “imagen segmentation “) OR (“Graphics Processing Unit” OR “multiprocessor”) AND (“precision agriculture”))	(Images Processing) and (automatic detection) and (Graphics Processing Unit) and (precision agriculture)	((((Images Processing or automatic detection or imagen segmentation)) AND (Graphics Processing Unit)) AND precision agriculture)	(Images Processing or automatic detection or Image segmentation) and (Graphics Processing Unit) and (precision agriculture)
Total	89	18	62	26

Finally, in a third phase, introduction and conclusions sections, were reviewed to verify if the contained information, contributes and is related to the four proposed RQ. The documents obtained applying the three phases are presented in Table 3.

Table 3. SLR paper selection

Database	Phase I	Phase II	Phase III
IEEE Xplore	62	18	7
ScienceDirect	18	12	8
Scopus	89	21	7
SpringerLink	26	19	10
Total	195	70	32

The detail of the 32 finally selected papers, is presented in Table 4.

2.4 Relevant Data Extraction

The 32 finally selected papers (Table 4) were reviewed again by 3 members of the team, verifying the frequency and selection of the articles. In the case of articles that relate to more than one element of the RQ, they were classified into a single group for better results and discussion analysis and interpretation.

The data extracted from the articles were obtained according to the RQ, first the image analysis and processing techniques were reviewed, then the systematized tasks, the heterogeneous architectures in the image processing, and finally the techniques and tools that are used.

Table 4. SLR finally selected papers

Code and tittle	Database	Year	Country
A1. Special issue on design and architectures of real-time image processing in embedded systems	Springer	2014	France
A2. An automatic object-based method for optimal thresholding in UAV images: Application for vegetation detection in herbaceous crops	ScienceDirect	2014	Spain
A3. CT and MRI image compression using wavelet-based contourlet transform and binary array technique	Springer	2014	India
A4. Real-time multi-resolution edge detection with pattern analysis on graphics processing unit	Springer	2014	China
A5. Potential Applications of Computer Vision in Quality Inspection of Rice: A Review	Springer	2015	Iran
A6. Design and implementation of a computer vision-guided greenhouse crop diagnostics system	Springer	2015	USA
A7. Procesamiento de imágenes para reconocimiento de daños causados por plagas en el cultivo de Begonia semperflorens (flor de azúcar)	IEEE	2015	Colombia
A8. Acceleration techniques and evaluation on multi-core CPU, GPU and FPGA for image processing and super-resolution	Scopus	2016	Greece
A9. Efficient parallelization on GPU of an image smoothing method based on a variational model	Springer	2016	Portugal
A10. A novel global methodology to analyze the embeddability of real-time image processing algorithms	Springer	2016	France
A11. Approximate georeferencing and automatic blurred image detection to reduce the costs of UAV use in environmental and agricultural applications	ScienceDirect	2016	Spain
A12. A Survey on Detection and Classification of Rice Plant Diseases	IEEE	2016	India
A13. Wavelet-based multicomponent denoising on GPU to improve the classification of hyperspectral images	Scopus	2017	Spain
A14. Research on Intelligent Acquisition of Smart Agricultural Big Data	IEEE	2017	China
A15. An Autonomous Multi-Sensor UAV System for Reduced-Input Precision Agriculture Applications	IEEE	2017	Bulgaria
A16. Convolution Neural Network in Precision Agriculture for Plant Image Recognition and Classification	IEEE	2017	UK
A17. Real-time parallel image processing applications on multicore CPUs with OpenMP and GPGPU with CUDA	Springer	2017	Turkey

(continued)

Table 4. (*continued*)

Code and tittle	Database	Year	Country
A18. Design and implementation of WSN for precision agriculture in white cabbage crops	IEEE	2017	Colombia
A19. Computer vision and artificial intelligence in precision agriculture for grain crops: A systematic review	ScienceDirect	2018	Brazil
A20. Digital mapping of soil available phosphorus supported by AI Technology for precision agriculture	IEEE	2018	China
A21. Unsupervised detection of vineyards by 3D point-cloud UAV photogrammetry for precision agriculture	ScienceDirect	2018	Italy
A22. Deep leaning approach with colorimetric spaces and vegetation indices for vine diseases detection in UAV images	Scopus	2018	France
A23. Detection of stored-grain insects using deep learning	ScienceDirect	2018	Canada
A24. Failure Detection in Row Crops from UAV Images Using Morphological Operators	Scopus	2018	Brazil
A25. The recognition of rice images by UAV based on capsule network	Springer	2018	China
A26. An Image Processing Method Based on Features Selection for Crop Plants and Weeds Discrimination Using RGB Images	Springer	2018	France
A27. On-line crop/weed discrimination through the Mahalanobis distance from images in maize fields	ScienceDirect	2018	Ecuador
A28. A review on weed detection using ground-based machine vision and image processing techniques	Scopus	2019	China
A29. CropDeep: The Crop Vision Dataset for Deep-Learning-Based Classification and Detection in Precision Agriculture	ScienceDirect	2019	China
A30. A UAV Guidance System Using Crop Row Detection and Line Follower Algorithms	Scopus	2019	Brazil
A31. Development and evaluation of a low-cost and smart technology for precision weed management utilizing artificial intelligence	Scopus	2019	USA
A32. Automatic detection of cereal rows by means of pattern recognition techniques	ScienceDirect	2019	Finland

3 Results

The results obtained in each of the four research questions (Table 1) proposed for this study are listed below:

3.1 RQ1 Which Techniques of Image Processing and Analysis have been used in Precision Agriculture?

The 32 articles previously selected (Table 4) addressed image processing and analysis techniques (i.e. 100%). Some of them deal with one or more techniques, which classifies it in the one with the greatest contribution to RQ. The techniques are summarized below:

Vegetation Indices-Based Techniques. Vegetation indices allow the extraction of spectral characteristics by combining two or more spectral bands, based on the properties of the reflectance produced by vegetation [16]. In the review, the images obtained by unmanned aerial vehicles (UAV) and the use of global positioning systems (GPS) were considered. The works in this category are explained below:

In [28] a system of autonomous multi-sensor UAV images was developed, which is designed to provide spectral information related to water management for an orchard. Water and vegetation stress indices were derived from multispectral and thermal data collected simultaneously from the developed system, then used as indicators to determine water stress and crop health condition.

In [29], infected areas of grapes using UAV were identified, and with the data obtained they proposed a method based on a convolutional neural network (CNN), to detect symptoms of diseases in crops. The results of the CNN are compared using different color spaces, vegetation indices, as well as the combination of both information.

In [30] an autonomous system of detection and monitoring of plants guided by computerized vision was built to continuously monitor the temporal, morphological and spectral characteristics of a lettuce crop in a hydroponic system of nutrient film technique (NFT).

In [31] a network of capsules (CapsNet) was designed using images of rice crops captured by UAV. The images were previously processed using the histogram equalization method in grayscale images and through the superpixels algorithm, a segmentation of super pixels was obtained. The two results are sent to CapsNet for training and it predicts the output vector according to the routing protocol. The study was used to monitor the growth of rice and prevent diseases and pests.

In [32] a precision agriculture system was implemented that uses wireless sensor networks supported by GPS to determine soil quality in white cabbage crops. In [33] a complete guidance system for use in UAV (hardware and software) was proposed based on image processing techniques for selective spray of agrochemicals.

In [34] an innovative unsupervised algorithm was presented for vineyard detection and row grape characteristics evaluation, based on 3D point cloud maps processing, in order to determine the number of rows and the number of missing plants in the rows.

In [35], the authors conducted a study on a methodology to reduce the cost of generating geomatic products by automatically detecting blurry images, in a set of images captured with a UAV, by establishing an indicator number, that describes the level of blur in the images for a specific camera setting.

Thresholding Based Techniques. This technique provides a threshold automatically, with appropriate behaviors, even under changing or adverse lighting conditions [2].

Some literature reviews were carried out considering processes of preprocessing, segmentation, characteristic extraction and classification, applied to the weed and crop detection [36]; grain production [37]; classification and detection of species [38]; rice quality review [14]; rice plants decease detection [19].

Segmentation is the method used by several authors in different applications [39], where images are used to parallelize heterogeneous technologies in real time. In [40] a method for the automatic extraction of geoparcels of images was proposed to determine soil nutrients. In [41] an algorithm (OBIA) was developed that uses segmentation for weed detection and germination monitoring. In [42] a highly efficient algorithm was implemented for smoothing noise in damaged or altered images of large dimensions. In [43] a method to detect missing plants in coffee crops was implemented using image segmentation. In [44] a model was developed to determine the appropriate treatment plans for different types of crops and different regions by compiling a database of images through remote sensors.

In [45] an image processing algorithm was implemented for the representation of edges based on the classification of the types of edges in four categories: ramp, impulse, step and sigmoid (RISS). In addition, the proposed algorithm performs a connectivity analysis (CA) on the edge map to ensure that small and disconnected edges are removed. Performance analysis in experiments allows the multiple resolution edge detection algorithm, proposed with edge pattern analysis, lead to more effective edge detection and localization with improved accuracies.

Learning-Based Techniques. These techniques are based on both supervised and unsupervised learning processes [2]. In [38] a data set was presented containing 31,147 images with more than 49,000 registered instances in 31 different ways, applying deep learning to classify and detect different species in real time. In [46] a detection and identification method for stored grain insects was developed, by applying a deep neural network. In [47], several heuristic methods of prediction were proposed for image characteristic extraction in different agricultural processes. In [48] a segmentation method was used with many restrictions to overcome light acquisition conditions and combined it with an appropriate morphological filtering, in order to eliminate segmented images to discriminate crops and weeds in RGB images. In [49] an automatic method was developed for the identification of interspersed weeds within the rows of corn crops, by applying a minimum distance criterion based on Mahalanobis distances.

Wavelet-Based Techniques. In agricultural images processing, it is sometimes necessary to highlight the details (edges) and detect the textures, analyzing the image from different angles. Thus, two categories can be established: (1) analysis of the content of the frequency, considered as a signal to separate the low (smooth variations in color) and high frequencies (edges that provide details) [2]. In [50] a high-frequency sub-band method was obtained from the wavelet transform that decomposes into multiple directional sub-bands through a bank of directional filters to diagnose people's diseases.

Table 5 summarizes the image processing and analysis techniques used for precision agriculture activities.

Table 5. Processing and image analysis used techniques

Technique	Selected papers	Number of papers
Segmentation	A1, A2, A4, A5, A7, A9, A15, A17, A20, A22, A24, A26, A28, A32	14
Vegetation indices	A6, A11, A14, A18, A21, A25, A30, A31	8
Learning	A8, A10, A12, A16, A23, A27, A29	7
Wavelet	A3, A13, A19	3
Total		**32**

3.2 RQ2 Which Agricultural Tasks have been Automated Using Image Processing and Analysis Techniques?

In total, 21 articles presented the systematization of agricultural processes through image analysis (65%). Some of the articles deal with one or more processes, which are summarized below:

Weed Detection. In five articles (A2, A26, A27, A28, A31) were presented methods of image processing for the detection of weeds in agricultural crops. Technological advances are most frequently used in the detection of weeds by image processing [36]. In [51] an intelligent sprayer was developed using vision and artificial intelligence to distinguish weeds from vegetable crops and accurately spray weeds. In [41], an algorithm was designed that processes ultra-high-resolution UAV images to discriminate weeds in the germination process and detect the specific places where they are found for further treatment. In [48] an RGB imaging system was presented for crop and weed discrimination.

Crop Monitoring. In total 3 articles (A7, A21, A25) present studies related to crop monitoring through image processing. The monitoring of vineyards was obtained from point clouds 3D maps, generated by multispectral UAV images [34]. Monitoring of the germination of corn, sunflower and wheat crops [41]; rice growth and prevent diseases and pests through multispectral UAV imaging [31]; monitoring of lettuce crops [30], and the possible attack of pests in the begonia crop [52].

Disease Detection. Two articles (A12, A22) detailed a research on disease detection for agricultural crops: identification of infected areas using UAV images to prevent the massive spread of diseases in grape crops [29]; and detection of plants with diseases by means of image processing techniques and automatic learning in rice crops [19].

Plague Detection. Two articles (A19, A23) are related to studies on plague detection. In [46] a method was developed for detecting and identifying insects in stored grains using a neural network with RGB images. In [37] a literature review was conducted on tools and techniques of image processing with artificial vision and intelligence for different approaches related to the detection of diseases, grain quality and phenotyping with the aim of improving the production of five grains: corn, rice, wheat, soybeans and barley.

Plant/Aliment Counting. Three articles (A5, A24, A32) presented studies on plants and food counting, for different types of crops. In [14] the theoretical and technical principles of computer vision was indicated for quality inspection and rice monitoring where the best grain for human consumption is selected. In [43], a method was tested for the detection of coffee plants missing in the sowing, which improves agricultural management from aerial images, obtained by UAV.

Fumigation. An article (A30) presents an investigation about the use of agricultural image processing for crop fumigation. The selective pulverization of agrochemicals is essential to maintain high productivity and quality of agricultural products [33], and the precise selection of weeds, reducing negative impacts related to costs and application of agrochemicals [51].

Soil Characteristics. Three articles (A6, A15, A20) detailed research on obtaining soil characteristics in different types of agricultural crops. The distribution of soil nutrients is a key step for the application of precision agriculture and digital mapping of soil is an effective technology [40]. The management of water and the distribution of soil nutrients in pomegranate crops is very important to obtain optimal agricultural profits through the processing of UAV images [28]. The determination of the stress of the plants and determination of nutrients in the soil by means of spectral images for lettuce crops is important to improve the quality of the products and obtain good prices in the market [30].

Table 6 summarizes the agricultural tasks that have been automated using image processing and analysis techniques.

Table 6. Automated agricultural tasks

Activity	Selected papers	Number of papers
Crop monitoring	A7, A21, A25	3
Disease detection	A12, A22	2
Weed detection	A2, A26, A27, A28, A31	5
Plague detection	A19, A23	2
Plant/food counting	A5, A32, A24	3
Fumigation	A30	1
Soil characteristics	A6, A20, A15	3
Total		**19**

3.3 RQ3 Which Heterogeneous Architectures are used in Precision Agriculture?

Eight articles (25%) that propose the use of heterogeneous architectures applying image processing and artificial intelligence techniques for different application contexts, were analyzed. In [37], the literary review identified great opportunities and the authors recommend the exploitation of GPUs and advanced artificial intelligence techniques, such as DBN (Deep Belief Networks) in the construction of solid methods of computer vision applied to precision agriculture. The multiprocessor architectures found are detailed below:

GPU. A GPU is adequate for addressing problems that can be expressed as parallel calculations of data with a high arithmetic intensity. Experiments with GPUs showed that by increasing the size of the fragment in the images, the execution time decreases approximately four times compared to serial computing [39]. The use of GPU in image processing presents great opportunities to systematize agricultural activities using high resolution images and large sizes [1]. Parallel processing techniques with GPU accelerate the reconstruction of definition content with ultra-high-resolution images, achieving three times faster than the normal process of image processing [53]. Algorithms for image preprocessing with GPUs can improve damaged images, especially in the elimination and smoothing of noise [28]. The processing of images with GPU and the CUDA platform showed that increasing the number of cores reduces response times [54]. The parallel implementation of image processing on a GPU shows scalable acceleration as the resolution of an image increases [45].

Multicore. A multicore is a processor that combines two or more independent microprocessors in a single package, often a single integrated circuit. Applications that use image processing for agricultural tasks over time require better image resolutions (cameras) and the integration of processing algorithms with heterogeneous systems [47]. The inclusion of multiprocessing and multi-core technologies allows the segmentation of many images in parallel, covering surfaces with several objects to be used in different agricultural applications [39]. The use of multicore architectures with the OpenMP platform [11], showed that when the size of the images increases, the execution time decreases approximately four times in comparison with serial computation [39, 54].

FPGA. Field-Programmable Gate Array is a programmable device that contains logic blocks whose interconnection and functionality can be configured at the moment, through a specialized description language. The processing and analysis of images have great application in the agricultural context demanding computing power for the automation of different processes [1]. The use of FPGA architectures in the reconstruction of high-resolution images shows a reduction in execution times four times faster on the low-end Xilinx Virtex 5 devices and 69 times faster on Virtex 2000t [53].

Table 7 summarizes the heterogeneous architectures used by image processing for various activities, including the agricultural field.

Table 7. Heterogeneous architectures used in image processing

Architecture	Selected papers	Number of papers
GPU	A4, A9, A19, A31	4
Multicore	A10, A13, A17	3
FGPA	A1, A8	2
Total		**9**

3.4 RQ4 Which are the Challenges and Trends in Precision Agriculture?

Most of the articles reviewed, a total of 29 (90.6%) presented research on current trends in the applications of processing and analysis of images in the agricultural context, being the following:

Real Time. Real-time image processing applications have been implemented using multiprocessing and multi-core technologies to improve response times [1, 39, 45, 47, 50, 53, 54]. In addition, artificial vision combined with image processing techniques has become a promising tool for accurate weed detection in real time [36].

UAV Images. Nowadays, unmanned devices intervene in agricultural activities, reducing operating costs [35], avoiding the compaction of soil and reducing resource waste in crops [33]. The exact detection of plants from high resolution images [31], multispectral [34, 43, 52], and ultra-high resolution [41] help improve agricultural management by preventing pests and diseases. The adoption of precision agriculture techniques and localized treatments is important to avoid soil contamination [28].

Artificial Intelligence Techniques. Artificial intelligence has been considered the biggest challenge to promote the economic potential and efficiency in precision agriculture. The integration of algorithms and technologies for image processing promises a propitious future to solve agricultural problems with better results, several techniques and technological advances have been studied and implemented in the articles analyzed, such as Machine Learning [19, 48, 51]; Deep Learning [36, 38, 44]; Neural networks [29, 31, 46]; Big Data and Internet of Things (IOT) [38, 55] for detection of plants, weeds, diseases, pests, quality of grains and vegetables.

Table 8 summarizes recent trends for image processing used to solve precision agriculture problems.

Table 8. Recent technology trends of image processing applied on agriculture

Technology trends	Selected papers	Number of papers
Real time	A3, A4, A8, A9, A10, A13, A17, A27	8
UAV images	A2, A7, A11, A15, A21, A24, A30, A32	8
Artificial intelligence techniques	A1, A12, A14, A16, A18, A19, A22, A23, A25, A26, A28, A29, A31	13
Total		**29**

4 Discussion

Most of the articles analyzed in the literary review, refer to image processing for agricultural tasks using algorithms and architectures that improve the levels of productivity in various types of crops. In each study, depending on the problem to be solved, the selection of the most appropriate techniques, algorithms and architectures is required.

The segmentation technique is one of the most used currently, while the wavelet-based technique has been the least used, as shown in Fig. 2, from Table 5.

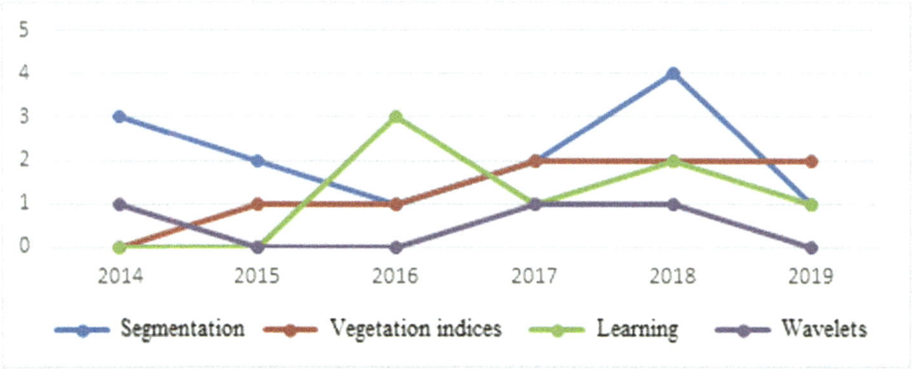

Fig. 2. Used image processing techniques

The integration of computer vision and artificial intelligence has given good results in different agricultural tasks, such as detection of diseases, detection of weeds, detection of pests, among others. The reviewed works solve real problems adapting to the needs of producers and buyers, obtaining better profits with agricultural products of good quality.

The detection of weeds in different crops is the most studied agricultural task from 2014 to the present, as shown in Fig. 3, from Table 6.

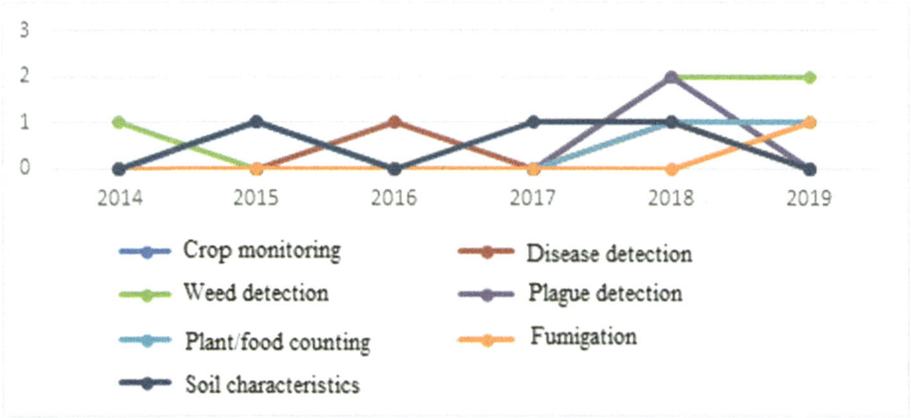

Fig. 3. Precision agriculture tasks

Related to heterogeneous technologies such as Multicore, GPU and FGPA, in the studies reviewed, they have not been used in precision agriculture activities. GPU and FGPA are used in the processing of images in other contexts, especially medicine, achieving important benefits in speed of image processing with different resolutions and large sizes. Here, the images come from different sources of the electromagnetic spectrum such as X and gamma rays, ultrasound, radio waves, electron beam and visible band. Programming languages such as CUDA and OpenMP are used in the parallelization of image processing and analysis algorithms with successful results in decreasing response times and processing speed.

The GPU architecture is the one that has received the most attention in the studies reviewed, the same one that integrating with artificial intelligence techniques has solved problems with large amounts of images in very low times. Figure 4 shows a summary of the architectures used (see Table 7).

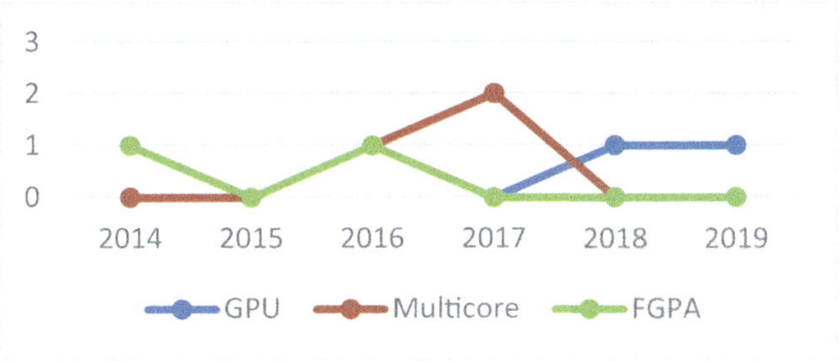

Fig. 4. Heterogeneous architectures used in image processing

As mentioned before, some artificial intelligence techniques have been used in the processing of agricultural images using images obtained from UAV in real time, achieving good results and reducing response times. Technological advances facilitate the application of new techniques such as machine learning and deep neural networks to increase the capacity of image processing in pattern recognition tasks and image classification applications with high accuracy rates.

Real-time image processing applications from UAV and using Deep Learning techniques are the most used in the reviewed studies as shown in Fig. 5 (and Table 8). The images acquired with UAV in real time can be exploited by new algorithms, programming languages, processing techniques and image analysis, heterogeneous architectures to create agricultural solutions that provide benefits in quality and production, both for producers and consumers of products agricultural.

Some limitations in this study, mentioned in Sect. 2.3, the following are considered: (i) use of four bibliographic databases, (ii) documents in the English language of the last 5 years, (iii) articles, reviews and conferences evaluated by pairs. Therefore, future work can be done considering these and other elements that were excluded in the SLR.

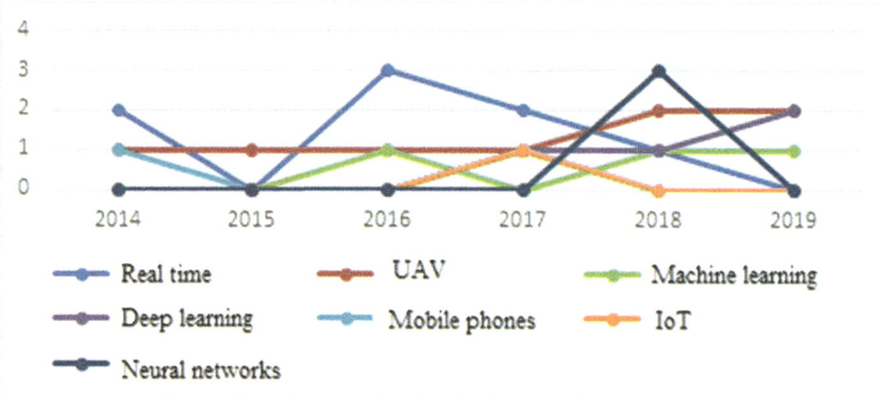

Fig. 5. Technology trends in image processing

5 Conclusions

Most articles and reviews analyzed focus on the processing and analysis of images using image processing and analysis techniques (segmentation, vegetation indices, learning, wavelets) (Table 5 and Fig. 2) to systematize agricultural tasks of different crops. like corn, coffee, lettuce, passion fruit, rice, potatoes. The use of artificial vision techniques has determined great benefits for producers and consumers of agricultural products. Through the processing of images, it is feasible to automate difficult agricultural tasks, minimizing time and resources, avoiding soil erosion and environmental contamination.

The detection of plants, pests, insects, diseases, soil quality and crop monitoring (Table 6 and Fig. 3) by means of image processing techniques is an efficient process that facilitates the systematization of agricultural tasks, which implies the reduction of work manual and physical effort, improving the quality of the products and the quality of life of the producers and consumers.

The use of heterogeneous architectures (Table 7 and Fig. 4) together with advanced artificial intelligence techniques are also promising alternatives for future work in real-time image processing applications, which can increase the capacity of image processing with high resolution and large sizes (Table 8 and Fig. 5), which will allow the reduction of time, complexity and computational costs in different disciplines using heterogeneous architectures and high-performance programming languages for parallel processing. Currently, heterogeneous architectures are used only in the systematization of agricultural tasks (Table 7 and Fig. 4), so it is an opportunity to create new and efficient applications using image analysis techniques.

References

1. Chillet, D., Hübner, M.: Special issue on design and architectures of real-time image processing in embedded systems. J. Real-Time Image Process. **9**(1), 1–3 (2014). https://doi.org/10.1007/s11554-014-0401-6
2. García-Santillán, I., Pusdá, M., Pajares, G.: Identificación automática de vegetación utilizando imágenes agrícolas: una revisión de métodos. In: Imbaquingo, D. (Ed.) Tecnologías Aplicadas a la Ingeniería, pp. 155–162. Universidad Técnica del Norte, Ibarra (2017)
3. Guerrero, J.M., Pajares, G., Montalvo, M., Romeo, J., Guijarro, M.: Support vector machines for crop/weeds identification in maize fields. Expert Syst. Appl. **39**(12), 11149–11155 (2012). https://doi.org/10.1016/j.eswa.2012.03.040
4. Jiménez-Brenes, F.M., López-Granados, F., Torres-Sánchez, J., Peña, J.M., Ramírez, P., Castillejo-González, I.L., et al.: Automatic UAV-based detection of Cynodon dactylon for site-specific vineyard management. PLoS ONE **14**(6), 1–21 (2019). https://doi.org/10.1371/journal.pone.0218132
5. Roldán-Serrato, K.L., Escalante-Estrada, J.A., Rodríguez-González, M.T.: Automatic pest detection on bean and potato crops by applying neural classifiers. Eng. Agric. Environ. Food **11**(4), 245–255 (2018). https://doi.org/10.1016/j.eaef.2018.08.003
6. Campos, Y., Sossa, H., Pajares, G.: Spatio-temporal analysis for obstacle detection in agricultural videos. Applied Soft Comput. J. **45**, 86–97 (2016). https://doi.org/10.1016/j.asoc.2016.03.016
7. Barbedo, J.G.: A review on the main challenges in automatic plant disease identification based on visible range images. Biosyst. Eng. **144**, 52–60 (2016). https://doi.org/10.1016/j.biosystemseng.2016.01.017
8. Leibovich, F., Chichizola, F., De Giusti, L., Naiouf, M., Tirado Fernández, F., De Giusti, A.: Programación híbrida en clusters de multicore. Análisis del impacto de la jerarquía de memoria. In: Jornadas - III-LIDI – Unlp, 2nd edn. La Plata (2012)
9. HajiRassouliha, A., Taberner, A.J., Nash, M.P., Nielsen, P.M.F.: Suitability of recent hardware accelerators (DSPs, FPGAs, and GPUs) for computer vision and image processing algorithms. Sig. Process. Image Commun. **68**, 101–119 (2018). https://doi.org/10.1016/j.image.2018.07.007
10. Weinstock, J.A., Murillo, L.G., Leupers, R.: Parallel system C simulation for ESL design. ACM Trans. Embed. Comput. Syst. **16**(1), 25 (2016). https://doi.org/10.1145/2987374
11. Murilo, B.: Modelos paralelos para la resolución de problemas de ingeniería agrícola. Universidad Politécnica de Valencia, España (2015). https://dialnet.unirioja.es/servlet/tesis?codigo=95661
12. Manuel López, M.: Procesos en paralelo en la PC a través de CUDA. Unocero. https://www.unocero.com/noticias/procesos-en-paralelo-en-la-pc-a-traves-de-cuda/. Accessed 13 July 2019
13. Liao, S.W., Kuang, S.Y., Kao, C.L., Tu, C.H.: Halide-based synergistic computing framework for heterogeneous systems. J. Sig. Process. Syst. **91**, 2019–233 (2019). https://doi.org/10.1007/s11265-017-1283-1
14. Zareiforoush, H., Minaei, S., Alizadeh, M.R., Banakar, A.: Potential applications of computer vision in quality inspection of Rice: a review. Food Eng. Rev. **7**(3), 321–345 (2015). https://doi.org/10.1007/s12393-014-9101-z
15. Vithu, P., Moses, J.A.: Machine vision system for food grain quality evaluation: a review. Trends Food Sci. Technol. **56**, 13–20 (2016). https://doi.org/10.1016/j.tifs.2016.07.011

16. Andrade, F., Taboada, M., Lema, D., Maceira, N., Echeverría, H., Posse, G., et al.: Los desafíos de la agricultura argentina, 1st edn. Ediciones INTA, Buenos Aires (2017)
17. García-Santillán, I., Peluffo-Ordoñez, D., Caranqui, V., Pusdá, M., Garrido, F., Granda, P.: Computer vision-based method for automatic detection of crop rows in potato fields. Adv. Intell. Syst. Comput. **721**, 355–366 (2018). https://doi.org/10.1007/978-3-319-73450-7
18. García-Santillán, I.D., Montalvo, M., Guerrero, J.M., Pajares, G.: Automatic detection of curved and straight crop rows from images in maize fields. Biosyst. Eng. **156**, 61–79 (2017). https://doi.org/10.1016/j.biosystemseng.2017.01.013
19. Shah, J.P., Prajapati, H.B., Dabhi, V.K.: A survey on detection and classification of rice plant diseases. In: 2016 IEEE International Conference on Current Trends in Advanced Computing, ICCTAC 2016, pp. 1–8 (2016). https://doi.org/10.1109/ICCTAC.2016.7567333
20. Kamilaris, A., Prenafeta-Boldú, F.X.: Deep learning in agriculture: a survey. Comput. Electron. Agric. **147**, 70–90 (2018). https://doi.org/10.1016/j.compag.2018.02.016
21. Kitchenham, B., Charters, S.: Guidelines for performing systematic literature reviews in software engineering. Tech. Rep. **5**, 40–44 (2007). https://userpages.uni-koblenz.de/~laemmel/esecourse/slides/slr.pdf
22. Fernandez, A., Insfran, E., Abrahão, S.: Usability evaluation methods for the web: a systematic mapping study. Inf. Softw. Technol. **53**(8), 789–817 (2011). https://doi.org/10.1016/j.infsof.2011.02.007
23. Velthuis, M.G.: Métodos de investigación en ingeniería del software, 1st edn. Grupo Editorial RA-MA, Madrid (2014)
24. Institute of Electrical and Electronic Engineers. IEEE Xplore Digital Library. https://ieeexplore.ieee.org/Xplore/home.jsp. Accessed 21 Mar 2019
25. ScienceDirect: ScienceDirect.com | Science, health and medical. https://www.sciencedirect.com. Accessed 21 Mar 2019. Journals, full text articles and books
26. Scopus: Scopus preview - Scopus - Welcome to Scopus. https://www.scopus.com/home.uri. Accessed 21 Mar 2019
27. Springer: Springer Link. https://link.springer.com. Accessed 21 Mar 2019
28. Katsigiannis, P., Misopolinos, L., Liakopoulos, V., Alexandridis, T.K., Zalidis, G.: An autonomous multi-sensor UAV system for reduced-input precision agriculture applications. In: 24th Mediterranean Conference on Control and Automation (MED), pp. 60–64. Athens (2016). https://doi.org/10.1109/MED.2016.7535938
29. Kerkech, M., Hafiane, A., Canals, R.: Deep leaning approach with colorimetric spaces and vegetation indices for vine diseases detection in UAV images. Comput. Electron. Agric. **155**, 237–243 (2018). https://doi.org/10.1016/j.compag.2018.10.006
30. Story, D., Kacira, M.: Design and implementation of a computer vision-guided greenhouse crop diagnostics system. Mach. Vis. Appl. **26**(4), 495–506 (2015). https://doi.org/10.1007/s00138-015-0670-5
31. Li, Y., Meiyu, Q., Pengfeng, L., Qian, C., Xiaoying, L., Junwen, G., et al.: The recognition of rice area images by UAV based on deep learning. In: MATEC Web of Conferences, vol. 232, p. 02057 (2018). https://doi.org/10.1051/matecconf/201823202057
32. Juan Núñez, V.M., Faruk Fonthal, R., Yasmín Quezada, L.M.: Design and implementation of WSN for precision agriculture in white cabbage crops. In: 2018 IEEE ANDESCON, ANDESCON 2018 - Conference Proceedings, pp. 1–4 (2018). https://doi.org/10.1109/ANDESCON.2018.8564674
33. Basso, M., Pignaton de Freitas, E. A.: UAV guidance system using crop row detection and line follower algorithms. J. Intell. Robotic Syst. Theory Appl 1–17 (2019). https://doi.org/10.1007/s10846-019-01006-0

34. Comba, L., Biglia, A., Ricauda Aimonino, D., Gay, P.: Unsupervised detection of vineyards by 3D point-cloud UAV photogrammetry for precision agriculture. Comput. Electron. Agric. **155**, 84–95 (2018). https://doi.org/10.1016/j.compag.2018.10.005
35. Ribeiro-Gomes, K., Hernandez-Lopez, D., Ballesteros, R., Moreno, M.A.: Approximate georeferencing and automatic blurred image detection to reduce the costs of UAV use in environmental and agricultural applications. Biosyst. Eng. **151**, 308–332 (2016). https://doi.org/10.1016/j.biosystemseng.2016.09.014
36. Wang, A., Zhang, W., Wei, X.: A review on weed detection using ground-based machine vision and image processing techniques. Comput. Electron. Agric. **158**, 226–240 (2019). https://doi.org/10.1016/j.compag.2019.02.005
37. Patrício, D.I., Rieder, R.: Computer vision and artificial intelligence in precision agriculture for grain crops: a systematic review. Comput. Electron. Agric. **153**, 69–81 (2018). https://doi.org/10.1016/j.compag.2018.08.001
38. Zheng, Y.Y., Kong, J.L., Jin, X.B., Wang, X.Y., Zuo, M.: CropDeep: the crop vision dataset for deep-learning-based classification and detection in precision agriculture. Sensors **19**(5), 1058 (2019). https://doi.org/10.3390/s19051058
39. Samet, R., Bay, O.F., Aydin, S., Tural, S., Bayram, A.: Real-time image processing applications on multicore CPUs and GPGPU. In: International Conference on Parallel and Distributed Processing Techniques and Applications, vol. 1, pp. 116–122 (2015)
40. Dong, W., Wu, T., Sun, Y., Luo, J.: Digital mapping of soil available phosphorus supported by AI technology for precision agriculture. In: 2018 7th International Conference on Agro-Geoinformatics (2018). https://doi.org/10.1109/Agro-Geoinformatics.2018.8476007
41. Torres-Sánchez, J., López-Granados, F., Peña, J.M.: An automatic object-based method for optimal thresholding in UAV images. Comput. Electron. Agric. **114**, 43–52 (2015). https://doi.org/10.1016/j.compag.2015.03.019
42. Gulo, C.A., de Arruda, H.F., de Araujo, A.F., Sementille, A.C., Tavares, J.M.: Efficient parallelization on GPU of an image smoothing method based on a variational model. J. Real-Time Image Process. 1–13 (2016). https://doi.org/10.1007/s11554-016-0623-x
43. Oliveira, H.C., Guizilini, V.C., Nunes, I.P., Souza, J.R.: Failure detection in row crops from UAV images using morphological operators. IEEE Geosci. Remote Sens. Lett. **15**(7), 991–995 (2018). https://doi.org/10.1109/LGRS.2018.2819944
44. Abdullahi, H.S., Sheriff, R.E., Mahieddine, F.: Convolution neural network in precision agriculture for plant image recognition and classification. In: 2017 Seventh International Conference on Innovative Computing Technology (INTECH), pp. 1–3. Luton (2017). https://doi.org/10.1109/intech.2017.8102436
45. Jiang, B.: Real-time multi-resolution edge detection with pattern analysis on graphics processing unit. J. Real-Time Image Process. **14**(2), 293–321 (2018). https://doi.org/10.1007/s11554-014-0450-x
46. Shen, Y., Zhou, H., Li, J., Jian, F., Jayas, D.S.: Detection of stored-grain insects using deep learning. Comput. Electron. Agric. **145**, 319–325 (2018). https://doi.org/10.1016/j.compag.2017.11.039
47. Saussard, R., Bouzid, B., Vasiliu, M., Reynaud, R.: A novel global methodology to analyze the embeddability of real-time image processing algorithms. J. Real-Time Image Process. **14**(3), 565–583 (2018). https://doi.org/10.1007/s11554-017-0686-3
48. Ahmad, A., Guyonneau, R., Mercier, F., Belin, É.: An image processing method based on features selection for crop plants and weeds discrimination using RGB images. In: International Conference on Image and Signal Processing ICISP 2018, pp. 3–10 (2018). https://doi.org/10.1007/978-3-319-94211-7_1

49. García-Santillán, I., Pajares, G.: On-line crop/weed discrimination through the Mahalanobis distance from images in maize fields. Biosyst. Eng. **166**, 28–43 (2018). https://doi.org/10.1016/j.biosystemseng.2017.11.003
50. Uma Vetri Selvi, G., Nadarajan, R.: CT and MRI image compression using wavelet-based contourlet transform and binary array technique. J. Real-Time Image Process. **13**(2), 261–272 (2017). https://doi.org/10.1007/s11554-014-0400-7
51. Partel, V., Charan Kakarla, S., Ampatzidis, Y.: Development and evaluation of a low-cost and smart technology for precision weed management utilizing artificial intelligence. Comput. Electron. Agric. **157**, 339–350 (2019). https://doi.org/10.1016/j.compag.2018.12.048
52. Cáceres, C.A., Amaya, D., Ramos, O.L.: Procesamiento de imágenes para reconocimiento de daños causados por plagas en el cultivo de Begonia Semperflorens (Flor de Azúcar). Acta Agron. **64**(3), 272–279 (2015). https://doi.org/10.15446/acag.v64n3.42657
53. Georgis, G., Lentaris, G., Reisis, D.: Acceleration techniques and evaluation on multi-core CPU, GPU and FPGA for image processing and super-resolution. J. Real-Time Image Process. **1**, 1–28 (2016). https://doi.org/10.1007/s11554-016-0619-6
54. Aydin, S., Samet, R., Bay, O.F.: Real-time parallel image processing applications on multicore CPUs with OpenMP and GPGPU with CUDA. J. Supercomput. **74**(6), 2255–2275 (2018). https://doi.org/10.1007/s11227-017-2168-6
55. Wu, Q., Liang, Y., Li, Y., Liang, Y.: Research on intelligent acquisition of smart agricultural big data. In: 25th International Conference on Geoinformatics. IEEE, Buffalo (2017). https://doi.org/10.1109/GEOINFORMATICS.2017.8090913

Educational Robot Using Lego Mindstorms and Mobile Device

José Varela-Aldás[1](✉), Oswaldo Miranda-Quintana[1], Cesar Guevara[1], Franklin Castillo[1], and Guillermo Palacios-Navarro[2]

[1] SISAu Research Group, Universidad Tecnológica Indoamérica, 180103 Ambato, Ecuador
{josevarela,cesarguevara, franklincastillo}@uti.edu.ec, oswamiranda1991@gmail.com

[2] Department of Electronic Engineering and Communications, University of Zaragoza, 44003 Zaragoza, Spain
guillermo.palacios@unizar.es

Abstract. Modern education at different levels has integrated the use of technology as a teaching-learning assistant, especially in the initial stage of training because it motivates curiosity and helps the abstraction of knowledge. This paper presents the development of an educational robot using components of the Lego Mindstorms EV3 kit and a mobile app to interact with the student; the robot is build using two motors and two sensors, one to differentiate colours and the other one to measure proximity. The objective is to teach the different colours and explain the spatial position to children under five years old, while the screen of the mobile device interacts with the user. The results show the finished product, the data generated by a learning test, and a usability test, concluding that the robot fulfils its purpose, but requires certain improvements.

Keywords: Educational robot · Lego Mindstorms · Mobile app · Test

1 Introduction

Robotics is currently considered one of the most prestigious fields in the world, focusing on the study of robots, these are reprogrammable machines capable of facilitating human work [1, 2]. Nowadays, several advances in robotics can be appreciated, new studies are made so that a robot is able to simulate human attitudes. The robot can make decisions outside its capabilities using techniques, statistical probability methods and behaviour patterns. The most commonly used robotic equipment are mobile robots, which are electromechanical devices capable of moving in a workspace according to their type, whether with wheels or legs. These robots are those that are being used and implemented by teachers for educational assistance [3, 4].

Educational assistance is a service which aims to help a large percentage of students improve their education, applying a temporary intervention in learning. In this area, studies have been carried out that expose the main characteristics of the low concentration

J. Nummenmaa et al. (Eds.): CSEI 2019, AISC 1078, pp. 71–82, 2020.
https://doi.org/10.1007/978-3-030-33614-1_5

level of students. For this reason, the STEM (Science, Technology, Engineering, and Mathematics) methodology is ideal for building knowledge and strengthening academic development. This methodology seeks to exploit the capabilities of students so that they are able to understand a new field, whether technology or other sciences [5].

The work described in [6] and [7] presents a methodology in the use of mobile applications with smartphones, tablets, computers, etc. Applications can be obtained in virtual stores depending on the operating system, such as Play Store if you have Android and App Store if it is iOS. These applications are developed by companies specialized in mobile programming, with the main objective of giving a help to the user by providing information, entertainment, and so on. In [8], an educational robot built with Lego pieces and controlled by a mobile application is presented, with the aim of being used as a learning tool for primary school children, mixing pedagogy with science and technology, using programming lines and algorithms to solve any problem, achieving in them curiosity and approach in the understanding of science, turning robotics into a point of attention, visualizing, exploring and testing the concepts of reasoning and in turn opening new fields that allow students have a new vision to the professional life.

In the study developed in [9], it is established that robotics in education has as its main objective the development of new skills and competences that will allow children to work in a tangible and playful way, their system is composed of a robot, an application for mobile devices, a set of programming cards and a web page. In [10] it is indicated that the use of robotics in the classroom as a learning tool generates multi-disciplinary learning environments that allow students to strengthen their learning process while developing different skills that will allow them to face the challenges of the current society.

According to [11], robotics in early childhood education is to achieve an adaptation of students to current production processes, where it plays a very important role in the world of technology, however, robotics is considered a system that goes beyond a work application, for the development is used as research instruments an experiential test and a drawing test that have been adapted to the evolutive characteristics of the children of Early Childhood Education. The webQDA software has been used to support the qualitative data analysis.

In [12], the knowledge about the teaching of technology is analyzed as a pedagogical tool of great utility in the academic formation of boys and girls equally, in addition, when the student is involved in this type of academic process, is fostering creativity and motivation, which will allow you to develop cognitive and manual skills later on. In [13] says that the use of technology is presented in education as a tool that is oriented in an appropriate way, benefit and stimulate teaching-learning processes, in addition to having generated a significant change in methodologies in teaching, not be considered as teaching tools, but also as an axis of study.

On the other hand, [14] establishes that mobile applications are a feasible learning alternative, also says that technology in recent years has been getting more involved in mobile devices and mobile applications in the educational process. In [15] it is mentioned that mobile applications is the innovation for study and learning that combines education, pedagogy and technology. Furthermore, the use of mobile devices for education is an opportunity to expand the possibility that students can collaboratively build their knowledge through these ICT (Information and Communications Technology).

Finally, in [16] an educational manipulation robot is built for the development of children's learning that allows experimental tests to measure the effectiveness and efficiency of skills that are developed when working with assembly figures in the construction of mobile robots such as spatial reasoning. The results of the first developed phase of the robot research project, clearly show the need to use software and hardware tools to teach basic robotics with Lego allowing the student to create and imagine a diversity of intelligent and autonomous devices that can be applied in areas of industry, medicine or routines of the daily life of the human being. In this way, several studies demonstrate the usefulness of technology and robotics in education [17].

This work realizes the implementation of an educational robot using Lego components and a mobile device, the objective is to interact with the learner to achieve his attention, teaching him the colors and spatial position. The following sections contain: Sect. 2 details the design of the proposal; Sect. 3 describes the implementation of the educational robot; Sect. 4 presents the results obtained, finally, Sect. 5 presents the final conclusions.

2 Design of the Proposal

The school classroom requires didactic elements that facilitate students' learning and motivate the use of new technologies from early ages, for which purpose the design of a technological learning tool is proposed, using easily existing components in the market, which allows the proposal to be replicated in a short time.

The first users of the educational robot are children between 3 and 5 years old and the project helps to learn of spatial position and the recognition of colours through the use of a robot, in order to motivate and facilitate the learning of the students. The design components of the educational robot are shown in Fig. 1, for the construction of

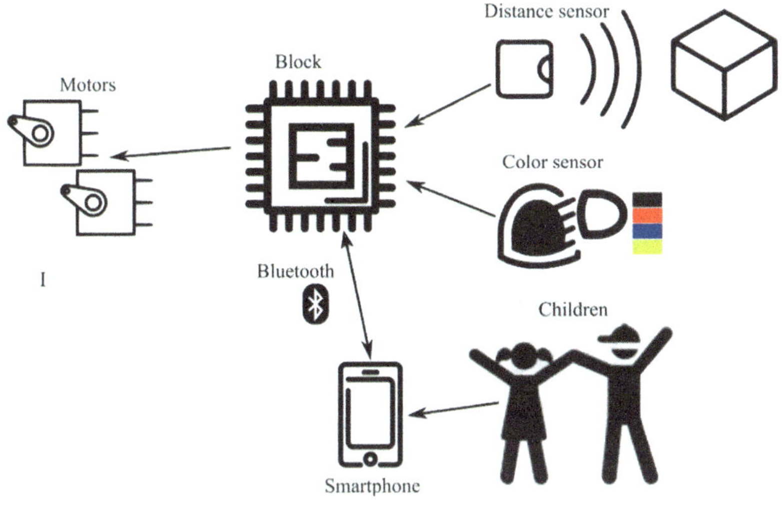

Fig. 1. Educational robot design.

the robot a central processor is required to perform the control of the elements of the system, to which the input and output peripherals are connected. The input components are the colour and proximity sensors, and the output elements are the motors that make up the mobile robot. In addition, the processor is connected to a mobile device via Bluetooth to control the movements of the robot and interact with the user.

Given that one of the main objectives of this work is to create an easy-to-access teaching tool, the robot is led for children in early childhood education, which can be used simply by students and teachers.

3 Development of the Proposal

3.1 System Connections

This section describes the stages of development of the robot that is responsible for carrying out certain processes automatically. With the use of input and output devices such as: servo motors, infrared sensor, and colour sensor, the robot allows children to be instructed to differentiate colours and spatial orientation.

The connection used for the development of the educational application is shown in Fig. 2, where the block of the Lego Mindstorms EV3 kit is the central segment of the robot given that it contains all the logical and electronic components that allows the instructions for the robot (stored in an internal memory). In addition, an Android smartphone controls the robot, and to program a computer with the application App Inventor 2 is used.

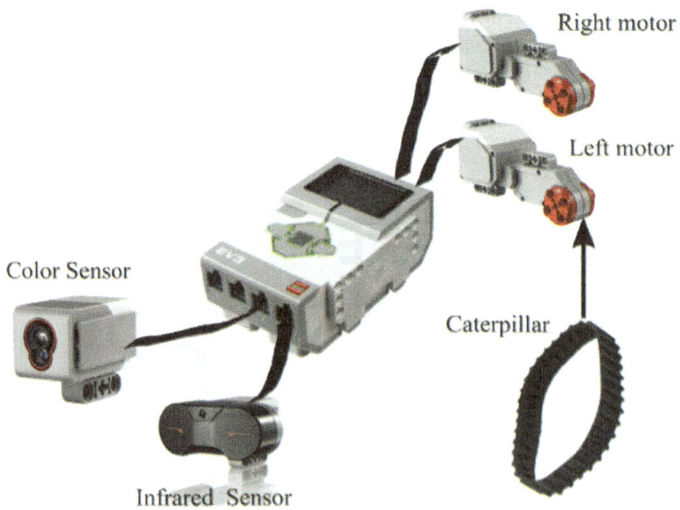

Fig. 2. System connection.

The prototype is assembled in the form of mobile robot with caterpillars with a unicycle-like behavior that allows two degrees of freedom (linear and angular movement). All the pieces that make up the Lego Mindstorms Ev3 kit allow to assemble each element of the robot. In addition, a support is designed on top of the mobile robot to charge the mobile device, so that the screen is located on the front of the robot, because it will show the robot's face. An important component in the Lego educational robot are the wheels, because they allow the movement of the logical block in a real space and interact with the environment that surrounds it. In order not to have inconvenience with the movement of the robot it is necessary to connect the sensors and motors of the Lego block in such a way that they do not interfere with the wheels, that is, that the robot can rotate freely, for which it is necessary to spin the cables around of different pieces of the robot and thus prevent them from blocking the movements.

The processor inputs are located at the top of the screen of the block, in which the infrared sensor and the color sensor are connected to receive information from the outside. Another sensor of the Lego EV3 kit allows detecting colors and intensity of light, this sensor can recognize up to seven colors: black, blue, green, yellow, red, white, and brown; while the infrared sensor allows detection of the proximity of objects. On the other hand, the outputs of the robot are located in the lower part of the screen of the block, where 2 servomotors are connected to provide movement to the mobile robot.

Finally, the Lego Mindstorms EV3 block has Bluetooth communication, which allows the connectivity, programming, and manipulation of the block from different devices, in this case a Smartphone was used. To establish the connection between the EV3 block and the mobile device, the configuration of the Bluetooth communication is made in each element.

3.2 Robot Operation

This section describes the programming code that controls the robot. This program is implemented in the mobile device because it controls all the inputs and outputs of the robot. The algorithm used is shown in Fig. 3; this program is stored on the mobile device and the application is developed in App Inventor 2. The robot starts its operation and provides a menu in which the user can choose the working mode, that is, if it requires proximity recognition, the color recognition or command the robot remotely.

When choosing the *Color mode*, the system starts to detect the colour of the object located next to the sensor, then indicates which color is by text and the reproduction of a pre-recorded audio. When selecting *Distance mode*, the robot maintains a fixed distance of 1 m (desired value) between the child and the robot, reproducing an audio when it is far or near, before performing distance correction by means of movements. The audio playback is in synchronization with the movements or gestures of the face shown on the Smartphone screen.

The Control mode shows a menu to command the robot from the mobile device, controlling the robot by linear movements (forward and backward) and angular movements (right and left), and contains a stop button to stop the robot movement.

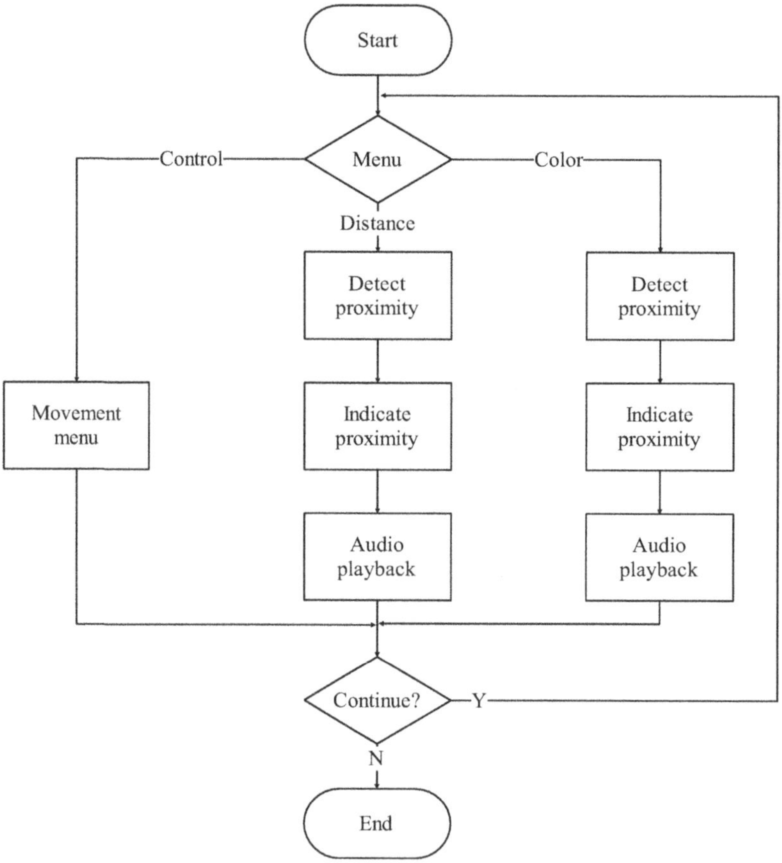

Fig. 3. Flow diagram of the program

The established algorithm meets the required requirements such as the continuous execution of different tasks: the measurement of external variables, the conservation of a distance, and finally the reproduction of an audio to interact with the child and teach the activities carried out.

3.3 Application Mobile

The user interfaces shown in Fig. 4 are developed using the blocks provided by the application (App Inventor 2), the control menu is implemented, the buttons are inserted and the required actions are configured, including the connection and disconnection functions via Bluetooth.

The animation of the face in the robot has been implemented by means of sequence of Gifs that simulate the movement of the lips and the eyes, these characteristics help the interaction between the robot and the child.

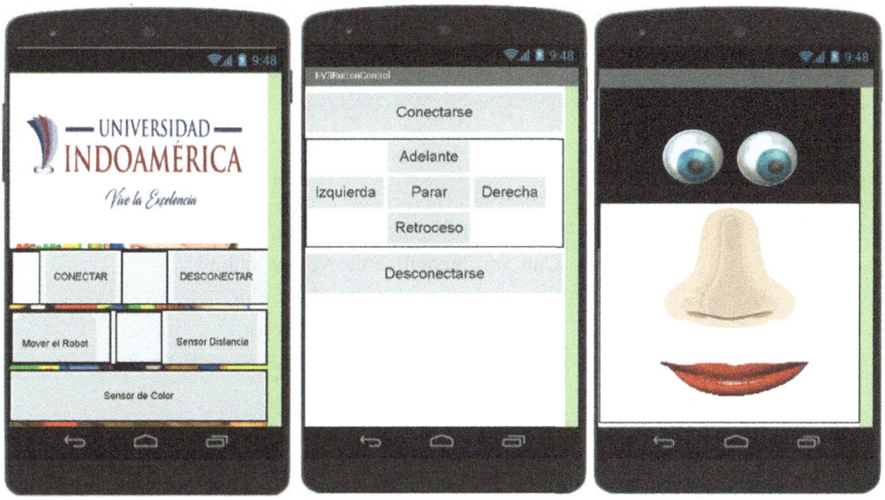

Fig. 4. Mobile application interfaces

4 Results

Following the procedures detailed in the previous section, the educational robot is assembled. Figure 5 presents the mobile robot implemented and the group of students that evaluates the robot's performance. This assembly with LEGO parts has not been implemented so that the user modifies its structure, also, the resistance of the robot is limited to the characteristics of the Mindstorms Ev3 kit.

Fig. 5. Educational robot implemented

4.1 Previous Analysis

Prior to the use of the educational robot in preschool children, a basic knowledge test is applied as detailed in Table 1. The children evaluated belong to the educational unit

"La Gran Muralla" in Ambato city, Ecuador, and its average age is 4 years old. The questions are used to know how many children know colours and recognize their spatial position.

Table 1. Test questions

Number of question	Question
1	Can you recognize the green color?
2	Can you recognize the yellow color?
3	Can you recognize the blue color?
4	Can you recognize the red color?
5	Can recognize if the robot is less than 1 m?

Figure 6 presents the data resulting from the test of the Pretest of the five questions, which were carried out on 16 students. The result of the first question has given as correct a 56.25% of children, the second question gives us a result of 75%, the third question gives us a percentage of 68.75%, in the fourth question it gives us a result of 75%. In the fifth question he gave us a result of 43.75%.

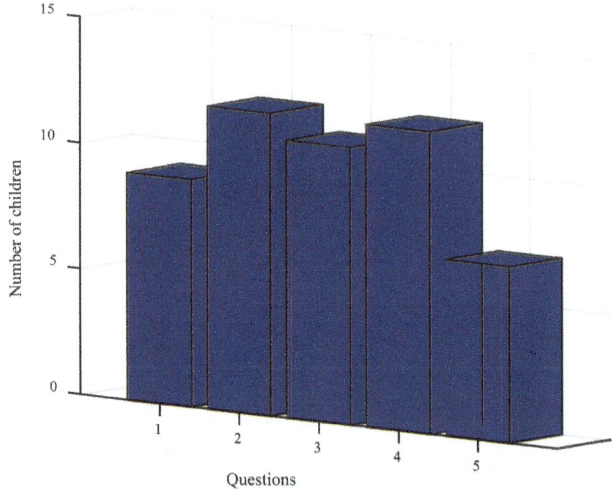

Fig. 6. Performance of the first test result

Although some children differentiate certain colors, not all correctly identify all colors, in addition, children do not distinguish whether the robot is close (<1 m) or far (>1 m).

4.2 Post Analysis

Once the educational robot is used in the classroom, the test is performed again, in Fig. 7 the results obtained are shown. Giving as a deduction the following data: in the first question a percentage of 87.5% answered correctly, in the second question a percentage of 93.75%, in the third question a percentage of 87.5%, in the fourth question a percentage of 93.75% and in the fifth question a percentage of 100%.

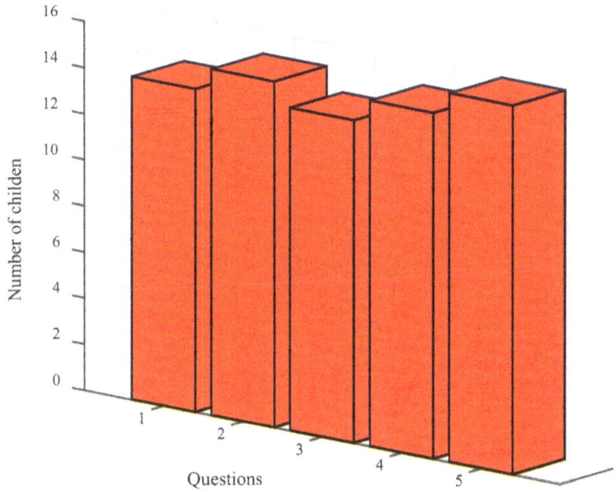

Fig. 7. Performance for the second test results

The results show that the children obtained knowledge about the names of the colors and recognize when they are at distances less than and greater than 1 m. In addition, children were observed attentive to the activities that the robot performed.

4.3 Usability Test

The robot's features are evaluated by users through a simple usability questionnaire designed by the authors for young children. The test contains five questions that allow to know the satisfaction of the child with the robot, the answers are presented in five faces that express emotions and represent a numerical value according to the position, so that the first face is equivalent to 1 and the last to 5. Table 2 presents the questions used in the usability test.

Applying the usability test in 16 children, the answers presented in Table 3 are obtained, calculating the mean of each question. The results indicate a good rating for the robot, but it can improve the ease of use and the characteristics.

Table 2. Usability survey

Number of question	Question	Options
1	Do you want to continue playing with the robot?	
2	Is it easy to play with the robot?	
3	The games that the robot has are enough?	
4	Do you think other children want to play with the robot?	
5	Do you feel safe playing with the robot?	

Table 3. Survey results

Number of question	Answers																Mean
1	5	5	5	5	5	5	5	3	5	5	5	5	5	5	5	5	4.9
2	4	4	5	4	4	4	4	2	4	3	3	4	4	3	3	4	3.7
3	4	5	5	3	3	5	5	3	3	4	4	5	5	3	3	3	3.9
4	5	5	5	5	5	5	5	4	5	5	5	5	5	5	5	5	4.9
5	4	5	5	3	4	5	5	5	4	3	4	5	5	4	4	5	4.4

5 Conclusions

Robotics has had an expansion to different areas in recent years, providing new possibilities in social areas. Education has seen the opportunity to rely on computer tools to maintain student attention and achieve more interactive learning. In this work an educational robot is implemented using components of the Lego Mindstorms Ev3 kit and a mobile application for Android, the purpose is to teach the colors and the spatial position to children between 3 and 5 years old, the proposal shows the elements used and the connections required for the operation of the robot.

The results show the robot assembled and in operation. To evaluate its contribution in teaching, a knowledge test of five questions is applied. The data obtained prior to the use of the robot in the classroom show deficiencies in the identification of colors and distances with references. After applying the teaching with the robot, the children repeat the test, observing an improvement in the evaluated knowledge. In addition, a usability test is applied that shows children's satisfaction with the robot, but need of improvements in ease of use and increase features. The development of this robot can serve as a basis for education in educational institutions, awakening interest in the school classroom and promoting STEM competencies from early ages.

References

1. Herrera, D., Roberti, F., Carelli, R., Andaluz, V., Varela, J., Ortiz, J.: Modeling and path-following control of a wheelchair in human-shared environments. Int. J. Humanoid Rob. **15**, 1–33 (2018). https://doi.org/10.1142/S021984361850010X
2. Ortiz, J.S., Molina, F., Andaluz, V.H., Varela, J., Morales, V.: Coordinated control of a omnidirectional double mobile manipulator. In: Lecture Notes in Electrical Engineering, pp. 278–286 (2017). https://doi.org/10.1007/978-981-10-6451-7
3. Zawieska, K., Duffy, B.R.: The social construction of creativity in educational robotics. In: Szewczyk, R., Zieliński, C., Kaliczyńska, M. (eds.) Progress in Automation, Robotics and Measuring Techniques. Advances in Intelligent Systems and Computing, vol. 351. Springer, Cham (2015). https://doi.org/10.1007/978-3-319-15847-1_32
4. Ospennikova, E., Ershov, M., Iljin, I.: Educational robotics as an inovative educational technology. Procedia Soc. Behav. Sci. **214**, 18–26 (2015). https://doi.org/10.1016/j.sbspro.2015.11.588
5. Arís, N., Orcos, L.: Educational robotics in the stage of secondary education: empirical study on motivation and STEM skills. Educ. Sci. **9**(2), 73 (2019). https://doi.org/10.3390/educsci9020073
6. Saura, J.R., Palos-Sanchez, P., Reyes-Menendez, A.: Marketing a través de aplicaciones móviles de turismo (m-tourism). Un estudio exploratorio. Int. J. World Tourism **4**(8), 45–56 (2017)
7. Contreras, I., Hernández, G.: Sistema de localización en tiempo real mediante un servidor web y aplicaciones móviles. Pistas Educativas **39**, 171–186 (2017)
8. Afari, E., Khine, M.S.: Robotics as an educational tool: impact of lego mindstorms. Int. J. Inf. Educ. Technol. **7**(6), 437–442 (2017). https://doi.org/10.18178/ijiet.2017.7.6.908
9. Sullivan, A., Bers, M.U.: Robotics in the early childhood classroom: learning outcomes from an 8-week robotics curriculum in pre-kindergarten through second grade. Int. J. Technol. Des. Educ. **26**(1), 3–20 (2016). https://doi.org/10.1007/s10798-015-9304-5
10. Emily Toh, L.P., Causo, A., Tzuo, P.-W., Chen, I.-M., Yeo, S.H.: A review on the use of robots in education and young children. J. Educ. Technol. Soc. **19**(2), 148–163 (2016)
11. Torres, N., González, R., Carvalho, J.: Roamer, un robot en el aula de Educación Infantil para el desarrollo de nociones espaciales básicas. RISTI - Revista Ibérica de Sistemas e Tecnologias de Informação **28**, 14–28 (2018)
12. Filippov, S., Ten, N., Shirokolobov, I., Fradkov, A.: Teaching robotics in secondary school. IFAC-PapersOnLine **50**(1), 12155–12160 (2017). https://doi.org/10.1016/j.ifacol.2017.08.2143
13. Keane, T., Keane, W.F., Blicblau, A.S.: Beyond traditional literacy: learning and transformative practices using ICT. Educ. Inf. Technol. **21**(4), 769–781 (2016). https://doi.org/10.1007/s10639-014-9353-5
14. Cuervo Gómez, W.O., Ballesteros-Ricaurte, J.A.: Framework para desarrollo de aplicaciones educativas móviles, basado en modelos de enseñanza. Praxis & Saber **8**(17), 125–153 (2017). https://doi.org/10.19053/22160159.v8.n17.2018.7204
15. Domingo, M.G., Garganté, A.B.: Exploring the use of educational technology in primary education: Teachers' perception of mobile technology learning impacts and applications' use in the classroom. Comput. Hum. Behav. **56**, 21–28 (2016). https://doi.org/10.1016/j.chb.2015.11.023

16. Lancheros, D.J.: Design and implementation of a didactic module for the learning in the construction, implementation and manipulation of robots. Formación Universitaria **3**(5), 3–8 (2010). https://doi.org/10.4067/S0718-50062010000500002
17. Preciado Alvarado, C.C.: Determinación e implementación de una metodología en robótica que potencialice los conocimientos de enseñanza aprendizaje en los estudiantes. Universidad Técnica de Machala, Ecuador (2016)

Use of E-Learning and Audio-Lingual Method for the Development of Listening Comprehension Skills

Ruth Viviana Barona-Oñate[1(✉)], Sonia de los Angeles López-Pérez[2], Jimmy P. López López[3], and Julio A. Mocha-Bonilla[2]

[1] Centro de Idiomas, Universidad Tecnica de Ambato, UTA, 180103 Ambato, Ecuador
rv.barona@uta.edu.ec

[2] Facultad de Ciencias Humanas y de la Educacion, Universidad Tecnica de Ambato, UTA, 180103 Ambato, Ecuador
{sda.lopez,ja.mocha}@uta.edu.ec

[3] Facultad de Contabilidad y Auditoria, Universidad Tecnica de Ambato, UTA, 180103 Ambato, Ecuador
jlopez9964@uta.edu.ec

Abstract. The ability of listening comprehension is fundamental in the learning process of any foreign language. The objective was the application of the audio-lingual method with the support of E-learning in the development of listening comprehension skills in English language teaching. The study was implemented with 58 students of English of level A2, divided equally, 29 for the experimental group where an intervention was carried out based on the method and use of E-learning, while 29 students were from the control group with a teaching traditional. The analysis of the academic performance was based on the scores obtained by the students in the diagnostic test, the mid-term exam and the final exam. The results showed that there were significant improvements in the experimental group, exceeding the control group by 1.6 points in the average scores of the final test, which demonstrates the usefulness of the method.

Keywords: Audio-lingual method · Receptive skills · Comprehension · Auditory skills · Communication

1 Introduction

Throughout the world, English is widely used for communicating in different areas such as, business, entertainment, technology and science. In a world that is provided with a rich variety of cultures, traditions, and especially languages, people have had the need to find something in common, so that they can be able to communicate. Now that technology has taken over, everything is faster.

J. Nummenmaa et al. (Eds.): CSEI 2019, AISC 1078, pp. 83–98, 2020.
https://doi.org/10.1007/978-3-030-33614-1_6

Needless to say, the world needs to have a common language that aims to make interaction quicker and more efficient.

In recent years, this language has been gaining great importance among Ecuadorians due to the fact that it is a prodigious tool that provides access to updated information in any field. Furthermore, according to Reglamento de Régimen Académico [4] in article 31, it establishes that higher education students must obtain a B2 level certification according to Common European Framework of Reference for Languages [16] as a requisite of graduation. However, the study called English Proficiency Index (EPI) published at the end of 2015 indicates that Ecuador is in position 35 in a list of 65 countries that were analyzed [14]. This means low proficiency level and evidently, the teaching and learning process of English must be improved.

Considering the previous facts, the institutions of higher education offer English courses, through languages centers where learners have free access to programs of English. Consequently, students can achieve the necessary level. Nowadays, the underlying purpose of requiring students to get a B2 level is to help them be able to communicate and face the challenges of understanding English and using it to sort out problems in their fields of study.

The implementation of methods in schools in Ecuador aim at improving the teaching and learning process of English as a foreign language in the institutions. The authorities have implemented through the curricula some alternatives in order to reduce the problems between the material that students of primary, elementary, secondary and higher education must study. Thus, in the implementation of the Curriculum for General Basic Education, its objective is to lead their learners to reach B1 level according to the Common European Framework of Reference for Languages by the time they finish 3rd year of Bachillerato General Unifi ado (BGU). This new curriculum has been implemented since the school year 2016–2017 for the Amazon and Highland Region and 2017–2018 for the Coast Region. Thus, it is difficult to evidence its results on the receptive skills and productive skills.

Meanwhile, Universidad Técnica de Ambato (UTA) in order to meet the requirement stated in Consejo de Educación Superior (CES), offers English and other foreign language courses to students through its Languages Center. According to the latest reform of it, higher education students must approve a B2 level which means that they must be able to communicate efficiently in the target language. Specially, they have to demonstrate that they have acquired the productive and receptive skills of the target the language.

Students that come from high schools are currently in false beginner levels. In other words, their language skills, specially listening comprehension skills are not developed and their understanding of instructions from their teachers and also the listening passages presented in course materials is deficient. Consequently, UTA Language Center has considered appropriate to offer five levels to comply with the level set for higher education institutions: A1 Starter, A2 Elementary, B1 Pre- intermediate, B1+ Intermediate and B2 Upper Intermediate. At the end of the program, students take an institutional proficiency test which is the final requirement to finish their language studies. The focus of UTA Language Center is to train students so that they acquire the abilities of the language for a successful use in real life.

Student have two options to join the language courses, they can start from the starter level or skip it to begin with the following course. Students who know that have developed more successfully their language skills can take an optional examination, if they score an 80% correct they can skip the Starter level. Students who do not approve this optional exam have to start in level A1 which, according to the Evaluation Center of UTA Language Center, happens to most of them. Level A1 content covers the basic knowledge of English language which focuses on vocabulary and structures that will give students the base for the rest of the levels.

In A2 level, students must be able to comprehend basic relevant sentences and expressions that are frequently used in daily life. Therefore, they must possess good listening comprehension skills to understand the language used in routine errands that require a straight exchange of information as said by [5]. It has been observed that the ability with which A2 students have most difficulty with is Listening comprehension [10]. This was evidenced in diagnostic tests taken in a semester according to the archives from the Evaluation Center of UTA Language Center. The students' average in this skill showed that the 70% of them got scores below the passing grades which is seven out of ten (7/10). This phenomenon also appeared on listening comprehension quizzes and tasks done during the classes. It is said that listening comprehension is the most difficult skill to teach and learn [20].

That is why students need help to improve this skill which is relevant for progressing in a language learning process. Listening is the basic receptive skill necessary for comprehension then comes reading skill [2]. In general terms, this investigation is focused on the lack of use of the audiolingual method and its influence in the development of the listening comprehension skill English learners at level A2 at the Languages Center of Universidad Técnica de Ambato. This issue arises from the fact that English language has 24 consonant sounds, 12 vowel sounds, and 3 diphthongs, which makes it difficult for Spanish speaking learners of English to recognize sounds in the language due to the fact that their native language solely has five vowel sounds. Consequently, developing the appropriate listening skills may be difficult without the appropriate method [22]. Additionally, the limited information that learners have about pronunciation of sounds in English results in difficulties at the time of recognizing the sound in English, obviously they are relevant to develop the listening skills required for comprehension of spoken language [8].

Moreover, having ineffective activities for teaching the pronunciation of sounds in English influences two relevant aspects in the language teaching and learning process. Firstly, the process becomes ineffective and the listening comprehension is heavily compromised. Secondly, as a result of the lack of appropriate methodology low scores on these tests taken by learners appear.

Finally, the limited amount of time for proper English pronunciation activities in class causes teachers to postpone or neglect those activities in order to prioritize other parts of the content they have to teach such as grammar, vocabulary or reading comprehension skills. However, English language teaching requires the teaching of pronunciation to give students the correct sounds for them to understand conversations properly. Teaching a foreign language, specifically English, to university students require the implementation of a strategy that combines phonological and electronic learning characteristics. Especially within university education, the educational platforms are used, so that E-learning is a support in face-to-face teaching, as well as for lifelong learning. Initially, the phonological aspect requires face-to-face instruction so that English teaching and learning take place at the same time and place, which is beneficial for live monitoring and control purposes. Secondly, the technological component of E-learning provides a high level of influence on student participation in language learning due to the fact that it adapts to the population of young adults with whom this project works and will boost their education by motivating students to work autonomously.

On the other hand, the absence of an appropriate strategy results in imbalance in the teaching-learning process and lack of motivation in students since their academic performance may not be optimal [7]. Finding a methodology that meets the needs of students is very useful because it will benefit the mentioned students in two aspects: motivation and the better understanding of the mechanics of the language studied and to better develop the ability to listen and speak in English. This study is relevant because is focused on the development of the listening comprehension skills of students of skills of UTA Languages Center. According to CES, it determines that it is responsibility of the institutions of higher education to provide to people who go through any of the careers or programs with actual knowledge of their rights as citizens and the country's socio-economic, cultural and ecological reality as well as the mastery of a foreign language.

Moreover, it is important because of the fact that university students learn to communicate in a language other than their own. The students attending to universities in Ecuador are Spanish speakers who experience big difficulties at the time of listening to spoken information that are due to the differences existing between Spanish and English. This may cause interference in developing comprehension and oral expression in the English language, evidenced by the poor performance on listening comprehension tests administered.

This research is essential for students to improve all the skills but especially the listening comprehension skill since it is the process in which they actively receive input from which they subtract the meaning and then articulate a response. This is essential in communication and helps students to progress in the learning of a foreign language. The input received by listening focuses on selected aspects for constructing meaning and relate it to the knowledge that already exists. It is the base for speech and cognitive aspects to develop which makes it the most important of the skills. A reason that makes students have difficulties to perform well at this skill in life is that they are immerse mostly in a Spanish speaking environment which is more challenging since they are not constantly practicing the skill [23].

This study influences the methods used in EFL classes because it proposes the audio linguistic method as a factor of development of the listening comprehension skills with the support of E-leaning [25]. The audio-lingual method is defined as an instructional method that refers to English language teaching and centers in listening and speaking. There are various benefits of using audiolingual method. Firstly, this method instructs students in the use of foreign language and provide them the opportunity to be immerse in a context where the language is mainly used by real people and environment. Secondly, although it teaches grammar implicitly, it emphasizes grammatical structures which help students focus on communicating the message efficiently. Thirdly, it enhances listening comprehension and speaking skills which are the goal of learning a language. Moreover, it uses visuals to transmit the meaning of phrases which means that the time that would have been spent in translating can be used to, in fact, practice the language. Lastly, it makes structures that are related with the functions of language easy to learn because the focus is on repetition, correct pronunciation and real use of the language, rather than only theory, for students to internalize the knowledge.

It was selected the level A2 because students already have basic structures and vocabulary which make it possible to work with functions of the language rather than single words. On the other words, they can use the audio-lingual method to reinforce the activities through the implementation of E-leaning. Therefore, the application of the audiolingual method with the support of E-learning is proposed as an objective to encourage the development of listening comprehension skills.

2 State of the Art

Once checked other sources such as databases of international academic journals, there are similar research projects done about either the audiolingual method or the listening skill separately. However, investigations had done about the two variables interacting with each other seem to be scarce, here there are some:

Currently, an educational issue that it is urgent to solve is listening comprehension. It is important because is crucial in language teaching and learning for developing communicative competences. This article addresses the use of linguistic findings when teaching listening comprehension in a foreign language through the linguistic approach. It shows the connection between the speech perception process and the type of language which must be considered when designing comprehension teaching techniques and programs. It reported findings on English words and sentences peculiarities which have implications in teaching a foreign language [18].

A thorough study states that the audio-lingual method and the structural approach has a relative effect on the academic achievement of elementary school students in English language [17]. This experimental study done with a sample of two groups of 30 students, matched using intelligence tests. One was taught with audio-lingual method and the other with the structural approach resulting in equally significant contributions on the academic development of the children.

The theoretical and empirical research done by [13] suggests that metacognition and cognitive strategies improve L2 listening skill. This study was done at a university level French class through observation of five listening based sessions of a second-year class that had 26 students and their teacher during one semester. Data was collected using a teacher self-evaluation questionnaire and structured observation. Resultantly, there was evidence of cognitive work before, during and after listening teaching however, there was lack of indication of explicit metacognitive teaching and strategy assessment.

As stated by [26], the audiolingual method can improve the learning of vocabulary like plural nouns. This investigation was done in the first year class of SMP Advent Surakarta using classroom action research which contained planning, action, observation and reflection. Quantitative and qualitative data was collected subsequently; it was shown that the audiolingual method improved the learning of plural nouns evidenced by the scores of students in posts tests.

According to [11] stated that audio-lingual method was appropriated to teach speaking to students of second semester of English in the University of Riau Kepulua because the meaning of it motivated them to be more active in learning which links ages and the kinds of knowledge units. This method helped the learners to acquire the meaning of the words in English and developed the ability to react which could result in native like competence in English language. A study done in Ecuador by [21] claimed that English was easy to learn if the institutions implemented methodologies that allows students to develop it. The audio-lingual method contributed to improve the level of the students in the target language through the correct pronunciation and grammar as well as accurate and quick responses in conversational situations [6].

The audio-lingual method was used in this classroom action research to provide clear input of vocabulary along with repetition, memorization and drills through media. Observations and tests were used to identify the fact that spoken words were identified through phonetic symbols which was useful on spoken texts. With the audio-lingual method, 30 of the 40 students researched scored the passing grade which was 75. The most relevant similar studies were done about this topic and presented in this section, showed that this investigation could be done in order to gather more information about the correlation between audio-lingual method and listening comprehension skills. This will benefit the teaching and language process of English.

This study is based on two theories which are: Firstly, the Behaviorism theory, which is the base of the creation of Audio-Lingual method [3], was designed by Skinner (1977) who argued that feelings are internal processes that should be studied by the usual scientific methods, emphasizing controlled experiments with both animals and humans. His theory states that human behavior reacts to external stimuli by means of which a behavior is reinforced. This has the intention to make it repeated or exterminated according to the consequences that the stimulus entails. Skinner's theory is based on the idea that events related to learning change or modify our behavior and our ways of acting according to certain circumstances. These changes are the result of the individual response to the stimuli a person experience. The response comes according to the stimulus which are received. These could be positive or negative [19].

Secondly, the Socio Cognitive theory, which took the basis on the behavioral learning and then moved forward, was proposed by [1] who accepts that humans acquire skills and behaviors in an operative and instrumental way, rejecting that our learning is carried out according to the behavioral model. It emphasizes how cognitive factors intervene between observation and imitation which helps the subject to decide whether the pattern is going to be imitated or not. In addition, through a significant social model an unviable behavior is acquired using only instrumental learning.

In Ecuador, higher education students who are enrolled in any major have to obtain a B2 level certificate according to the Common European Framework of Reference for Languages (2002) as stated in Article 31 of CES. In the mentioned article it is also claimed that the level B2 is considered as Sufficiency and students have to acquire it before they enroll in the last academic period of the major chosen which is important for them to continue their studies. That is to say that students need to get the certification as soon as they can, otherwise they will not be able to get their degree. One of the components of the sufficiency exam is the listening comprehension section which is worth 20% of the test. It is important for university students to develop the mentioned skill.

Moreover, Plan Nacional de Desarrollo 2017–2021 of Ecuador, in objective 1 claims that the government will guarantee a dignified life with equal opportunities for all people. That is why the English learning and its improvement as well as use in real life is important so that more people can have the same opportunities.

3 Methodology

The following study was based on the quantitative and qualitative approaches in cooperation and in two theories. The first is Skinner's Behaviorism which highlights positive or negative reinforcement to a certain type of behavior as a way of learning. The second is Bandura's Socio Cognitive theory which states that people not only learn skills through behavior conditioning but also emphasizes that cognitive and social factors intervene in observation and imitation.

It was quantitative because it determined the cause and effect of the hypothesis, which was specified at the beginning of the research, it collected data in an objective way, which represented and summarized in numbers and it will indicate samples to signify the population.

It was qualitative because the research included describing a continuous process which was the teaching and learning process of English. Furthermore, certain pieces of data were summarized in narrative forms that needed to be described.

Finally, this research focused on the study of the behavior in its natural setting because it analyzed cultural aspect influencing an EFL class [9].

The information was acquired directly from the place where the facts were found, in our study it was in UTA Language Center by a direct interaction between the researcher and the students that were taking the A2 level of English. This research examined the statistical relationship or correlation between independent and dependent variables through the use of tests in order to get quantitative data from the study subjects for this reason, the correlational study was chosen.

According to [12] who mentioned that this type of investigative modality "has the purpose of evaluating the relationship that exists between two or more variables" (p. 122). It is therefore proposed the application of the audiolingual method with E-learning support in the development of listening comprehension skills during the teaching and learning process of the English language.

3.1 Subjects

In order to determine the sample of our research, the study subjects were taken into consideration, in this case the students who studied level A2 of English, applied a probabilistic cluster sampling, which was detailed in the following table (See Table 1).

Table 1. Study sample

Sample	
English level A2 students, Control Group	29
English level A2 students, Experimental Group	29
Total	58

This research was conducted at UTA Languages Center. During the academic period March-August 2018. This study took in consideration 58 A2 students.

3.2 Distribution

This research considered two groups for the study. The first was the control group which did not experience the method proposed. This group had 29 participants who were studying A2 level in the regular program at UTA Languages Center and their age ranged from 19 to 21 years old. On the other hand, there were 29 participants who experimented the method and had similar characteristics to the first group.

3.3 Instruments

Application of E-learning. Another aspect that this project included is the application of electronic technologies to access the content taught using the Audio-Lingual method. Nowadays, the electronic component that has been enhanced by the use of Internet has been widely accepted by people of all ages. By using E-Learning, the students can manage their own pace at education [24]. Listening exercise can be practiced at home by students. In this way the learning and teaching process becomes more productive because students can practice independent [15], the main advantages offered by virtual education are the reduction of costs and the attention to more participants, which is not achieved only with the traditional method in a classroom. In addition, the flexibility of schedules is a very important factor because it allows the student to schedule the course in the best possible way. Another interesting advantage is the interaction that the courses generate awakening the interest of the student and helping those who are shy to be the most active in classes through discussion forums and other means of participation.

The experimental group used the platform Cambridgelms. This platform contains three parts which are Assessment, Online Workbook and Extension Activities. The Assessment section contains one progress test per unit which was activated just for the experimental group due to the fact the it contains audios of the functions taught in class which students were assigned as homework. Regarding the Online Workbook, both of the groups were assigned because it contains activities to practice grammar and vocabulary. On the other hand, the extension activities were assigned to the experimental group only as it has listening comprehension activities which students developed on their own at home. The progress of these activities was checked online in order to control the amount of practice the students were getting at home (See Fig. 1).

A digital tool known as Forms from Office 365 was used in order to create the online version of the questionnaires. This tool not only facilitated the process of application of the survey, but also the addition of multimedia to make it more explanatory and the data tabulation.

3.4 Protocols

Regarding the academic performance percentages of student, two classes from the Languages Center were taken for analysis of their scores. The first class was one that underwent a textbook piloting process of UTA Languages Center in the semester March-August 2018. This class used a material that included several exercises of repetition and substitution drills in order to make students memorize a dialogue and, after that, apply the expressions learnt in a new context. On the other hand, a second class was chosen for the study. Although the data was collected during the same period of time, this class did not undergo the textbook piloting process in the mentioned center.

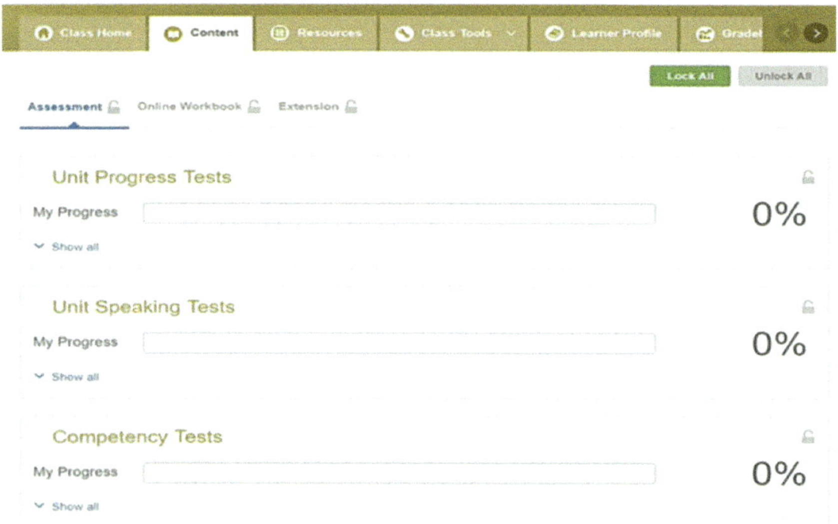

Fig. 1. Image of the support of the platform Cambridgelms

3.5 Data Processing and Analysis Plan

Once the data has been collected the following steps were shadowed:

1. A comparative analysis of the academic performance registers was done which allowed us to delimit the dimensions of the study
2. A deep analysis was done to obtain the most possible knowledge of the data, with the intention of defining all the dimensions of the study which was reached by using the software SPSS Version 21
3. Later on, the answers of the test were classified according to the variables studied considering the objective of the research. The complete set of responses given in the test was considered as data, and the answers given to each item of the questionnaire were considered as analysis unit.
4. Data analysis. By reading the responses to the test, information needed to each dimension, observing the frequency in which each answer appears.

Which lead to:

* Codified the information gathered.
* Group quantitative data in percentages and frequencies.
* Created table to synthesize information.
* Defined the significate categories that constituted the variables.
* Did an analytic study by calculating the performance indicators with statistical figures.
* Interpreted data for the studied context.
* Draw conclusions.

The described process permitted to organize and analyze all the relevant information to test the hypothesis: Using the audio-lingual method with E-learning support develops listening comprehension skills.

4 Results

A comparative analysis was done to determine the effectiveness of the method applied in the experimental group. A diagnostic test was given at the beginning of the semester was considered as the starting point. The midterm and final exams were compared to get full understanding of the progress the students were making in the experimental as well as the control group.

The following charts present the data obtained by control group without the application of the method.

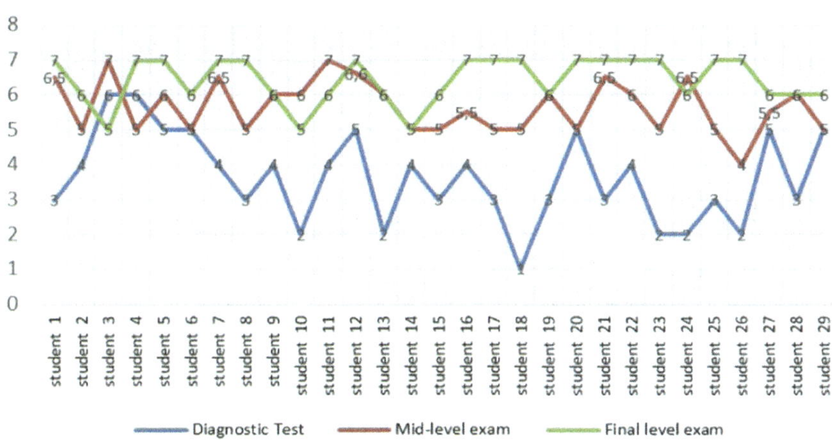

Fig. 2. Performance results without the method

According to the data obtained (See Fig. 2), the academic performance of this control group of students show little improvement. The scores of the listening comprehension evaluation in the diagnostic test implied a poor performance on understanding oral discourse with the highest score being five out of ten (5/10) and the lowest, one out of ten (1/10); whereas the mode score, which showed the score that occurred more frequently, is four out of ten (4/10). Clearly, students at the beginning of the semester demonstrated they had little development in listening comprehension skill due to a variety of facts that may mean that appropriate methods were not used previously.

Regarding the midterm test, it was applied two months and two weeks after the diagnostic test. The scores of the listening comprehension evaluation in the midterm test imply a little improvement compared to the diagnostic test. The highest score turned out to be seven out of ten (7/10) whereas the lowest one was four out of ten (4/10). The mode score, which showed the score that occurred more frequently, was five out of ten (5/10) which showed that most students got a score below the passing grade which was seven (7/10) in Universidad Técnica de Ambato. On the other hand, the average score, that was five point six out of ten (5.6/10), showed an improvement of two point four points (2.4) considering that the average score of the listening comprehension evaluation in the diagnostic test was three point two out of ten (3.2/10).

Concerning the final test, it was applied two months and three weeks after the midterm test. The scores of the listening comprehension evaluation in the final test suggest a slight improvement compared to the midterm test. The highest score was seven out of ten (7/10) whereas the lowest one was five out of ten (5/10). The mode score, which showed the score that occurs more frequently, was seven out of ten (7/10) which showed that most students got a score just on the passing grade which was seven in this university. In contrast, the average score, which was six point four out of ten (6.4/10), displays an improvement of zero point eight points (0.8) considering that the average score of the listening comprehension evaluation in the midterm test was five point six out of ten (5.6/10).

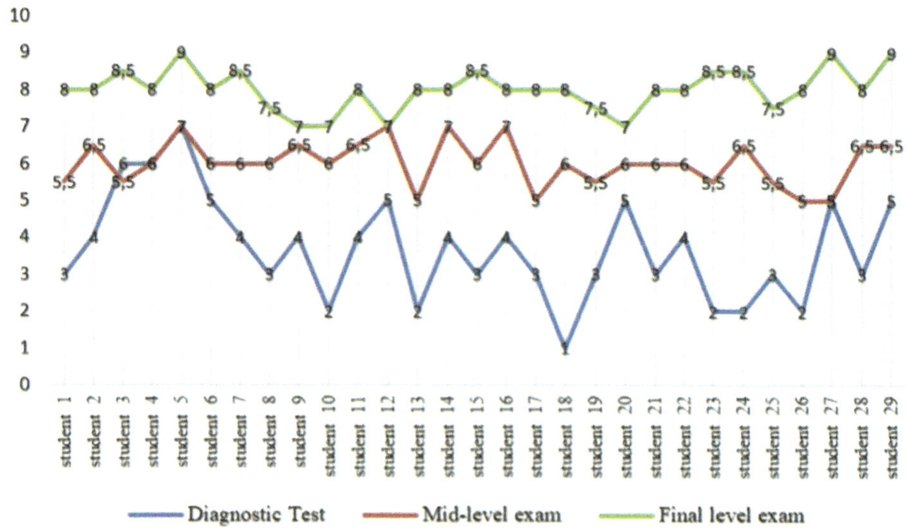

Fig. 3. Performance results without the method

According to the data gathered, the academic performance of this experimental group of students showed a significant improvement. The tests applied contain evidence on this fact (See Fig. 3).

The scores of the listening comprehension evaluation in the diagnostic test implied a poor performance on understanding oral discourse with the highest score being six out of ten (6/10) and the lowest, one out of ten (1/10); whereas the mode score, which showed the score that occurred more frequently, is three out of ten (3/10). Clearly, students at the beginning of the semester demonstrated they had little development in listening comprehension skills due to a variety of facts that may mean that appropriate methods were not used previously.

Regarding the midterm test, it was applied two months and two weeks after the diagnostic test. The scores of the listening comprehension evaluation in the midterm test imply an improvement compared to the diagnostic test. The highest score turned out to be seven out of ten (7/10) whereas the lowest one was five out of ten (5/10). The mode score, which showed the score that occurs more frequently, was six out of ten (6/10) which showed that most students got a score a little below the passing grade which was seven out of ten (7/10) in Universidad Técnica de Ambato. On the other hand, the average score, which was six out of ten (6/10), showed an improvement of two point three points (2.3) considering that the average score of the listening comprehension evaluation in the diagnostic test was three point seven out of ten (3.7/10).

Concerning the final test, it was applied two months and three weeks after the midterm test. The scores of the listening comprehension evaluation in the final test suggested a significant improvement compared to the midterm test. The highest score was nine out of ten (9/10) whereas the lowest one was seven out of ten (7/10). The mode score, which shows the score that occurred more frequently, was eight out of ten (8/10) which showed that most students got a score above the passing grade which was seven (7/10) in this university. In the same way, the average score, which was eight out of ten (8/10), displayed a noteworthy improvement of two points (2) considering that the average score of the listening comprehension evaluation in the midterm test was six out of ten (6/10).

4.1 Statistical Results of the Intervention

(See Figs. 4, 5 and 6).

	Groups	N	Media	Typical Deviation	Typical. Error. Of the mean
Listening Comprehension	Control	**29**	**6,4138**	,68229	,12670
	Experi-mental	**29**	**8,0000**	,56695	,10528

Fig. 4. Difference test

	Kolmogorov-Smirnov[a]			Shapiro-Wilk		
	Statistic	gl	Sig.	Statistic	Gl	Sig.
Listening Comprehension	,178	58	,000	,930	58	,003

Fig. 5. Difference test

		T test for means equality		
		Sig. (bilateral)	Means difference	Tip error of the difference
Listening comprehension	Equal variances have been assumed			
	No equal variances have been assumed	,000	-1,58621	,16473

Fig. 6. Difference test

4.2 Final Decision

After the intervention, it was evaluated whether the samples, of the experimental group that underwent the application of the audiolingual method, presented significant improvements or not. For which the t-test for independent samples was used; first, the normality of the data was established, the second step was to apply a t test for independent samples and compare the results of the group. In this sense, the p value of the experimental group shows a value of 0.000 there was a significant difference of the sample of the experimental group. Therefore, with a Sig. (Bilateral) of, 000 it was concluded: The intervention of E-learning and audio-lingual method obtained significant improvements in the listening comprehension of the A2 students of English of the Languages Center-UTA.

5 Conclusions

According to the analyzed data, it was determined that after the application of the audio-lingual method with the E-learning support, the students of the experimental group obtained better scores with average values of 8.0/10.0; while the students of the control group obtained average values of 6.4/10.0. In conclusion, there is a significant increase with the implementation of audiovisual resources to develop pronunciation.

With regard to the listening comprehension skill, it can be concluded that the repetitions in each class and at home through E-learning exercises help students to recognize how words in phrases sound together and thus strengthen the vocabulary and pronunciation. This fact makes effective communication stronger, which allows students to express their ideas clearly and fluently.

On the other hand, the information collected from the Languages Center students established that the E-Learning implementation and the audio-lingual method works as a support instrument in the process of teaching listening comprehension.

References

1. Bandura, A., Walters, R.H.: Social Learning Theory, vol. 1. Prentice-hall, Englewood Cliffs (1977)
2. Cigdem, H., Ozturk, M., Topcu, A.: Vocational college students' acceptance of web-based summative listening comprehension test in an EFL course. Comput. Hum. Behav. **61**, 522–531 (2016). https://doi.org/10.1016/j.chb.2016.03.070. https://linkinghub.elsevier.com/retrieve/pii/S0747563216302436
3. Condori, E.: La comunicación oral como clave para el aprendizaje del español en la escuela sueca.: El empleo del método audio-lingual en las lecciones de español del åk 6, 7, 8 y 9 de dos escuelas suecas (2019)
4. Consejo de Educacion Superior: Reglamento de Regimen Academico, vol. 51. Quito, Ecuador (2017)
5. Council of Europe. Council for Cultural Co-operation. Education Committee. Modern Languages Division. Common European Framework of Reference for Languages: learning, teaching, assessment. Cambridge University Press (2001)
6. Foote, J.A., Trofimovich, P., Collins, L., Urzúa, F.S.: Pronunciation teaching practices in communicative second language classes. Lang. Learn. J. **44**(2), 181–196 (2016). https://doi.org/10.1080/09571736.2013.784345. http://www.tandfonline.com/doi/full/10.1080/09571736.2013.784345
7. Garcia, C., Naranjo, J., Alvarez-M, E., Garcia, M.: Training virtual environment for teaching simulation and control of pneumatic systems. LNCS, vol. 11613, pp. 91–104 (2019). https://doi.org/10.1007/978-3-030-25965-5_8
8. Gilakjani, A.P., Sabouri, N.B.: Learners. listening comprehension difficulties in English language learning: a literature review. English Lang. Teach. **9**(6), 123 (2016). https://doi.org/10.5539/elt.v9n6p123
9. Harooni, M., Pourdana, N.: Politeness and indirect request speech acts: gender-oriented listening comprehension in Asian EFL context. Int. J. Appl. Linguist. Engl. Lit. **6**(2), 214–220 (2017). https://doi.org/10.7575/aiac.ijalel.v.6n.2p.214
10. Hasanvand, Z.M., Mehrdad, A.G., Tabar, M.M.A.E.: The effect of teaching prosodic features on pre-intermediate EFL learners'listening comprehension. Mod. J. Lang. Teach. Methods **6**(1), 475 (2016)
11. Herawati, N.: Audio-lingual method as method in improving speaking ability of second semester of english department students of unrika. J. DIMENSI. **1**(3) (2016)
12. Hernández, R., Fernández, C., Baptista, M.: Metodología de la Investigación. McGraw-Hill, México (2014)
13. Hernandez Wilson, J., Izquierdo, J.: Metacognicion y comprension oral en l2: Observacion de la practica docente en nivel universitario. Rev. Electronica de Invest. Educativa **18**, 39–52 (2016)
14. EF English Proficiency Index: EF English proficiency index trends. Technical Report (2015). http://www.ef.com/epi
15. Karimi, F., Chalak, A., Biria, R.: Pedagogical utility of pre-listening activities for improving iranian elementary EFL learners' listening comprehension. Int. J. Instr. **12**(1), 1127–1140 (2019)

16. Little, D.: The common European framework of reference for languages: content, purpose, origin, reception and impact. Lang. Teach. **39**(3), 167–190 (2006)
17. Mallick, M., Bhushan, S.: Academic achievement in English language: relative effect of audio-lingual method and structural approach. Man India **96**, 1323–1331 (2016)
18. Masalimova, A., Porchesku, G., Liakhnovitch, T.: Linguistic foundation of foreign language listening comprehension. Math. Educ. **11**(1), 123–131 (2016). https://doi.org/10.12973/iser.2016.21012a. https://www.scopus.com/inward/record.uri?eid=2-s2.0-84957559889&doi=10.12973%2fiser.2016.21012a&partnerID=40&md5=bf8363981f5084188199a6af5c90d24d
19. Morales, Y.A.: Revisión teórica sobre la evolución de las teorías del aprendizaje. Revista Vinculando (2018)
20. Ngo, N.T.H.: The impact of listening strategy instruction on listening comprehension: a study in an English as a foreign language context. Electron. J. Foreign Lang. Teach. **13**(2), 245–259 (2016)
21. Ramos Cuadrado, C.M., Cuadrado, R., del Rocìo, M.: The use of audio lingual method to develop listening and speaking skills in the students of 10th C at Chambo high school during the school year 2016-2017. B.S. thesis, Riobamba, UNACH 2016 (2016)
22. Samawiyah, Z., Saifuddin, M.: Phonetic symbols through audiolingual method to improve the students' listening skill. Dinamika Ilmu **16**(1), 35–46 (2016)
23. Schneider, B.A., Avivi-Reich, M., Daneman, M.: How spoken language comprehension is achieved by older listeners in difficult listening situations. Exp. Aging Res. **42**(1), 31–49 (2016). https://doi.org/10.1080/0361073X.2016.1108749. http://www.tandfonline.com/doi/full/10.1080/0361073X.2016.1108749
24. Singh, R.N., Hurley, D.: The effectiveness of teaching and learning process in online education as perceived by University faculty and instructional technology professionals. J. Teach. Learn. Technol. **6**(1), 65–75 (2017). https://doi.org/10.14434/jotlt.v6.n1.19528. https://scholarworks.iu.edu/journals/index.php/jotlt/article/view/19528
25. Tahir, I.: Teachers beliefs in balancing linguistic competence and teaching performance in EFL classrooms. BRAIN. Broad Res. Artif. Intell. Neurosci. **9**(1), 50–58 (2018)
26. Utomo, A.A.P.: Improving students achievement in learning plural noun through the using of audio–lingual method.(a classroom action research of the first year of smp advent surakarta in the academic of 2013/2014). Karya Ilmiah Mahasiswa Progdi Pendidikan Bahasa Inggris FKIP **2**(2) (2016)

Experiments on a Mashup Web-Based Platform for Increasing e-Participation and Improving the Decision-Making Process in the University

Víctor Peñafiel(✉) and Hernando Buenaño

Technical University of Ambato, Ambato, Ecuador
{vi.penafiel, edwinhbuenaniov}@uta.edu.ec

Abstract. Nowadays, there are two important techno-social confluent phenomena: the increasing interest of citizenship to participate on the public decision-making process and the rising use of ICTs in all vital areas. For this reason, there are many participation instruments being developed and implemented using ICTs and Internet. This paper aims: to deploy a generic and flexible ICT architecture for supporting e-participation in the university and to present the results of several experiments carried out with this ICT tool. This web-based platform supports many tasks such as debate, dissemination of information, vote, clarifying issues concerning the university community, use of questionnaires, among others. The stakeholder decides how to manage each task. The use of ICTs and Internet will allow to take advantage of the necessary technical hardware in order to enhance the university processes and tasks. The experiments were carried out in the Technical University of Ambato, in four different scenarios of the decision-making process. The obtained results may be used as reference by other universities or groups needing to improve the quality of the e-participation and decision-making processes with the use of ICTs.

Keywords: Participatory tasks · e-participation · e-Democracy

1 Introduction

The mission of universities in society is not only to train researchers or professionals, but also to form citizens that will participate later in the democratic life and in the decision-making processes. In this respect, the ICTs provide several tools that help in democratic practices.

We can mention not only the Internet-related technologies but also some fields like Artificial Intelligence [1], information security and group decision-making techniques [2]. A great number of participation instruments can be included among these tools. These instruments, being used with the proper electronic support, can increase and improve not only the participation of the university community but also the quality of the decision-making process. We must consider that these instruments are a combination of basic tasks of participation. Likewise, it has been mentioned that the participation methods implemented in the universities are still traditional [3]. So far, these

J. Nummenmaa et al. (Eds.): CSEI 2019, AISC 1078, pp. 99–110, 2020.
https://doi.org/10.1007/978-3-030-33614-1_7

participation tasks have been carried out through onsite meetings. In this direction, ICTs and Internet use has allowed to take advantage of the existent web support and enhance the participation tasks process by setting up different types of web architectures for participation. The use of these platforms has resulted in a better structured communication, more debate and sharing of ideas, more participation of the university community, a web consensus and a rise on the ICTs usage, among other positive aspects.

Some examples of the above-mentioned participation techniques based on ICTs are: PARBUD [4] and MyUniversity. However, due to the high cost involved in the design and implementation of this sort of architectures, it is necessary to check if they are really useful and could help in achieving the desired results. For this reason, taking advantage of web services implementation growth, the establishment of new technologies as mashup [5], and considering that there is free software available on the network, we propose an experimental web-based architecture for online participation.

This platform has an "ad hoc" [6] methodology for specific problems in the university scenario. This participation instrument, which is a hybrid application importing content from various web applications [7] to create something new, can reduce or even eliminate the costs and can also increase the efficiency of participation tasks in the university. Moreover, this mashup gives the option to be used in different ways according to the analysis cycle of decisions and it also allows to add new methods for supporting groups. At the same time, it is generic and flexible which results in a cutting-edge participative methodology leading to a more mobile and electronic-based decision-making process. In addition, the information volumes and procedural requirements have grown exponentially.

The aim of this study is to obtain an experimental evaluation of the usability of the mashup platform built to provide a more effective and efficient support to the participation tasks in the Technical University of Ambato (UTA). In general, the participation tasks carried out in the UTA are the same of any educational unit. Finally, this participative methodology not only facilitates the decision-making process in a group, but it can also improve the quality of the participation of the university community.

2 Materials and Methods

The e-democracyUTA platform can provide a method for creating and supporting the participation instruments in real and specific problems. Its architecture defines a web system that gives support to the group decision-making process and it could be suitable for any university (see Fig. 1). Same ways, it achieves a fluid communication between the obtained web services, which represent the participative tasks. The architecture is designed to add more web applications. It uses the Decision Analysis Cycle [8] for a better understanding of a given problem and for organizing correctly the thoughts and communication inside the group. In this sense, new methods are added for supporting groups. Therefore, with the resulting web services used by e-democracyUTA and following the Decision Analysis Cycle, four experiments were carried out. For using this experimental architecture, feasible participative tasks of the university were chosen in order to complete their execution on the platform. This mashup platform can use various web applications corresponding to many participative tasks such as debate, vote, use of questionnaires, dissemination of information, explanations to the university community [9], election of representatives, preparation of final documents among others.

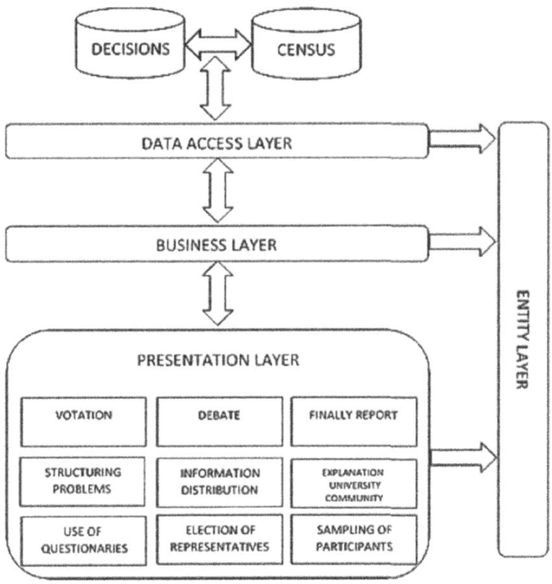

Fig. 1. Architecture of e-DemocracyUTA.

The e-DemocracyUTA platform is based on a layered object-oriented architecture [10]. This way, the code is better organized and structured in three layers: view layer, business layer, and data model layer. Therefore, the scalability of the software is increased. Two databases, both implemented in MySql [11], are deployed in Soomee. This cloud service assists the communication between the cloud and the e-DemocracyUTA platform and its databases. The data access layer oversees the interaction and communication with the actual database for processing data. The business layer oversees the update of data considering the business rules. The business layer is an intermediary between the presentation layer and the data model layer. The interfaces for the interaction with end users are implemented in the presentation layer. This layer represents the web applications created for each functionalities of the proposed architecture.

The operativity of the platform is characterized by the smooth engage of different technologies. These technologies include web sockets, Java Script (JQuery in particular) and Bootstrap. The web sockets technology supports the client-server communication. This technology is particularly important in the real time interaction between participants. JQuery is used in the validation of data forms. Together with Bootstrap, a friendly and adaptative interface is presented to the end user. JQuery is used together with the web sockets too. This technology allows to send asynchronous requests to the server and json responses are naturally instantiated and processed in the client layer of the application. The response time that this technology achieved makes it an enriched application in which the complexity of the interaction with the server layer of the application is hidden to the end user.

The extension of the style sheets by using Bootstrap facilitates the construction of a professionally aesthetic experience to the user. The web applications are encoded in ASP.NET by using Visual Studio. Finally, form the security point of view, it is important to mention that user passwords are encoded and stored as in the database as a MD5 hash. This way, the credentials of users are safe from ill-intentioned people in the TCP/IP channel.

2.1 e-DemocracyUTA Use at the Technical University of Ambato

For the experiment on the debate process, the participative tasks used were dissemination of information, debate, vote, explanations to the university community. Meanwhile, for experimenting on information display, the participative tasks that were used are dissemination of information, debate, vote, preparation of final documents and explanation to the university community. Concerning the election of representatives, the implicit participative tasks were dissemination of information, debate, vote and explanations to the university community. Finally, for the experiment on the use of questionnaires, the tasks used were the following: use of questionnaires online, dissemination and/or display of information and preparation of final documents.

For controlling the frontend and the right performance of the e-DemocracyUTA platform, the technical staff worked directly on the backend of the platform. In addition, all the processes executed with e-DemocracyUTA were developed using the client/server model. Also, all participants in the study could interact with the platform from any device connected to the Internet.

3 Experiments and Results

Four experiments were carried out with the e-DemocracyUTA platform in the UTA to check the effectiveness of the tasks (see Table 1). There were a total of 2558 participants among professors, students, administrative personnel and employees as members of the university community. After the experiments, a satisfaction survey was conducted regarding the usability of the platform.

3.1 Experiment 1: Debate and Vote

The e-democracyUTA platform includes web services of online debate forums for allowing the participants in the study to post their opinions and comments from any connected device and at any time. Moreover, the forums history is downloadable in order to be analyzed through text mining if necessary. The e-democracyUTA platform makes it easier to manage and store distribution lists, it allows debating online, consulting information and disseminating documents through the different web applications contained in the platform.

Table 1. Experiments performed with e-Democracy UTA

Task	Participants	Number of participants	Module to test
Debate and vote about installing another gate in the campus Ingahurco	Students Professors Administrative employees	2541	Debate and vote
Information analysis in the academic board in the Faculty of Information Technology, Telecommunications and Industrial Engineering	Professors	7	Information display
Election of representatives in a group of the Law Faculty	Students	10	Vote
Survey	Students Professors Administrative employees	2558	Use of questionnaires

For the task concerning the debate and vote, 2541 participants took part in the experiment. The population considered for the study was a campus of the UTA located in Ingahurco, where professors, students, administrative personnel and employees participated. The question that initiated the debate was the following: "Do you agree to have another access gate in the Campus of Ingahurco for the use of the university community?" Likewise, this question was used to carry out the vote. The forum was open for 15 days. After the participation of the university community members and when the deadline was met, the results were published in the website of the e-DemocracyUTA platform. The administrator played the role of moderator and stakeholder. Depending on the debate topic, the task could also be delegated to a participant.

Next, the process involved in the debate and vote experiment is detailed. In the first instance, the moderator created the corresponding group. Then, the participants received a user and password via email in order to access the platform. Once the group members logged in, they located the debate topic and the related forum links. By using a debate thread and the appropriate web application, the participants published their opinions. Afterwards, a vote was held using a web service with simple majority system. Finally, the vote results were displayed on the platform website keeping all participants posted (see Fig. 2).

For executing the task related to explanations to the university commgunity, according to the results published by the e-DemocracyUTA platform, the majority of participants were in favor of the motion. That is, 95.99% agreed to have another campus access gate in Ingahurco, while 2.24% was against and 1,77% were uncertain.

Finally, after having the vote participative task completed, another e-task was consequently deployed: explanations to the university community concerning the new access to the campus.

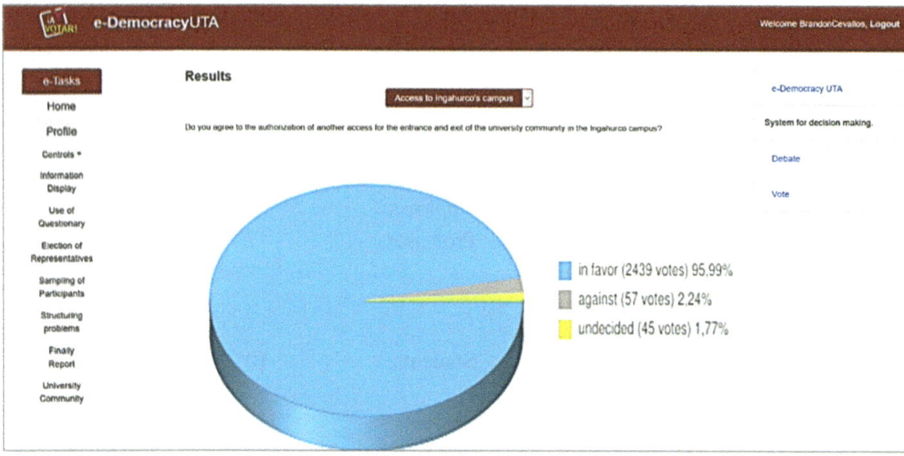

Fig. 2. Vote results

3.2 Experiment 2: Information Display for Decision-Making in the Academic Board of the IT, Telecommunications and Industrial Engineering Faculty (FISEI)

The second experiment consisted on previously uploading a document to the platform, having its content analyzed by the Academic Board of FISEI in the UTA. The document was related to an authorization granted by the Academic Board for a student to change his/her career. The seven members of the Academic Board checked the document using the platform and shared their comments and opinions. Later, a face-to-face meeting of the Board was held to decide on this case. By executing the vote task, the request of this student was approved. The experiment results were that, on one hand, due to a prior exchange of opinions between the group members, the sessions resulted to be more agile and shorter. On the other hand, time was saved by reviewing complex documents in advance. So the face-to-face meeting became more enriching and consequently, the decision-making process a lot easier. In addition, the information stored on the platform can be downloaded more easily, anytime it is needed in a face-to-face session. Once the Academic Board meeting ended, the final decision made was published through the platform (see Fig. 3).

For carrying out this experiment, firstly, the forum moderator, that is, the platform administrator, created a corresponding group, as it was made in the task of debate and vote. Secondly, the document was uploaded to the platform. To become users of the platform, the Academic Board members received a user and a password to sign in. In the main page of the platform, through the forums, the group members selected the corresponding topic and downloaded the document that they had to analyze. For sharing opinions on this document, creating a conversation thread with any Academic Board member or with the forum moderator, the procedure was the same that in the debate and vote experiment. Then, the moderator collected the opinions published on the platform. Finally, the decisions made by the Academic Board were displayed on the

Fig. 3. Results of decision-making process by the Academic Board of FISEI

e-democracyUTA platform keeping every participant posted. As we can see, this task originated other e-tasks such as debate, alternative generation, preparation of final documents and explanations to citizens.

3.3 Experiment 3: Election of Representatives in a Group of the Law Faculty

For the third experiment, one representative of presidents was elected in a group of the Law Faculty in the UTA. The ten candidates that intervened in this election represented the presidents of each group of each semester. These ten students were asked to participate through the platform. Next, by using the open source software applications available on e-democracyUTA, the election of the president was carried out. When the voting process ended, all participants had been able to cast their votes, being candidate 1 the winner. This result was immediately displayed as shown in Fig. 4.

The election was held though the following process. First, the invitation to the participants was sent via email. Once the invitation was accepted, the participants used the link in the email massage to access the platform. Using a web service of the platform, the participants wrote their comments and preferences about the candidates for the election. After selecting the candidates, the students proceeded to vote by using one of the web applications on the e-democracyUTA platform. Lastly, the results were displayed on the platform in order to keep the participants posted. At the same time, with the execution of this participation instrument, other participative e- tasks emerged, for example the debate, modelling preferences, vote, explanations to the university community and preparation of final documents.

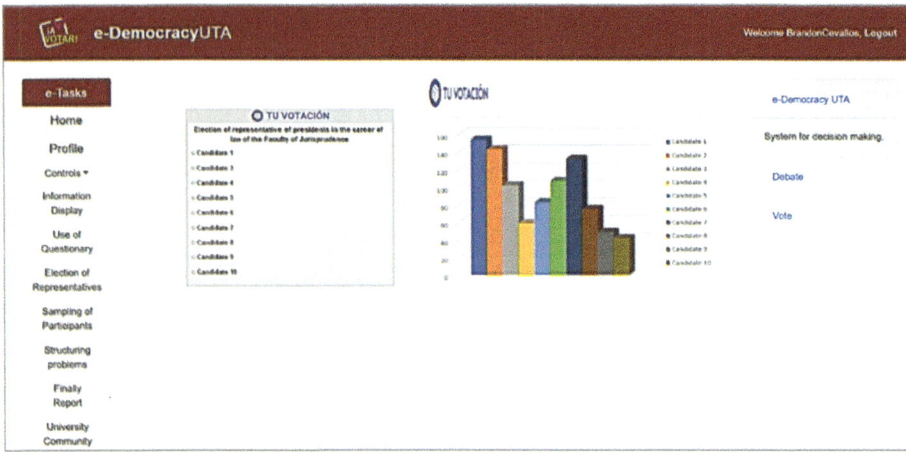

Fig. 4. Results of the election.

3.4 Experiment 4: Use of Questionnaires in the Satisfaction Survey Related to the Use of e-DemocracyUTA

This e-task helps to identify the main interests of the participants. At the end of the experiments and for knowing the user's satisfaction with the use of the e-democracyUTA, a survey was conducted among the 2558 participants. In general, the questions included referred to the use of virtual platforms, functionality and platform security, increasing participation and decision-making through e-democracyUTA, as well as the level of acceptance in the use of the architecture. Seven questions were included in the survey, six with a Yes or No format and one with Likert scale [12].

The detailed description of the experiment is the following: first, by using the email of all participants in the experiments, an invitation to do the survey was sent to them. Second, once the invitation was accepted, the users used the link included in the email message to enter the platform. Third, the participants finished the satisfaction survey through the corresponding web application of the platform. Finally, this information was stored for its further analysis.

On the other hand, concerning the first survey question, whose results are displayed in Table 2, related to the use of other platforms, it is shown that approximately three quarters of the participants had already used other e-participation platforms. In this sense, another experimental platform called MyUniversity was previously used in the UTA. This previous experience contributes to continue promoting electronic participation and it is in correspondence with the growing digital culture present in the universities

Regarding the third question, whose results are shown in Table 3, it is shown that most participants consider that the e-participation can be increased with the use of e-democracyUTA. We believe that this is because the university community is mainly composed of students, also called digital natives that belong to the technological era.

The fourth question was related to the ease of use of the platform. The results of the survey show that most participants think that the platform is fast and easy to use.

Table 2. Have you used an e-participation platform before?

Options		Frequency	Percentage
Valid	Yes	1916	74.91
	No	642	25.09
Total		2558	100.00

Table 3. Do you think that the e-democracyUTA platform promotes e-participation in the university community?

Options		Frequency	Percentage
Valid	Yes	1723	67.36
	No	364	14.23
	IDK	471	18.41
Total		2558	100.00

Nowadays, users demand instant response and mobility on the Internet due to the evolution of the web. In fact, our architecture is conceived for that web environment. Regarding the other participants who answered *no* to the question, it is possibly due to an incorrect usage of the platform (Table 4).

Table 4. Do you think that the e-democracyUTA platform is easy to use?

Options		Frequency	Percentage
Valid	Yes	2398	93.75
	No	89	3.48
	IDK	71	2.77
Total		2558	100.00

Concerning the fifth question, which was about the security and reliability of the platform, the results shown in Table 5 make us consider that most participants consider the architecture as reliable. The reason to this is because the participants are aware that now, most of the information systems are strongly secured with the use of commonly used information security technologies. With respect to the rest of participants, we think they didn't trust the platform because they do not rely on doing internet transactions or any other online activity, maybe due to the overabundance of information and software piracy.

The sixth question was related to whether e-democracyUTA facilitates the decision-making process or not. The results are shown in Table 6. We see that most of the university community is in favor. It is undeniable that every aspect of life is evolving with technology and that for people, especially for the digital natives, Internet dependency is an issue. This is the case of the university community of the UTA, mainly formed by young students born in the technologic era. That is why, they prefer

Table 5. Do you think that the e-democracy UTA plataform is reliable?

Options		Frequency	Percentage
Valid	Yes	1691	66.11
	No	277	10.83
	IDK	590	23.06
Total		2558	100.00

that important decisions are made online, through the different web applications available. On the contrary, regarding the rest of participants who answered no, perhaps they lack knowledge about of the new methodologies used in the present for decision analysis.

Table 6. Do you think that the e-democracyUTA facilitates the decision-making process?

Options		Frequency	Percentage
Valid	Yes	2256	88.19
	No	94	3.68
	IDK	208	8.13
Total		2558	100.00

From the results of the seventh question, we can observe in Table 7 that about 50% of the participants think that the platform does affect the human relationships in the university community. We believe that this is due to the overabundance of information and to the network technologies, leading to an overwhelming and drowning sensation among the users having to deal with an enormous amount of information. Nowadays, young people value more a social media post or WhatsApp message than a good conversation. University community is not exempt of this reality, so this community can successfully interact with the suggested platform.

Table 7. Do you think that the usage of the platform leads to a less personal interaction within the members of the university community?

Options		Frequency	Percentage
Valid	Yes	1352	52.25
	No	851	32.88
	IDK	385	14.87
Total		2588	100.00

Finally, concerning the results of the eighth question, displayed in Table 8, we observe that there is a high level of satisfaction, as most of the participants found that e-democracyUTA is a reliable tool, secure and useful for e-participation and decision-making in the UTA.

Table 8. What is your satisfaction level in the use of e-democracyUTA?

Options	Frequency	Percentage
Totally satisfied	1875	72.45
Satisfied	692	26.74
Little satisfied	12	0.46
Not satisfied	9	0.35
Total	2588	100.00

4 Conclusions

There is a countless number of participative tasks which already have a solution on the web but the one that is possible most known is the voting system. The proposed platform supports a better administration of the public politics at the university by encouraging the e-participation of the university community and by supporting the decision making in the institution.

The experiments that were carried out proved that the support of the Web 2.0 turns out to be less expensive and more efficient for the construction of a participatory instrument [13]. The platform e-democracyUTA is perceived as an attractive tool for the e-participation. The distributed environment has been exploited for supporting the decision making. This way, the platform supports the increase, not only of the transparency of decision-making processes in the university community, but also the participants' satisfaction. The scalability of the platform makes it possible to integrate a diverse amount of web applications for addressing real issues of the university community.

As a result of this study, three user roles were identified. These roles include the stakeholder, the participant-each member of the university community- and, the manager-member of the technical personnel. By knowing the participative tasks allows to solve the participation problem with the most proper instrument.

Acknowledgement. To the Technical University of Ambato, for its support to carry out these experiments in campus Ingahurco and campus Huachi Chico.

References

1. Azar, M.A., Tapia, M., García, J.L., Pérez, A.J.M.: Inteligencia artificial de las cosas. In: XXI Workshop de Investigadores en Ciencias de la Computación, Universidad Nacional de San Juan, Argentina (2019)
2. León, Y.O.L., Pravia, M.C.P., Delgado, F.M.: Procedimiento para la selección de la Comunidad de Expertos con técnicas multicriterio. Ciencias Holguín **22**(1), 34–49 (2016)
3. Peñafiel, V., Lavín, J.M.: Improving university decision making through e-Participation. In: 2018 International Conference on eDemocracy & eGovernment (ICEDEG), pp. 392–396. IEEE, Ecuador (2018)
4. Peñafiel, V.F., Guerrero, J.F., Lavín, J.M.: Experiments with a university e-Participation platform. Soc. Networking **6**(03), 239 (2017)

5. Ghiani, G., Paternò, F., Spano, L.D., Pintori, G.: An environment for end-user development of web mashups. Int. J. Hum. Comput. Stud. **87**, 38–64 (2016)
6. Rásury, R., Xavier, M.: Desarrollo de una herramienta de soporte metodológico a los procesos de e-participación. Master thesis (2019)
7. Haro, E., Guarda, T., Peñaherrera, A.O.Z., Quiña, G.N.: Backend development for web applications, Restful Web Services: Node. js vs Spring Boot. Revista Ibérica de Sistemas e Tecnologias de Informação (E17), pp. 309–321 (2019)
8. Capitán, Á.J.O.: Guía para el análisis de problemas y toma de decisiones, 2nd edn. ESIC Editorial, Madrid (2018)
9. Peña, R.M., Gómez, M.D.C.F., Barreiro, L.T., Rodríguez, K.V., Vera, M.A.V., Vera, A.L. V.: University social responsibility in the current knowledge society: a necessary approach. MediSur **15**(6), 786–791 (2017)
10. García-Holgado, A., García-Peñalvo, F.J.: Introduction to object-oriented análisis. Grupo GRIAL, Universidad de Salamanca (2018)
11. Arias, M.Á.: Aprende Programación Web con PHP y MySQL: 2nd edn. IT campus Academy, Colombia (2017)
12. Matas, A.: Diseño del formato de escalas tipo Likert: un estado de la cuestión. Revista electrónica de investigación educativa **20**(1), 38–47 (2018)
13. Cano, D.R., Gómez, J.I.A., Moro, F.J.G.: Metodologías colaborativas en la Web 2.0. El reto educativo de la Universidad. REDU: Revista de Docencia Universitaria **17**(1), 5 (2019)

GLORIA: A Genetic Algorithms Approach to Tetris

Diana Patricia Quintero Lorza(✉), Néstor Darío Duque Méndez, and Jacobo Andrés Gómez Soto

Universidad Nacional de Colombia,
Carrera 27 #64-60, Manizales, Caldas, Colombia
{dpquinterol,ndduqueme,jaagomezso}@unal.edu.co

Abstract. Tetris is a popular videogame developed by Alexey Pajinov, where the player will never win. This special feature makes it a popular reason to be used in Artificial Intelligence and Computational Intelligence techniques research to study the improvement in a game's performance. In this work, a genetic algorithms based approach will be presented, applying it to a game engine with some of the modern Tetris' gameplay features, where the objective is to observe the game agent performance in various test scenarios. The obtained results show the feasibility of using AI as a player in a game of Tetris.

Keywords: Genetic algorithms · Tetris · Artificial Intelligence

1 Introduction

The usage of intelligent systems techniques for solving various problems has become a popular practice; implementing these techniques in several game engines, either programmed in-house or with more complex engines such as [1–4] is no exception. This is a trend in Artificial Intelligence research, often making use of simplified versions of games with simple interactions and less details [5, 6]. Those games are mainly used to study the evolving performance that those game engines can show. In this work we show an evolutionary algorithm based on genetic algorithms, which learns how to play a game of Tetris with specific features. Tetris was chosen because as [7] states, the solution to games in Tetris can only be validated with a trial and error process.

Tetris is a popular computer game for one player, invented by Alexey Pajinov in 1985. As Font et al. state, it consists in the placement of tetramino pieces which are falling sequentially at an increasing speed in a board which usually has 10 columns and 20 rows. In this conditions, only two types of movements are allowed, a rotation and a horizontal movement before they reach the end of the game board [8]. The main goal of the game is to generate and clear the most amount of horizontal lines created with the tetraminos. The game ends when the number of pieces reach the height of the game board. The Fig. 1 shows the tetraminos and their rotations.

J. Nummenmaa et al. (Eds.): CSEI 2019, AISC 1078, pp. 111–126, 2020.
https://doi.org/10.1007/978-3-030-33614-1_8

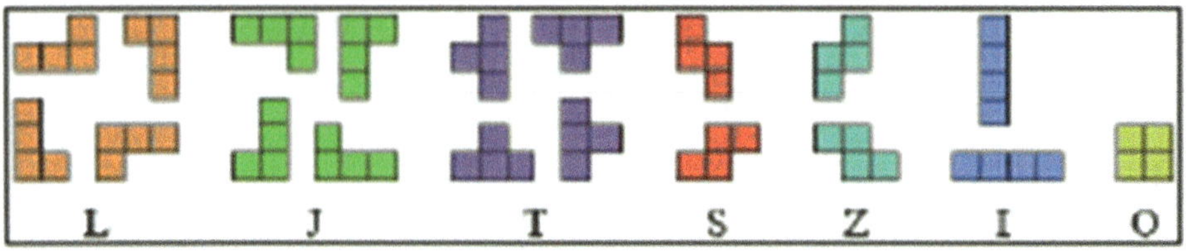

Fig. 1. Tetraminos and their rotations [8].

In [9], the authors state that Tetris is a game against nature because the piece selection for each game is generated randomly. Another of the game's main features, as presented by [10] is that due to the inexistence of a limit for the duration of a game, the player will always loses, no matter which method is used.

In this paper, as previously mentioned, we present the utilization of a genetic algorithm for studying the evolutionary efficiency on a Tetris game engine. First, we show a section focused in the literature review about the use of different intelligent systems techniques on the game of Tetris, specifically on using genetic algorithms (related works). After that, we explain, in a detailed manner, the methodology use for this research; in top of that, we included a subsection about the tests done. Following this line, we continue presenting a discussion in order to interpret the results obtained from the tests (analysis and discussion of results). Finally, we arrive to conclusions about the work done and we present some ideas for future works in order to get better findings (conclusions and future work).

2 Related Works

Tetris is a game which has drawn attention from a lot of AI researchers, because it allows the application of various techniques for results improvement, those techniques can be novel due to their playstyle. These techniques are mainly focused in the usage of intelligent agents as presented by the following authors.

In [11] the authors propose the usage of least squares method applied to learning. In this work although the method is not presented in a detailed way in various study cases, the authors propose in one of them the utilization LSPI in a Tetris game, where the authors affirmed that due to the usage of these methods they reached a number of cleared lines between 100 and 3000. Although the game board size is never specified in [1], taking the number of possible states is safe to say that the game board had a size of 10×20.

In [12, 13] the authors present the utilization of Gaussian processes in order to train a Tetris game agent. The authors propose using these processes as a learning regression technique, because according to the authors, Gaussian processes can help to improve and compete using regression trees and instance-based regressions. For this particular case, in [12] the authors achieved with the Tetris game agent an average of 50 lines with a maximum increment of 120 lines.

In [14] the authors used Ant Colony Optimization algorithms. These algorithms are often used to find metaheuristic-based solutions. Although the authors did not specify a Tetris implementation, they proposed a game with very similar features to Tetris. A linear weighting heuristic was designed, and the authors state that they achieved a maximum number of 17586 cleared lines.

In [15] the authors show the usage of evolutionary algorithms in covariance matrices on a Tetris game. In this work the authors discuss the game's essence and the nature of the search space. The obtained results are the algorithms' ability to rediscover strategies presented in other works. In [16] it is show the usage of that very same method on the same game, and because of that the authors conclude that the primordial feature on the designing of a game agent of this kind is the choosing of the characteristic functions that the game engine has.

In [17] the authors present the usage of swarm-based algorithms on a neural network in order to achieve an optimization on the game feature's weights. The authors state that although this approach is exceeded by hand-made optimization algorithms the strategies are similar, because using this approach the authors reported a maximum number of 1286705 cleared lines.

[18] present the usage of a genetic algorithm focused in the developing of a solution for randomly generated sequences of tetraminos; where each chromosome represents actions and an acceptable game strategy, to identify contour patterns and the actions for the best games for each tetramino. Additionally, the authors applied association rules for pattern extraction in different games.

[19] use a genetic algorithm to find optimum weights using a pondered function in order to train a game agent, in this paper the authors used two different sized boards, the first one with the normal 10×20 size, where the training showed a performance of 859520 cleared lines. The second board with a 6×12 size showed in its best game reported by the authors a performance of 4007 cleared lines. The authors state that this score is due to the fact that the games in the second experiment are shorter. One of the points which draw the most attention is the training time, because it took 3 months the creation of 30 generations, even reaching the point of a sudden stop of the training.

In the literature review we identified that in most cases are issues with the game agent training time, not only because this situation can occur for the game ending condition, but also in the fact that the better the game agent gets, the longer it will take to finish a game. Some authors presented solutions to this issue, such as [18] that stablishes a perfect game when 50 tetraminos are placed filling the 200 spaces in the game board. [1] propose various ways of measuring the game agent's performance, such as counting placed pieces, cleared lines or defining points for special actions done by the game agent, i.e. clearing more than one line at a time to stablish a count of those points, putting a hard limit to define a perfect game. [7] proposes the establishment of a fixed number of generations and pieces in order to measure the maximum number of lines that the game agent can clear, in this particular case, the authors stablished for their experiments a number of 30 generations and a fixed number of pieces between 100 and 1000 respectively.

3 Methodology

In this section we present the used methodology for this work, which is divided in three main sections: AI, which describes the details of the genetic algorithm, Game Engine describes the game features where the GA will be applied, and last a Functional Tests section which presents a measurement of the game agent efficiency and its training with the genetic algorithm in various situations.

3.1 AI Technique

The used AI technique is Genetic Algorithm; this is coded in Python 3 for training the game engine. GA was chosen because we considered that the solution space for the problem is too big for using other techniques. In the same way, in the reviewed works Genetic Algorithms showed in most cases the most optimal solutions comparing it to other solutions. The developed GA is constituted by heuristics, a population, a set of genetic operators (selection, crossover, mutation), a strategy of replacement and fitness function.

A compilation of used heuristics by different authors in the utilization of AI techniques for Tetris games can be found in [1]. In this particular case, the GA has four heuristics which are:

- Line numbers: Number of cleared lines after making a move or placing a tetramino.
- Number of holes: Number of holes resulting from making a move or placing a tetramino.
- Maximum height: Height of the tallest column in the board.
- Board uniformity: It refers to the accumulated difference between adjacent columns in the game board.

The weights of each chromosome are associated to each one of those heuristics. The population is composed by n individuals, where the initial population is randomly generated and the next generations are generated by the genetic operators in the following way:

- Selection: it is performed by a simple tournament with 10% of the population, where the two fittest individuals are selected.
- Crossover: Each gen of the selected individuals is weighted against the result of the fitness function of its own chromosome, after that the genes in the same position are added to generate a third gen in order to generate an offspring. Because elitism is not implemented in this GA, this crossover allows information sharing between the parents, but the offspring chromosome is biased to the fittest parent.
- Mutation: a delta mutation is used, where if there is mutation only a random gen will mutate a quantity between ±0.2.

Replacement: the new population doesn't retain any of the parents, because the offspring will replace the entire population.

Fitness function is defined as the total cleared lines in X games played using Y randomly generated tetraminos per game.

3.2 Game Engine

The game engine based in GA was developed in Python3. This game engine doesn't have spin mechanics, that means that tetraminos can't be "spin" to be fit in a space where it wouldn't fit normally. Also, it doesn't have lock delay function, which refers to how many frames a tetramino waits while it is in the floor before locking into place. The game engine allows playing while knowing the upcoming piece (see Fig. 2). For that mechanic we used the seven random bag randomizer, because this randomizer guarantees that the maximum wait between two pieces of the same particular tetramino is at most 13 pieces. This system is used mainly in modern iterations of Tetris.

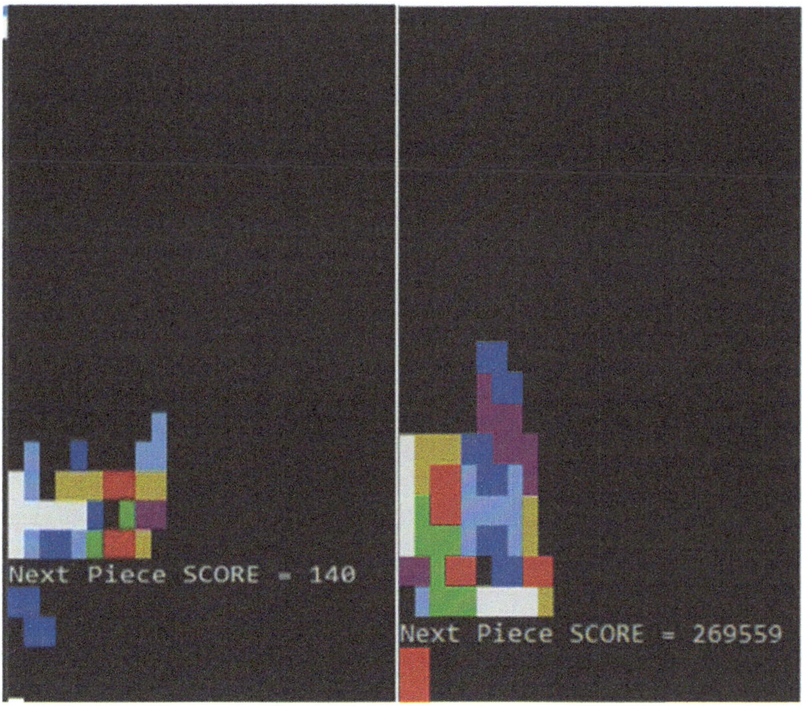

Fig. 2. Game engine playing with 200 tetraminos.

3.3 Functional Tests

One of the main sections in this work is the demonstration of the execution efficiency for the genetic algorithm in the developed game engine. For testing, 2 of the features considered the most important in the literature were changed and applied in the game engine, these features are the number of randomly generated tetraminos and the usage of the seven random bag randomizer function. Comparison between results was done using the fitness provided by the genetic algorithm as metric, in order to be able to compare between different generations and tests.

Following the theoretical proposal of [20] is considered for this practical case a perfect game to be able to eliminate two rows using 5 tetraminos generated in a random way, the calculated proportion would be 5 (random tetraminos)/2 (eliminated rows) = 2.5.

The first set of tests was done without the usage of the seven random bag function in order to assess the efficiency of the GA only by changing the number of tetraminos to be placed on each game. This was done in three tests: the first one using 200 tetraminos, the second one using 100 tetraminos, and the third one using 50 tetraminos. All of the test were ran for 50 generations. For each of the tests, a results comparison was done showing the improvement behavior at generation 1, 25, and 50, as shown in the Figs. 3 and 4.

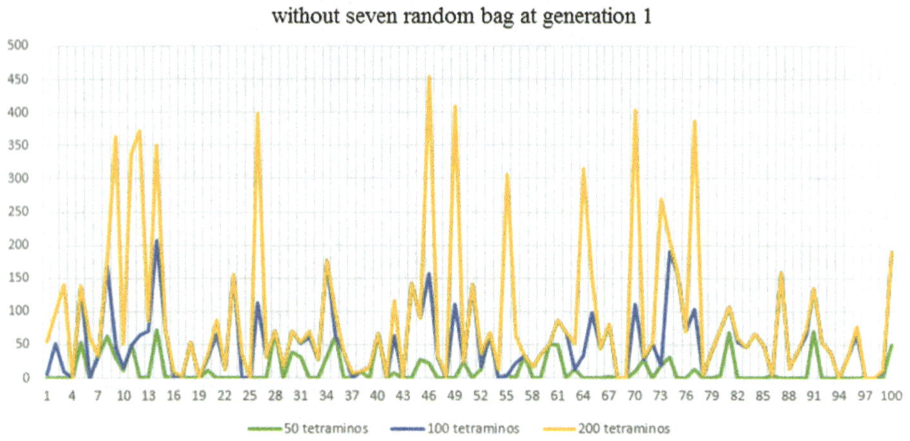

Fig. 3. Comparison between tests using 200, 100 and 50 tetraminos without seven random bag at generation 1.

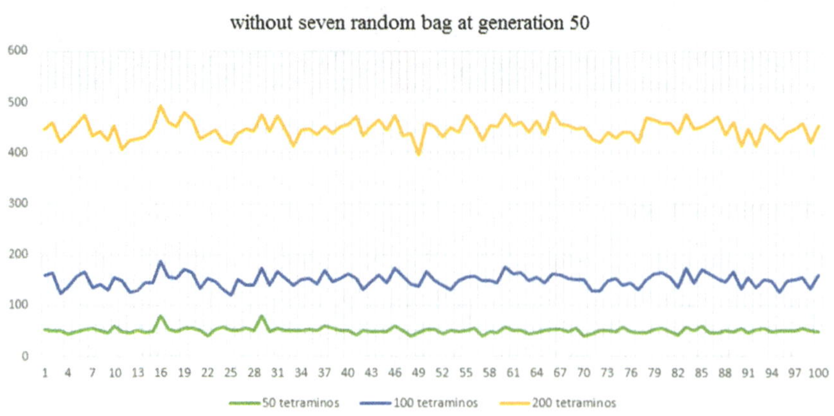

Fig. 4. Comparison between tests using 200, 100 and 50 tetraminos without seven random bag at generation 50.

We present a comparison between the results generated on each test at generation 1, 25 and 50 (see Fig. 3).

For the next set of tests, the importance of the seven random bag function was stated, and how having the information of the next coming piece can affect the game agent. In this set of tests, the number of tetraminos was also changed using 200, 100 and 50 tetraminos, but the number of generations was reduced to 25 for each test, as shown in the Figs. 5, 6 and 7, which present the behavior at generations 1, 13 and 25.

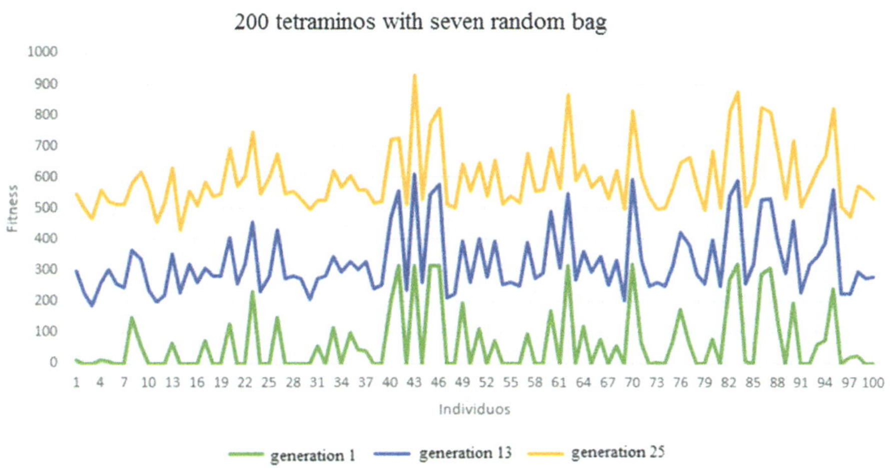

Fig. 5. Test with 200 tetraminos with seven random bag.

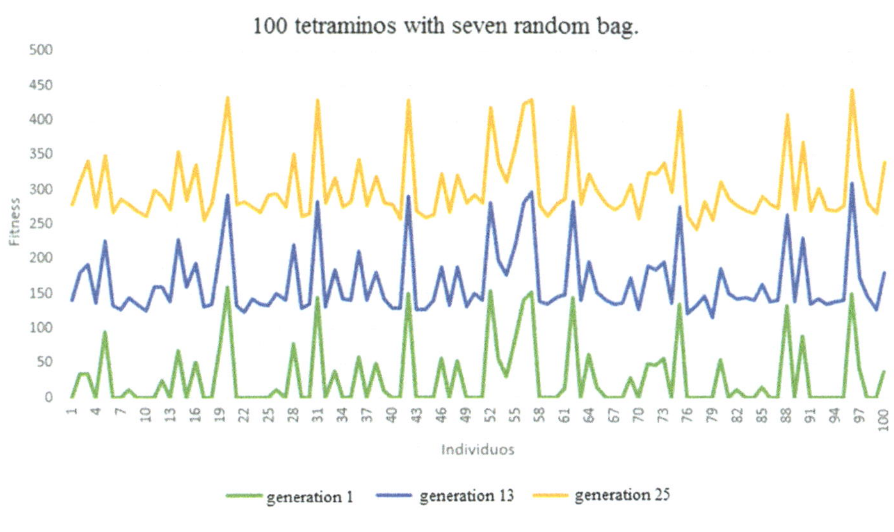

Fig. 6. Test with 100 tetraminos with seven random bag.

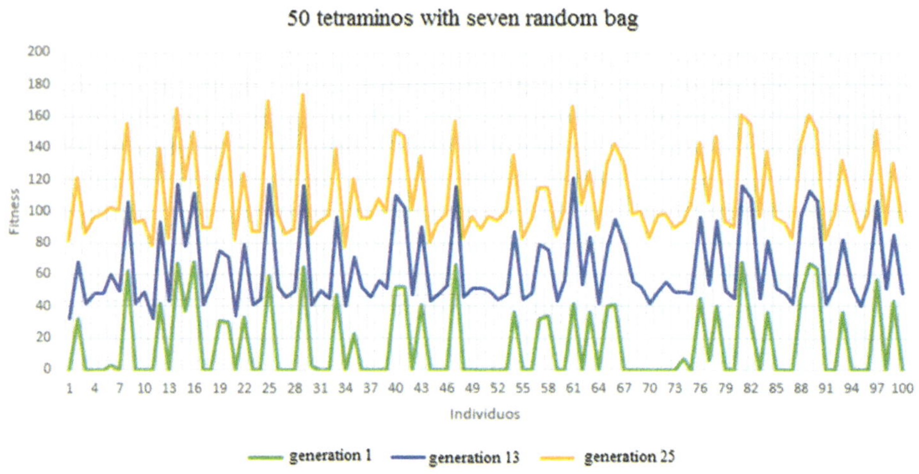

Fig. 7. Test with 50 tetraminos with seven random bag.

A comparison between results for each test in a generation was done in the same way as before (see Figs. 8, 9, 10, 11, 12, 13, 14, 15, 16, 17, 18 and 19).

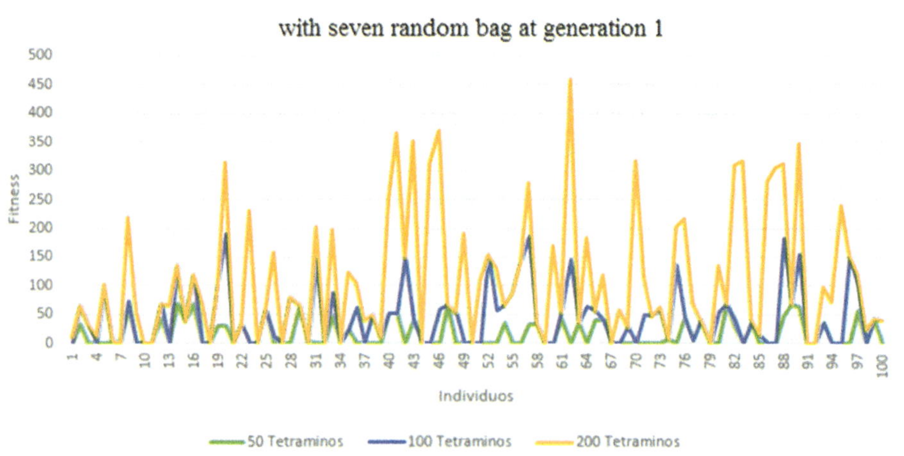

Fig. 8. Comparison between the 200, 100 and 50 tetraminos with seven random bag at generation 1.

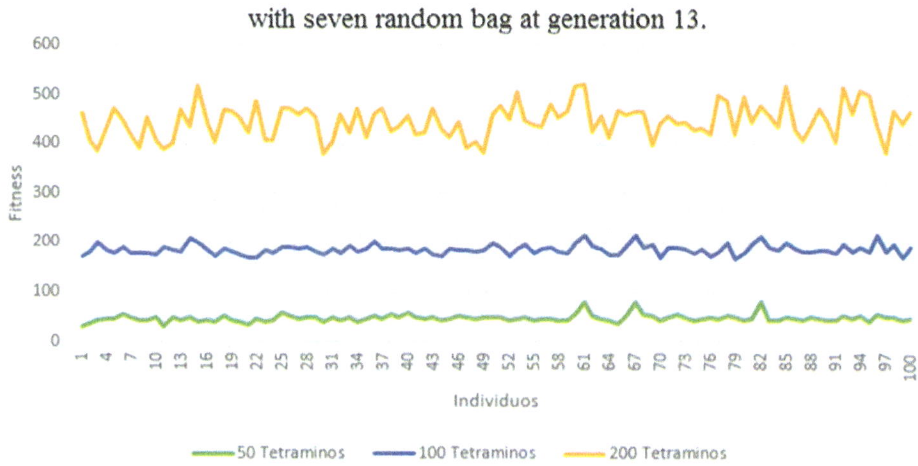

Fig. 9. Comparison between the 200, 100 and 50 tetraminos with seven random bag at generation 13.

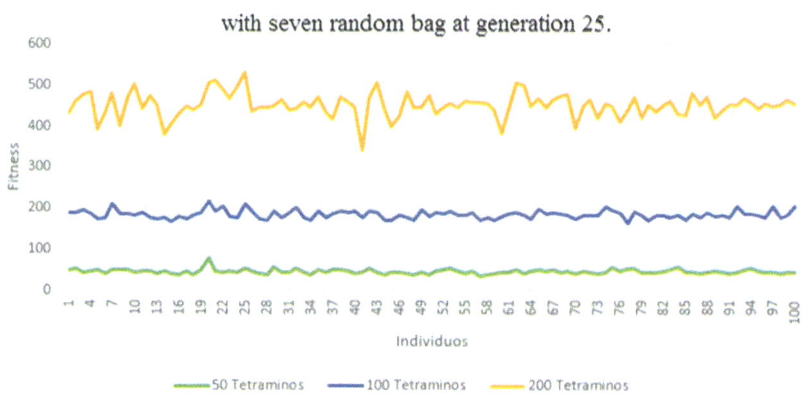

Fig. 10. Comparison between the 200, 100 and 50 tetraminos with seven random bag at generation 25.

At this point of the testing phase, a comparison between tests with and without seven random bag was done in order to better visualize the shown improvement. However, as the number of generation was not the same the comparison was done at generations 1, 13 and 25, because those numbers were the ones used in the second set of tests.

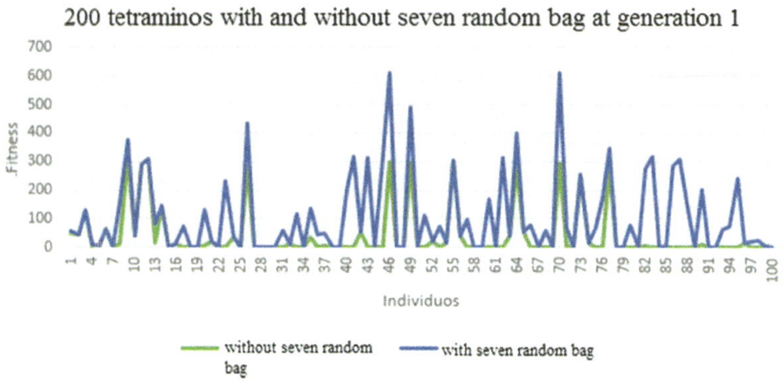

Fig. 11. Comparison between the tests using 200 tetraminos with and without seven random bag at generation 1.

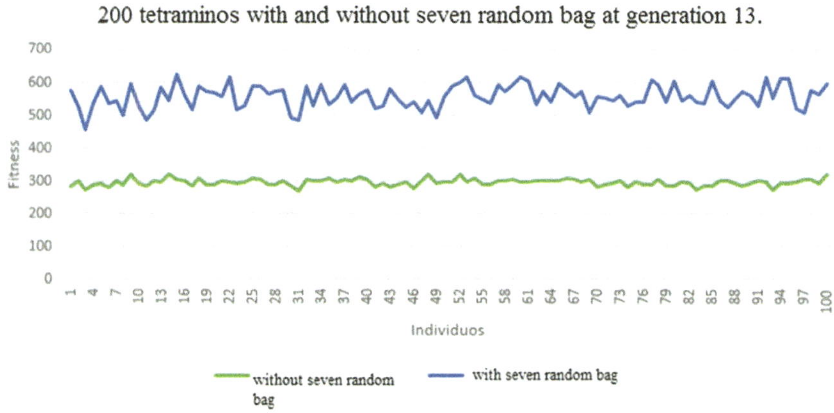

Fig. 12. Comparison between the tests using 200 tetraminos with and without seven random bag at generation 13.

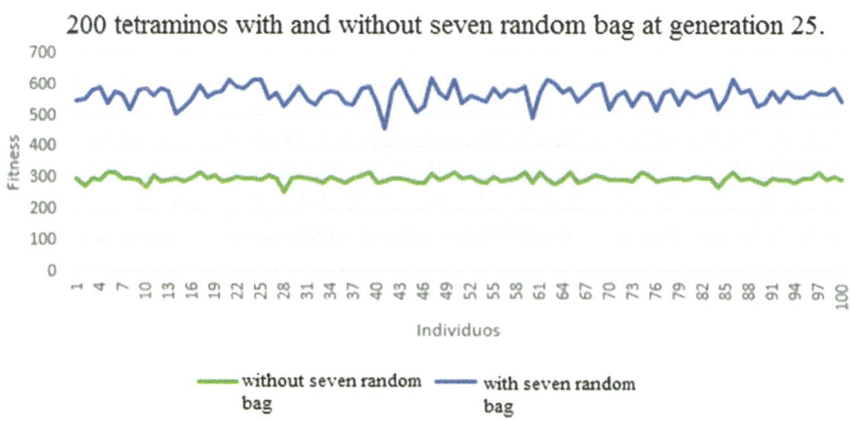

Fig. 13. Comparison between the tests using 200 tetraminos with and without seven random bag at generation 25.

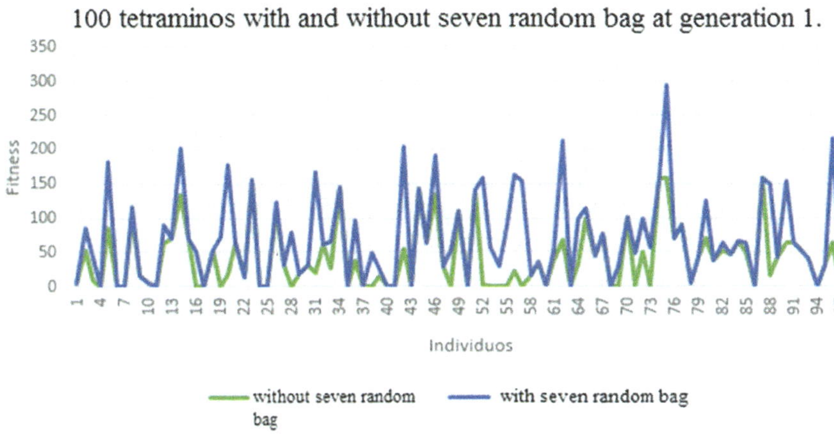

Fig. 14. Comparison between the tests using 100 tetraminos with and without seven random bag at generation 1.

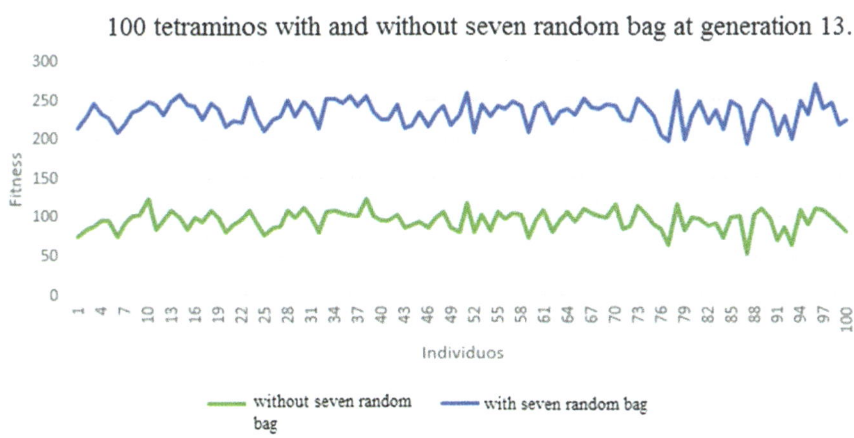

Fig. 15. Comparison between the tests using 100 tetraminos with and without seven random bag at generation 13.

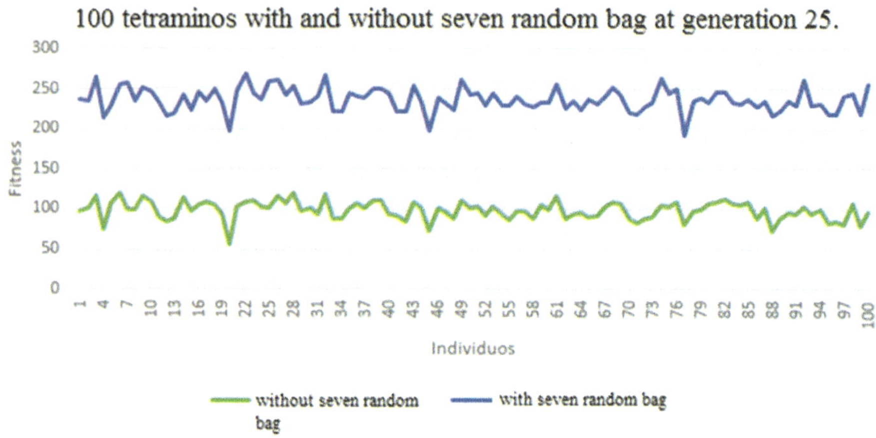

Fig. 16. Comparison between the tests using 100 tetraminos with and without seven random bag at generation 25.

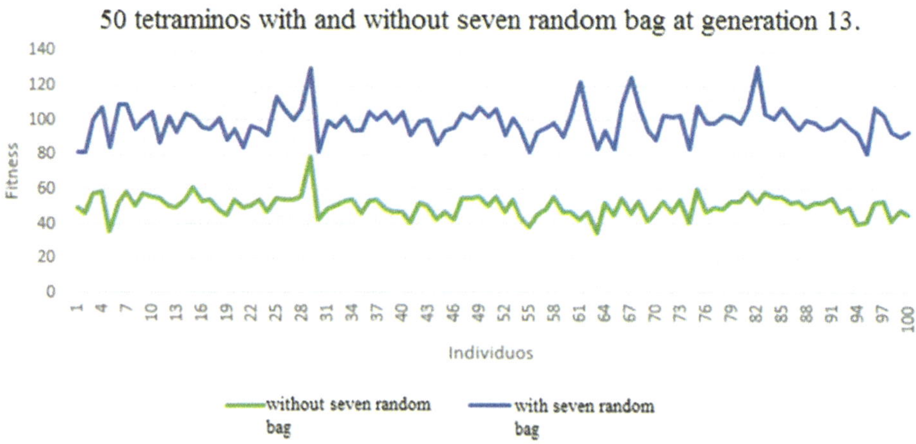

Fig. 17. Comparison between the tests using 50 tetraminos with and without seven random bag at generation 1.

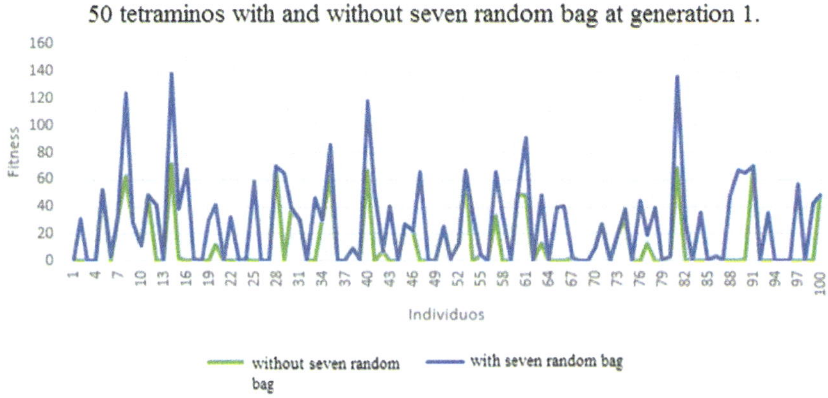

Fig. 18. Comparison between the tests using 50 tetraminos with and without seven random bag at generation 13.

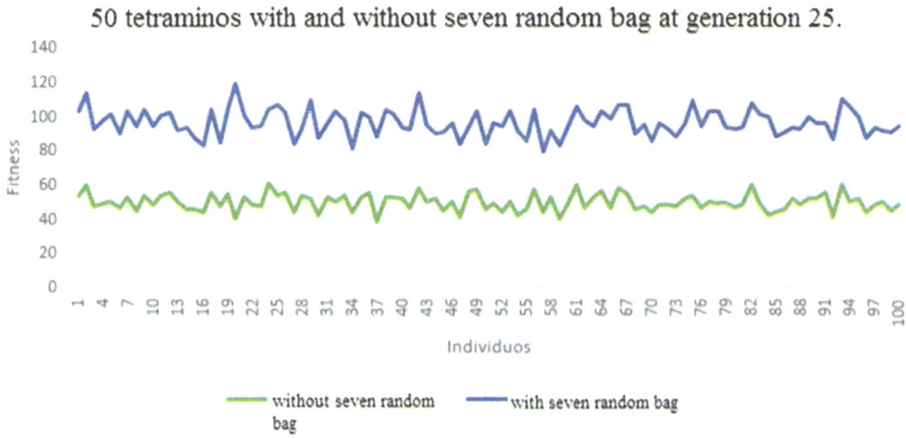

Fig. 19. Comparison between the tests using 50 tetraminos with and without seven random bag at generation 25.

4 Analysis and Results Discussion

In the first set of tests we can identify some main points; game behavior at generations 25 and 50 is very similar among the three games. The AI does produce an improvement result in the game performance, which is the search objective when using this kind of techniques. Another point to highlight in the experiments' results is the improvement between a game with more tetraminos and another one with less tetraminos, because we can observe if there are more tetraminos in play, a greater improvement can be seen from the first generation onwards, as shown in Fig. 5. Games played with 200 tetraminos have a better performance compared against games with 100 and 50 tetraminos, which show a less significant performance. Also, we can observe that the game behavior is less fluctuating as generations pass. Because of that, we can state that the AI training is more stable.

The second set of tests, which presents a popular function used in an actual Tetris game, doesn't show significant changes in behavior with different number of tetraminos, however, it does show that the games played with 200 tetraminos have better performance since generation 1, and the 100 and 50 games have a similar behavior compared against a 200 tetraminos game.

The third set of tests, where a comparison between the first and second experiments was done for each generation, we can show that, evidently, the second experiment has a significant improvement compared against the first experiment. Also, we can see that the behavior between generations is not similar, although there can be some degree of similarity between certain individuals. Another important point to highlight is the execution time for each experiment, as shown in Table 1.

Table 1. Time in hours of termination per experiment.

Experiment	Execution time (in hours)
200 tetraminos without seven random bag.	23
100 tetraminos without seven random bag	9
50 tetraminos without seven random bag	3
200 tetraminos with seven random bag	533
100 tetraminos with seven random bag.	300
50 tetraminos with seven random bag.	144

As it can be seen, the time difference between an experiment and another is really significant, even though both experiments have the same amount of tetraminos for the execution of the game.

5 Conclusions and Future Work

In this work, we present the utilization of a genetic algorithm on a Tetris game engine, where the making of certain experiments previously stated allowed us to conclude that the amount of tetraminos for a played game is not significant in order to study the game agent's behavior, however this feature is important at the execution time level, because the bigger the amount of tetraminos played, the longer the game time will be. From another perspective, we can conclude that using a function such as the seven random bag randomizer provides the game agent more possibilities of playing a perfect game, because after 11 days of continuous playing, the game agent did not lose a game, getting "perfect" games continuously.

Using the execution time results obtained from each experiment, we can observe that the training time is relative lower compared against works such as [8], which present training times longer than 3 months. Our approach presented a maximum training time of 533 h (22 days approximately). These faster training times can be attributed to the generation of game ending conditions, such as numbers of generations and ending a game where 2 lines are cleared using 5 tetraminos. The last condition is difficult to get using a random set of tetraminos while using the seven random bag function to serve the piece set.

For future works, some points can be polished such as the game engine, which can receive new features, i.e. spin mechanics and Lock Delay to get a similar experience like modern Tetris games offer. The piece randomizer function can be altered to the old Tetris' engines, which allows a repeated consecutive type of piece. The game agent performance can be compared between the two randomizer functions. For future works some points are proposed, such as using more heuristics or another set of heuristics, and changing the genetic operators' behavior.

References

1. Da Silva, R.S., Parpinelli, R.S.: Playing the original game boy tetris using a real coded genetic algorithm. In: Proceedings of the 2017 Brazilian Conference on Intelligent Systems, BRACIS, pp. 282–287. IEEE (2017)
2. Le, A., Arunmozhi, M., Veerajagadheswar, P., Ku, P.C., Minh, T.H., Sivanantham, V., et al.: Complete path planning for a tetris-inspired self-reconfigurable robot by the genetic algorithm of the traveling salesman problem. Electronics **7**(12), 344 (2018)
3. Lee, H., Shin, H., Chae, J.: Path planning for mobile agents using a genetic algorithm with a direction guided factor. Electronics **7**(10), 212 (2018)
4. Qi, L., Li, B., Chen, L., Wang, W., Dong, L., Jia, X., et al.: Ship target detection algorithm based on improved faster R-CNN. Electron **8**, 959 (2019)
5. Veerajagadheswar, P., Elara, M.R., Pathmakumar, T., Ayyalusami, V.: A tiling-theoretic approach to efficient area coverage in a tetris-inspired floor cleaning robot. IEEE **6**(35), 260–271 (2019)
6. Le, A., Prabakaran, V., Sivanantham, V., Mohan, R.: Modified a-star algorithm for efficient coverage path planning in tetris inspired self-reconfigurable robot with integrated laser sensor. Sensors **18**(8), 2585 (2018)
7. El Faddouli, N., El Falaki, B., Khalidi, M., Bennani, S.: Towards an adaptive competency-based learning system using assessment. IJCSI Int. J. Comput. Sci. Issues **8**(1), 265–274 (2011)
8. Font, J.M., Manrique, D., Larrodera, S., Criado, P.R.: Towards a hybrid neural and evolutionary heuristic approach for playing tile-matching puzzle games. In: 2017 IEEE Conference on Computational Intelligence and Games, pp. 76–79. IEEE (2017)
9. Papadimitriou, C.H.: Games against nature. J. Comput. Syst. Sci. **31**(2), 288–301 (1985)
10. Demaine, E.D., Hohenberger, S., Liben-Nowell, D.: Tetris is Hard, Even to Approximate. Cornell University, New York (2002)
11. Lagoudakis, M.G., Parr, R., Littman, M.L.: Least-squares methods in reinforcement learning for control. Second Hellenic Conference on AI, SETN 2002, pp. 249–260. Springer, Greece (2002)
12. Ramon, J., Driessens, K.: On the numeric stability of gaussian processes regression for relational reinforcement learning. In: ICML-2004 Workshop on Relational Reinforcement Learning, pp. 10–14. Springer, Canada (2004)
13. Driessens, K., Ramon, J.: Graph kernels and Gaussian processes for relational reinforcement learning. Mach. Learn. **64**(1–3), 91–119 (2006)

14. Esparcia-Alcázar, A.I., Mora, A.M., Agapitos, A., Burelli, P., Bush, W.S., Cagnoni, S., et al.: Preface. In: 17th European Conference on Applications of Evolutionary Computation. Lecture Notes in Computer Science, Spain, pp. 7–10 (2014)
15. Boumaza, A.: On the evolution of artificial Tetris players. Computational Intelligence and Games. CIG 2009, pp. 387–393. IEEE, Italy (2009)
16. Boumaza, A.: How to design good Tetris players. Hal Archives-ouvertes, hal-00926213 (2013)
17. Langenhoven, L., van Heerden, W.S., Engelbrecht, A.P.: Swarm tetris: applying particle swarm optimization to tetris. In: IEEE Congress on Evolutionary Computation, pp. 1–8. IEEE, Spain (2010)
18. Phon-Amnuaisuk, S.: Evolving and discovering Tetris gameplay strategies. Procedia Comput. Sci. **60**, 458–467 (2015)
19. Böhm, N., Kóokai, G., Mandl, S.: An evolutionary approach to Tetris. In: The Sixth Metaheuristics International Conference, pp. 137–48. Informs, Viena (2005)
20. Fahey, C.: Tetris. https://www.colinfahey.com/tetris/tetris.html. Accessed 13 July 2019

Electronics

Analysis and Determination of Minimum Requirements of an Autopilot for the Control of Unmanned Aerial Vehicles (UAV)

Hugo Loya[1], Víctor Enríquez[1], Franklin W. Salazar[2], Carlos Sánchez[2], Fernando Urrutia[2], and Jorge Buele[2,3](✉)

[1] Centro de Investigación y Desarrollo de la Fuerza Aérea Ecuatoriana, Ambato, Ecuador
giomh_1107@hotmail.com, venriquez@fae.mil.ec

[2] Universidad Técnica de Ambato, Ambato 180103, Ecuador
{fw.salazar, carloshsanchez, fernandourrutia}@uta.edu.ec

[3] SISAu Research Group, Universidad Tecnológica Indoamérica, Ambato, Ecuador
jorgebuele@uti.edu.ec

Abstract. Unmanned vehicles (UAVs) are technological tools whose application is carried out in the civil and military areas for surveillance, recognition and intelligence tasks. Given its versatility and features, in this paper the implementation of an autopilot is determined, after an analysis of commercial models. Said device must comply with technical requirements for the automatic flight control of a medium-sized UAV, considering the operation scenarios in the northern border of Ecuador. In addition, the parameters that intervene in the control loops and the relevant simulations are specified. The results obtained from the experimental flights of the aircraft are reflected in graphs, where the stability of the plant properly calibrated for autonomous flight is verified.

Keywords: Autopilots · Flight controller · Flight performance · UAV

1 Introduction

The Armed Forces of Ecuador (FFAA) within its action fields includes the development of air military power to ensure the national territory defense and contribute to security through Ecuadorian Air Force (FAE). Current institutional objectives have focused on the conflict zones control and drug trafficking, specifically on the northern border of the country [1]. National Defense Ministry in its prospective vision to develop the Defense Industry has commissioned to the Research and Development Center (CIDFAE) the design and construction of unmanned aerial vehicles (UAVs) with a high national added value [2]. This will achieve the aim of being used in support of surveillance, recognition and intelligence in operation zones (in the coastal, mountain and eastern regions) [3, 4].

J. Nummenmaa et al. (Eds.): CSEI 2019, AISC 1078, pp. 129–142, 2020.
https://doi.org/10.1007/978-3-030-33614-1_9

One of the greater development lines of research of this institution is dedicated to the design, development and integration of UAVs [5]. At the beginning they were remotely piloted aircraft, until the US Air Force, in alliance with the Boeing company, transform several F-16 fighters into drones (for training and later for national defense in military conflicts). The emergence of improved technological systems worldwide has allowed the emergence of revolutionary devices that adhere artificial intelligence, artificial vision, electronic nanotechnology, remote sensing satellite communications, etc. [6, 7]. That is why the International Civil Aviation Organization (ICAO) states that the operation of these vehicles must be as safe as manned aircraft, especially when any test could represent a danger to personnel on the ground [8, 9]. Its use includes civil and military applications, object detection, photogrammetry, surveillance and espionage in real time, recognition of danger zones, access to remote areas, etc. Its use and application leave open as a limitation the space and weight that UAV can support [10–12]. For correct functioning and fulfillment of planned tasks, it is required that the electronic device or autopilot should be selected with technical criteria [13]. This device constitutes the brain of the UAV and depends on its reliability, versatility and adaptability to embedded hardware [14].

2 State of Art

The lack of technological development of a system that allows UAVs (and other prototypes) to carry out missions autonomously, promotes the appearance of errors in flight parameters such as: speed, wind variations, altitude and direction. In addition, the lack of a mission zone recognition system, which sends real-time information to the ground base station to avoid instability in the navigation of the aircraft, generates the need to choose with technical criteria all the elements that make up an aircraft. Given this problem, the need arises for the joint work of the military air industry and the academy[1,2,3], as shown in the work presented below. The proposal of [15] presents the incorporation of a commercial electro-optical/infrared commercial camera used in UAVs tasks of recognition. Similarly [16] propose to increase the flight performance of a small unmanned aerial vehicle (UAV) by improving the autopilot system. Autopilot configuration parameters are modified using a stochastic optimization method and results obtained are used for simulations. In this context, this research project proposes to carry out the technical study for select the autopilot used to control the prototype of seven meters of spam, which has been called UAV "Gavilán". This has been done as part of Project Detection, Observation, Communication and Recognition (DOCR), in which it is linked to the FAE and the academy. Autopilot parameters are configured to obtain the best performance and for validation of this proposal, simulation and experimental tests are presented [17].

[1] http://repositorio.espe.edu.ec/jspui/handle/21000/9307.

[2] http://repositorio.uta.edu.ec/handle/123456789/19383.

[3] http://repositorio.puce.edu.ec/handle/22000/12152.

This paper is organized as follows: the introduction in Sects. 1, Sect. 2 state of art and Sect. 3 shows the guidance, navigation and control system of the UAV. Section 4 presents the control loops and their calibration. Results of the experimental tests carried out and the conclusions are described in Sects. 5 and 6 respectively.

3 Guidance, Navigation and Control System of the UAV

3.1 Autopilot

Based on the operational requirements for the design of the DOCR project prototypes, the main element and whose selection carries greater relevance in a UAV is the automatic pilot device. Among the main parameters to be considered are: dimensions and weight, cost and import permits, ability to follow waypoints, development of secondary flight routes, emergency patterns, etc. From the strategic point of view for control and monitoring of the aircraft, it is necessary to use two communication links. Primary link is typical of the autopilot and the secondary link is redundant and made through an external connection for a higher power radio [2]. Table 1 shows the comparison of the operational capabilities of 4 commercial brands.

Table 1. Comparison of operational capabilities

Item	MICROPI LOT (2128-3X)	PICCOLO II	KESTREL V2.4	APM 2.6
Operativity	Yes	Yes	Yes	Yes
Automatic Takeoff	Yes	Yes	Yes	Yes
Automatic Navigation	Yes	Yes	Yes	Yes
100% effectiveness in Automatic Landing	Yes	No	No	No
Simulations	Yes	Yes	Yes	Yes
Technical Support	Yes	Yes	Yes	Yes
Direct Interconnection with satellite communication system	Yes	No	No	No
Operation Modes	Manual Automatic	Total Manual Manual Assisted Automatic	Manual Automatic	Automatic
Plant adjustment in real time	Yes	Yes	Yes	Yes
Creation of Routes and flight patterns	Yes	Yes	Yes	Yes
Emergency handling	Yes	Yes	Yes	Yes
System and interface personalization	80%	60%	70%	90%
Laser altimeter	Yes	Yes	No	No
DGPS System	Yes	Yes	No	No

When performing the analysis in Table 1, it is determined that only the MicroPilot model 2128-3X system meets the requirements for the "Gavilán" UAV prototype. Additionally, it allows to have redundancy of hardware that admits to handle the sensors faults on board and sudden emergencies, as well as the integration of a differential GPS and laser altimeter that improve the measurements in critical flight stages: takeoff and landing.

3.2 Control System

The control system shown in Fig. 1, consists of electrical, electronic and mechanical components of advanced technology. As previously mentioned, autopilot system is the brain that controls each of the flight stages through coordinated movement of actuators, as well as the acquisition and processing of data from sensors integrated in the aircraft. Takeoff and landing stages have a high level of criticality, for which a DGPS (differential GPS) approach system and laser are incorporated, in order to comply with the safety regulations in a flight, in all its stages.

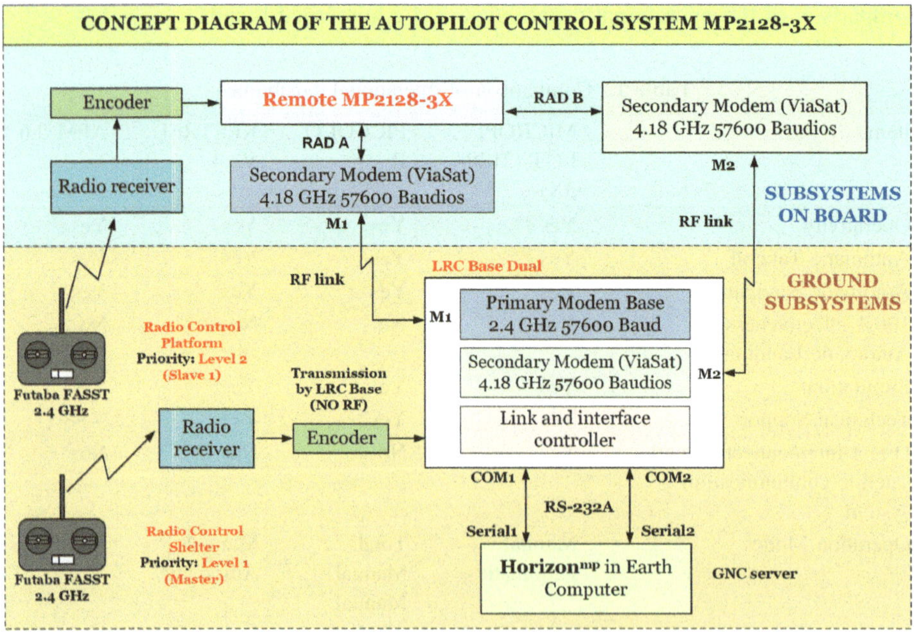

Fig. 1. Conceptual diagram of the UAV control system

For the automatic navigation stage, it is necessary to consider the UAV performance profile, velocity values, acceleration percentage, activation times of emergency routines and flight strategy. It is also needed to define the values that will serve to initialize the autopilot, where the different established PID control loops will allow a stable navigation that complies with the pre-programmed flight plan. Figure 2 shows the model of a trajectory and the reference points that make it up and in Table 2, nomenclature of both absolute and relative coordinates.

Table 2. Aerodynamical scenarios of simulation.

Absolute coordinates	Relative coordinates
(40:7.3986E, 90:30.96 N)	(4000, 4000)

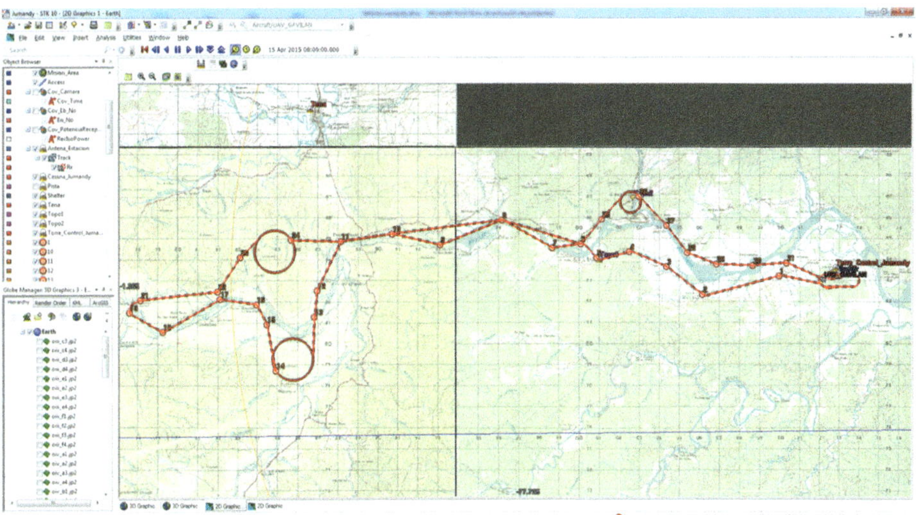

Fig. 2. Planned flight route, where the red line shows the communication link between Ground Control Station (GCS) and the aircraft.

4 Calibration and Control Loops

Calibration method for loops control is based on a series of gains established by default to use them as a starting point, they are stable gains for some aircraft models of similar characteristics to the UAV "Gavilán". Subsequently, depending on the aircraft behavior displayed in a simulator, modifications are made to obtain the most efficient gains, and then continue to adjust finely during the experimental flights using the adjustment technique by a heuristic method. The respective control loops are described in Table 3.

The first loops that must be calibrated are the aircraft stability and Pitch from Airspeed (Control the pitch to reduce the difference between the desired speed and the current one) and Roll from Heading (Controls the angle of the bank to reduce the difference between the desired course and the current one). This is done by evaluating the autopilot behavior while advancing by reference points following a previously set route (other control loops are calibrated in the same way). The performance evaluation of the parameters selected in the autopilot programming is done through a flight with different routes, ascents, descents, change of speed, height, etc. In Fig. 3 the interface that is available for calibration is shown and in Fig. 4 the loading of maps. Final gains values are presented in Table 4.

Table 3. Control loops description.

Item	Name	Control	Description
0	Aileron from roll	Aileron	Controls the ailerons to reduce the difference between the desired roll and the current one
			The gains are based on speed
1	Elevator from pitch	Elevator	Controls elevator plans to reduce between the desired pitch and the current pitch
			The gains are based on speed
2	Rudder from Y accelerometer	Rudder	Controls the rudder to reduce the difference between the Y accelerometer and the current one
			The gains are based on speed
3	Rudder from heading	Rudder	Controls the rudder to reduce the difference between the desired course and the current one. Used during takeoff
			The gains are based on GPS speed
4	Throttle from speed	Throttle	Controls the rudder to reduce the difference between the desired speed and the current speed. During the final approach
			The gains are based on current speed
5	Throttle from altitude	Throttle	Control the throttle to obtain the desired altitude
			The gains are based on current speed
6	Pitch from altitude	Desired pitch	Control the pitch to obtain the desired altitude
			The gains are based on current speed
7	Pitch from AGL	Desired pitch	Controls the pitch to reduce the difference between the desired altitude and the current one, as measured by the AGL
			The gains are based on current speed
8	Pitch from airspeed	Desired pitch	Controls the pitch to reduce the difference between the desired speed and the current one. It is activated during the ascent
			The gains are based on current speed
9	Roll from heading	Desired roll	Controls the angle of the bench to reduce the difference between the desired course and the current one. Active at any time
			The gains are based on GPS speed
10	Heading from crosstrack error	Desired heading	Control the desired course to minimize the distance of the MP and the next waypoint
			The gains are based on current speed

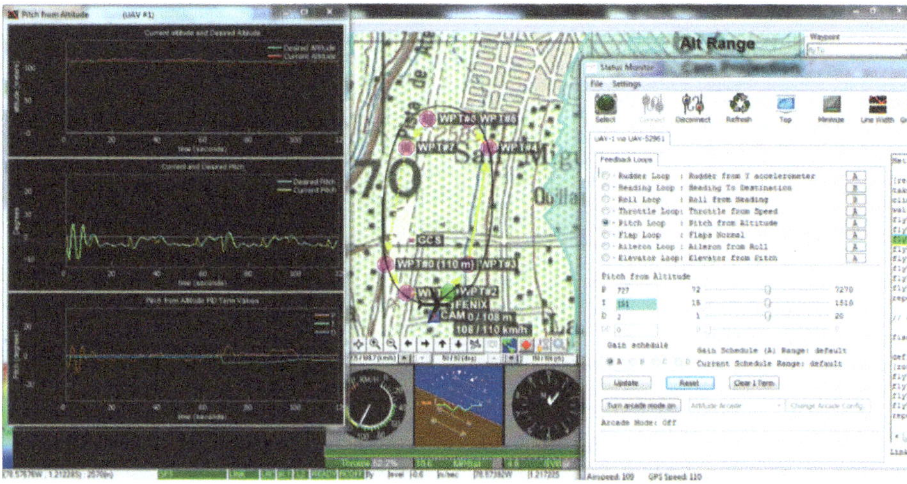

Fig. 3. Calibration interface of control loop gains in real time.

Table 4. Established final gains in UAV "Gavilán" prototype

Controller	Speed Range (Km/h)	Calibrated P constant	Calibrated I Constant	Calibrated D constant	Feeding
Aileron from Roll	0 a 120	−143000	−128	−3000	0
	120 a 219	−132000	−128	−1500	0
Elevator from Pitch	0 a 100	20000	10500	10000	−11626
	100 a 219	17265	10261	9565	−10657
Rudder from Y Acc.	0 a 219	138	0	40	−3750
Rudder from Heading	0 a 219	6000	2000	−20000	0
Throttle from Speed	0 a 219	−5000	−100	0	41
Pitch from Alt.	0 a 219	231	193	504	0
Pitch from Airspeed	0 a 219	13756	24	194	0
Roll from Heading	0 a 63	−250	0	−300	0
	63 a 91	−750	0	−350	0
	91 a 219	−1000	0	−400	0
Pitch from Descent	0 a 219	−1500	−150	−1719	0

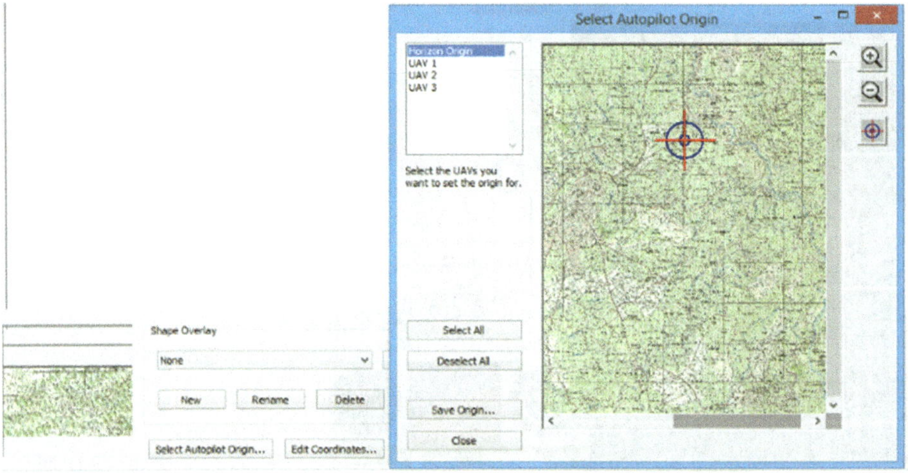

Fig. 4. Interface for loading respective maps.

4.1 Flight Substages

The sub-stages of flight are those in which the aircraft, after completing its main flight stage, has the capacity to achieve what the operator orders from the Ground Control and Control Station, in order to do this, the module destined to UAV control has a monitoring and control interface. In the current system used, flight sub-stages are activated by pressing buttons located on one side of the interface and are called secondary standards. These are programmed based on what is required by the operator, runway location and the type of mission to be carried out by the UAV. The display of the flight height range and the secure communications link are shown in Figs. 5 and 6 respectively.

Fig. 5. Sub-stage of flight annexed to the configuration of the UAV: range of flight height.

4.2 Simulations

The simulation is executed in the same graphic interface of a real flight, as shown in Fig. 7, which consists of the following elements:

- Flight plan.
- Emergency patterns, GPS loss, primary and secondary link, engine failure, power failure, partial or total control failure.
- Secondary patterns or routes.
- Adjustment of PID controllers.

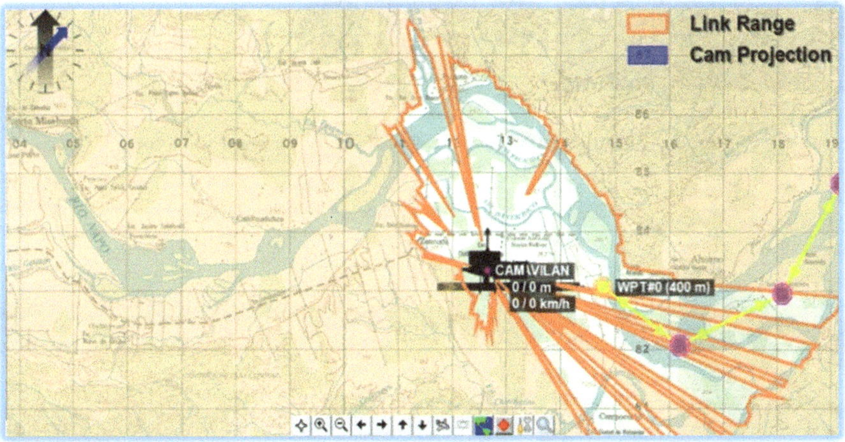

Fig. 6. Sub-step of flight annexed to the configuration of the UAV: communications link.

5 Experimental Tests

Among the test and validation flights of the "Gavilán" UAV, results of the flight mission developed at the airport of Jumandy Puerto Napo have been considered, where the runway has a height of 375 m above sea level, as can be seen in Fig. 8. At a cruising speed of 120 km/h and an altitude of 985 m above the takeoff point, the aircraft accomplished the mission in approximately 1 h and 18 min. Flight plan that was followed is composed of 24 reference points, although only 10 parameters of them have been placed in Table 5.

As can be seen in Figs. 9, 10 and 11, the aircraft presented the expected behavior during each of the mission phases. Height and speed were stable and each of the 24 programmed reference points in the autopilot system was met, without the loss of communications links.

Figure 12 shows that throughout the trajectory there is a line of sight, which is favorable for communications and their uninterrupted stability along the trajectory. Through this system it is possible to control the movements in azimuth and elevation, as well as autopilot communication through digital communication.

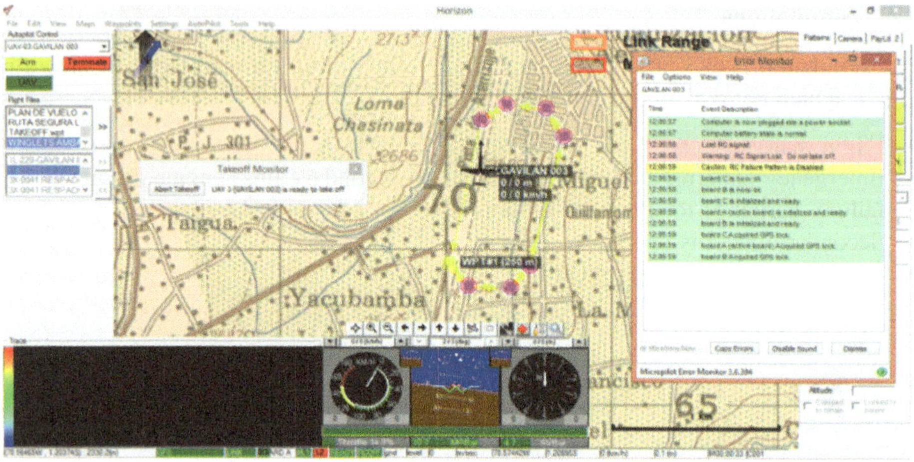

Fig. 7. Flight simulation with a previously planned trajectory.

Fig. 8. Experimental tests of the UAV "Gavilán" with the proposed system.

Table 5. Reference points that make up the flight route.

Controller	Latitude (Degrees)	Length (Degrees)	Altitude (MSL) (m)
1	−1.05866°	−77.6078°	389.784 m
2	−1.06665°	−77.642°	383.882 m
3	−1.05473°	−77.6581°	390.522 m
4	−1.04836°	−77.6747°	427.838 m
5	−1.05088°	−77.6895°	386.427 m
6	−1.04492°	−77.6965°	391.979 m
7	−1.04663°	−77.7092°	380.053 m
8	−1.03464°	−77.7316°	394.602 m
9	−1.04541°	−77.7591°	390.873 m
10	−1.04054°	−77.7806°	420.237 m

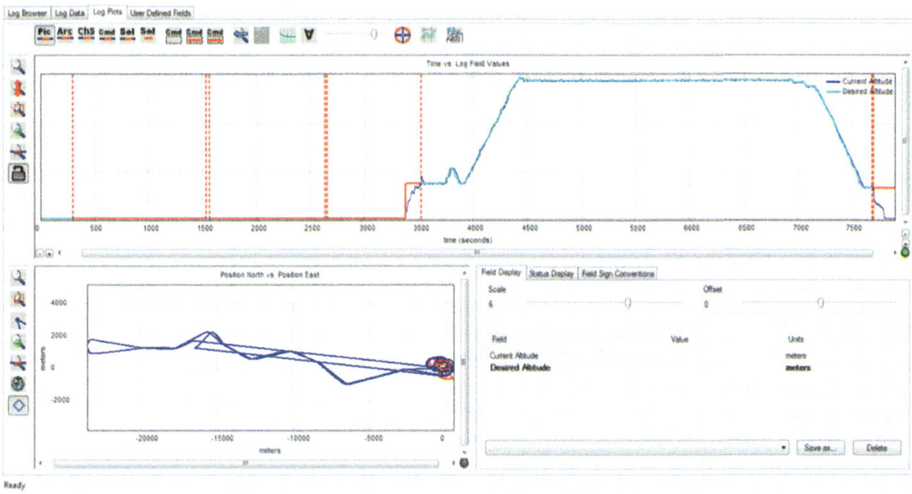

Fig. 9. Results of the UAV flight height control.

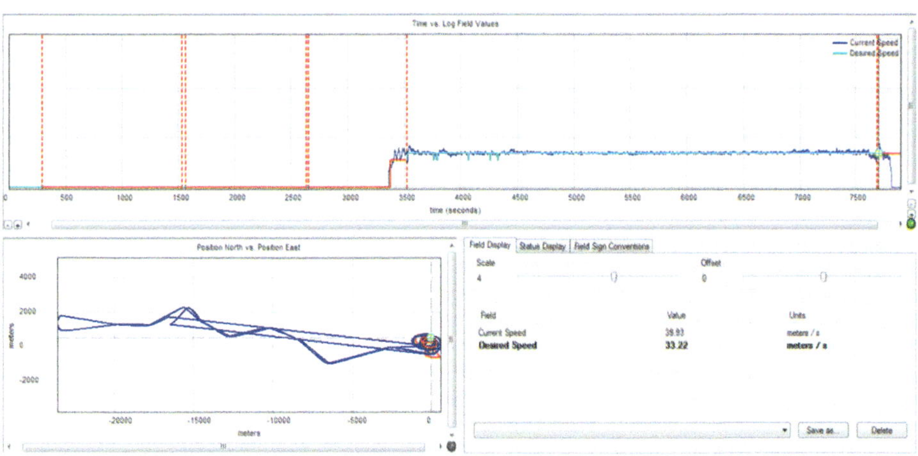

Fig. 10. Results of the UAV speed control.

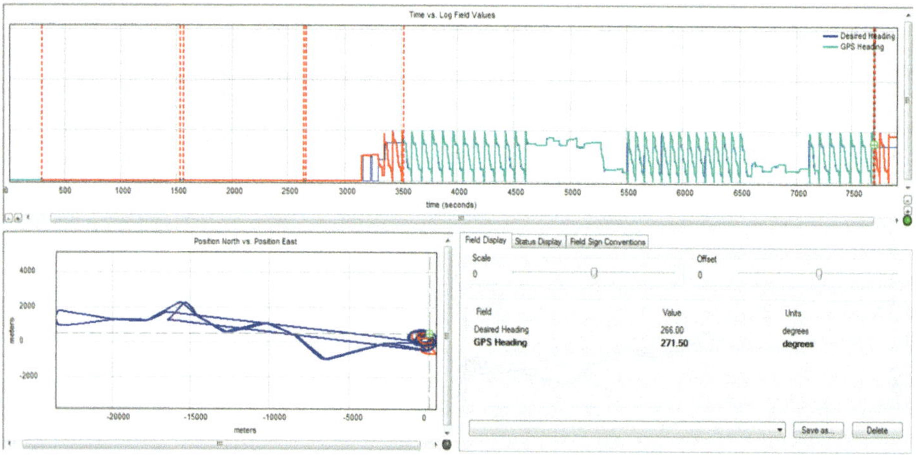

Fig. 11. Results of heading control of the UAV.

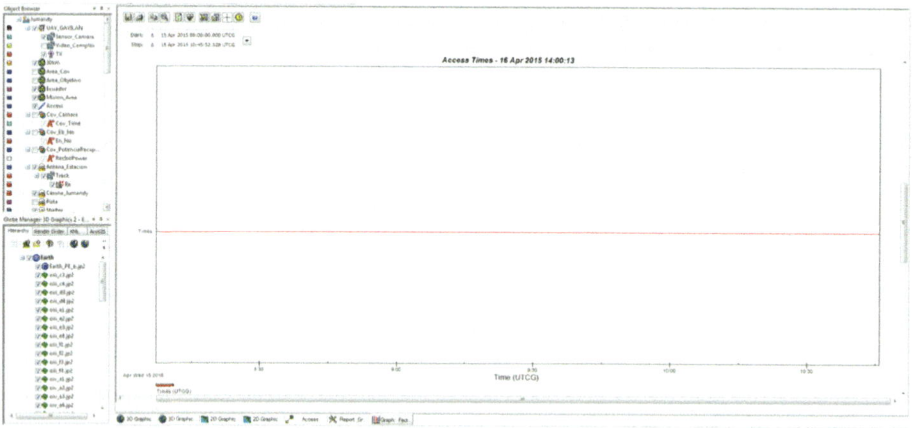

Fig. 12. Results of the existing line of sight.

6 Conclusions

Through the analysis carried out, an electronic autopilot system has been determined that complies with the technical requirements for the control of the "Gavilán" UAV prototype, in order to accomplish reconnaissance and surveillance missions. Initial adjustments of the control loops and simulations have been developed in the CIDFAE laboratories and in the Chachoán airport in the city of Ambato - Ecuador. All this in the search of contributing with the objective of promoting the Defense Industry, and in this way to promote the technological development in the aerospace field of the country, being evident the importance that CIDFAE should continue developing its capabilities together with the Academy to allocate resources in this field of research.

Results shown in the validation flights with the prototype have been reliable and satisfactory, this has taken the aircraft to its limits and sets a precedent for the search of continuous improvement. The stages of automatic take-off and landing have been tested on small prototypes, fulfilling satisfactorily. However, the calibration of these stages for the UAV Gavilán aircraft, as it is a larger and heavier aircraft, required the investment of more time, safety and personnel conditions.

The elaboration of this work constitutes a base for the continuous advance of the aerospace technology in Latin America. Therefore, searching for new scenarios and therefore a recalibration of parameters that can provide a stable and safe flight, in all its stages (take-off, flight and landing) is considered as future work.

Acknowledgments. This work was financed in part by Universidad Técnica de Ambato (UTA) and Dirección de Investigación y Desarrollo (DIDE) under project PFISEI 26. Special thanks to the support provided by Centro de Investigación y Desarrollo de la Fuerza Aérea Ecuatoriana (CIDFAE) and SISAu Research Group of the Universidad Tecnológica Indoamérica.

References

1. Mediavilla, J.R., Vélez, A.J.: Hacia la unificación de radares civiles y militares: Sistema de integración de la señal de radares de la Dirección General de Aviación Civil como apoyo a la Defensa y Seguridad del espacio aéreo ecuatoriano. RISTI - Revista Iberica de Sistemas e Tecnologias de Informacao **E15**, 66–75 (2018)
2. Medina-Pazmiño, W., Jara-Olmedo, A., Valencia-Redrován, D.: Analysis and determination of minimum requirements for a data link communication system for unmanned aerial vehicles-UAV's. In: 2016 IEEE Ecuador Technical Chapters Meeting (ETCM), pp. 1–6. IEEE (2016)
3. Stodola, P., Kozůbek, J., Drozd, J.: Using unmanned aerial systems in military operations for autonomous reconnaissance. In: International Conference on Modelling and Simulation for Autonomous Systems, pp. 514–529. Springer (2018)
4. Kaplan, C.: Air power's visual legacy: Operation Orchard and aerial reconnaissance imagery as ruses de guerre. Crit. Mil. Stud. **1**(1), 61–78 (2015)
5. Medina-Pazmiño, W., Jara-Olmedo, A., Tasiguano-Pozo, C., Lavín, J. M.: Analysis and implementation of ETL system for unmanned aerial vehicles (UAV). In: Advances in Intelligent Systems and Computing, vol. 721, pp. 653–662. Springer (2018)
6. Wild, G., Murray, J., Baxter, G.: Exploring civil drone accidents and incidents to help prevent potential air disasters. Aerospace **3**(3), 22 (2016)
7. Cunliffe, A.M., Anderson, K., DeBell, L., Duffy, J.P.: A UK Civil Aviation Authority (CAA)-approved operations manual for safe deployment of lightweight drones in research. Int. J. Remote Sens. **38**(8–10), 2737–2744 (2017)
8. Colefax, A.P., Butcher, P.A., Kelaher, B.P.: The potential for unmanned aerial vehicles (UAVs) to conduct marine fauna surveys in place of manned aircraft. ICES J. Mar. Sci. **75** (1), 1–8 (2017)
9. Muttin, F.: Umbilical deployment modeling for tethered UAV detecting oil pollution from ship. Appl. Ocean Res. **33**(4), 332–343 (2011)
10. Wu, Q., Zeng, Y., Zhang, R.: Joint trajectory and communication design for multi-UAV enabled wireless networks. IEEE Trans. Wireless Commun. **17**(3), 2109–2121 (2018)

11. Yang, D., Wu, Q., Zeng, Y., Zhang, R.: Energy tradeoff in ground-to-UAV communication via trajectory design. IEEE Trans. Veh. Technol. **67**(7), 6721–6726 (2018)
12. Zeng, Y., Zhang, R.: Energy-efficient UAV communication with trajectory optimization. IEEE Trans. Wireless Commun. **16**(6), 3747–3760 (2017)
13. Mammarella, M., Capello, E.: A robust MPC-based autopilot for mini UAVs. In: 2018 International Conference on Unmanned Aircraft Systems (ICUAS), pp. 1227–1235. IEEE (2018)
14. Liu, M., Egan, G.K., Santoso, F.: Modeling, autopilot design, and field tuning of a UAV with minimum control surfaces. IEEE Trans. Control Syst. Technol. **23**(6), 2353–2360 (2015)
15. Jara-Olmedo, A., Medina-Pazmiño, W., Mesías, R., Araujo-Villaroel, B., Aguilar, W.G., Pardo, J.A.: Interface of optimal electro-optical/infrared for unmanned aerial vehicles. Smart Innov. Syst. Technol. **94**, 372–380 (2018)
16. Oktay, T., Konar, M., Onay, M., Aydin, M., Mohamed, M.A.: Simultaneous small UAV and autopilot system design. Aircr. Eng. Aerosp. Technol. **88**(6), 818–834 (2016)
17. Bouzid, Y., Siguerdidjane, H., Bestaoui, Y., Zareb, M.: Energy based 3D autopilot for VTOL UAV under guidance & navigation constraints. J. Intell. Rob. Syst. **87**(2), 341–362 (2017)

Development and Analysis of a PID Controller and a Fuzzy PID

Morelva Saeteros(✉), Wilman Paucar, Cristian Molina, and Gustavo Caiza

Universidad Politécnica Salesiana, UPS, 170146 Quito, Ecuador
{csaeteros,Wpaucar,Cmolinac,gcaiza}@ups.edu.ec

Abstract. This document presents the development of a Proportional Integral Derivative (PID) and a Fuzzy-PID, for the control of the level of the MPS PA plant, which is constituted by industrial sensors and actuators. The mathematical model and the parameters of the system were obtained using the software the MATLAB environment using a STM32F4 card. The experimental results show the robustness of the Fuzzy PID, because it improves the response of the system and its parameters are tuned automatically according to the state of the process, thus enhancing the performance of the system. Besides, the conventional PID controller requires an adjustment of its parameters to operate in an optimal for each change of variable, and it does not respond efficiently to disturbances.

Keywords: Automation · PID · Fuzzy PID

1 Introduction

PID controllers are extensively utilized in the automation and control of industrial processes, because they are easy to implement and exhibit good response in linear systems. About 95% of the controllers are PID due to their simple structure, precision and reliability in the case of first order processes, and where the response times are not very critical [1]. For systems that present nonlinearities or disturbances, they exhibit a low performance which produces instability [2].

A conventional PID controller offers a good performance when the values of its parameters, namely the proportional (Kp), integral (Ki) and derivative (Kd) gains, are optimal; however, around 90% of the PID controllers are bad tuned at present [3]. Problems arise when the parameters are not well configured, and the performance of the system reduces thus producing losses at the industrial level. In the control of nonlinear complex and/or time-varying variables, the conventional PID does not yield good results [4, 5]. As a consequence, new intelligent control techniques have been developed to improve the response in this type of systems.

Automatic tuning of controllers is a very broad area, which is expanding with new control techniques, thus representing a great benefit for industries [3].

One of the techniques being utilized is a controller based on fuzzy logic to automatically tune the Kp, Ki, Kd parameters. The fuzzy logic technique provides a tool to improve the times of response, enabling to work with nonlinearities and uncertainties [1, 2].

J. Nummenmaa et al. (Eds.): CSEI 2019, AISC 1078, pp. 143–154, 2020.
https://doi.org/10.1007/978-3-030-33614-1_10

In 1974, Mandami developed a fuzzy deduction based on rules suggested by Zadeh, which was used to control dynamic plants [6]. Fuzzy-PID combines classic and fuzzy control, which continuously adjusts the parameters of the controller according to the variables, to achieve an optimal performance [7].

The purpose of this paper is to compare a conventional PID controller with a Fuzzy-PID controller. The simulation and programming were carried out in Matlab/Simulink. First, a conventional PID controller was realized, for which the transfer function of the plant was obtained, to further simulate it and determine the parameters Kp, Ki, Kd of the controller. Afterwards, based on experience and tests carried out in the plant, the algorithm for the Fuzzy-PID controller was realized. The control strategies were implemented on a MPS PA station, in which real processes may be implemented due to its structure and elements. The control is carried out using a STM32F4 card, which is programmed by means of the Matlab software, enabling the visualization of the state of the process in real time. In addition, the stability, time of response, overshoot and steady-state error may be analyzed, to observe the performance in the presence of disturbances and set-point changes.

The document is organized as follows: Sect. 2 describes the theoretical fundamentals and the methodology employed, Sect. 3 describes the identification of the plant, implementation and results and, at last, Sect. 4 includes the conclusions of the work.

2 Materials and Methods

2.1 PID

The PID controller and its variants: P, PI and PD, are used in the control of all types of processes in various industries such as: chemical and food industry, mining industry, mobile and aerospace industry among others. The algorithm depends on three parameters, namely: proportional, integral and derivative gains [8, 9]. In the automation and control of industrial processes, the conventional PID controller is extensively used due to its structure and performance [10]. The parameters of a PID controller are calculated by means of the equations:

$$u(t) = K_p e(t) + K_i \int e(t)dt + K_d \frac{d}{dt} e(t) \tag{1}$$

$$e(t) = r(t) - y(t) \tag{2}$$

Where Kp, Ki, Kd are the proportional, integral and derivative gains, respectively. These parameters constitute the control signal u(t) and affect the output of the system. Figure 1 shows the general structure of the controller.

The effect of the variation of parameters Kp, Ki, Kd of a PID controller on the response of the system, are summarized in Table 1.

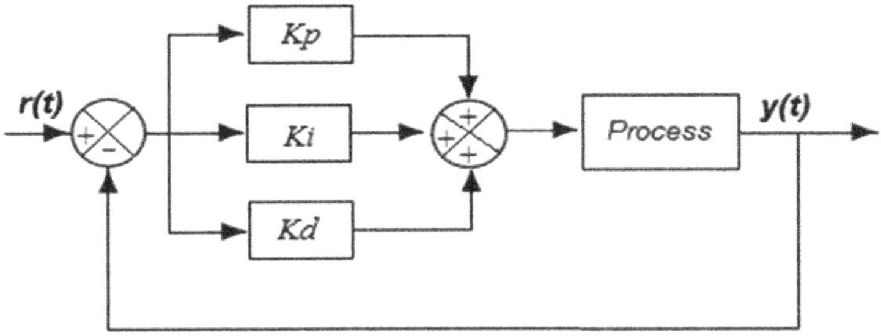

Fig. 1. Structure of a PID controller.

Table 1. Response of the system as a function of the PID parameters [11].

	Rise time	Overshoot	Settling time	Steady-state error
Kp	Reduces	Increases	Changes slightly	Reduces
Ki	Reduces	Increases	Increases	Eliminates
Kd	Changes slightly	Reduces	Reduces	Changes slightly

2.2 Fuzzy Controller

Fuzzy systems were developed as an alternative to classical control methods using, instead of analytical control theory, the logic of decision-making originated by artificial intelligence. Fuzzy control is a strategy based on the knowledge and dynamic of the system [2], which uses linguistic and imprecise rules based on the knowledge of experts [10], i.e. it is a method based on human experience and comprises 4 main parts: fuzzification, defuzzification, inference and rules [11, 12].

The parameters of the Fuzzy PID algorithm are automatically tuned using a fuzzy logic tuner [13, 14]. The inputs to the Fuzzy PID controller are the error and the derivative of the error, and the outputs are the parameters ΔKp, ΔKi, ΔKd, as can be seen in Fig. 2.

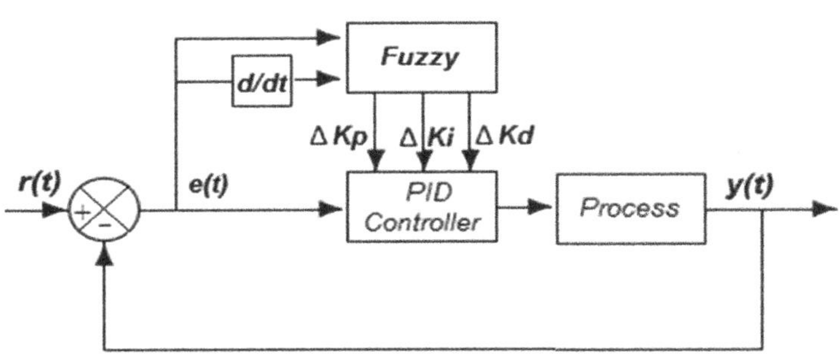

Fig. 2. Function block [13]

The Fuzzy PID control combines conventional PID control and Fuzzy techniques. Once the Fuzzy algorithm is realized, the calculated parameters compensate the parameters obtained in the classic PID controller, according to the equations:

$$K_p = K_{p0} + \Delta K_p \tag{3}$$

$$K_i = K_{i0} + \Delta K_i \tag{4}$$

$$K_d = K_{d0} + \Delta K_d \tag{5}$$

where K_{p0}, K_{i0} and K_{d0} are the parameters obtained for the conventional PID, and ΔK_p, ΔK_i and ΔK_d are the outputs of the Fuzzy controller, which change according to the behavior of the process.

2.3 Hardware

STM32F4
The microprocessor STM32F4 is based on a 32 bits processor with an ARM Cortex-M3 core, has 80 input and output pins, a maximum operating frequency of 72 MHz and 128 Kb flash memory. The main peripherals are: ADC Converter 12-bit, DAC 12-bit, Timers, USART7 UART for serial communication and SPI/I2C interfaces [15].

MPS Compact Station
The MPS PA (Modular Production System) is a compact workstation in which continuous and discontinuous systems can be implemented [16]. The module is designed to enable the implementation of real industrial applications, with the corresponding disturbances. Four variables, namely temperature, pressure, level and flow, can be controlled combining closed loops with sensors and analog and digital actuators, see Fig. 3.

3 Case Study

The process of level variation in the MPS workstation corresponds to a First Order Plus Dead Time (FOPDT) model, considering the controllers as a feedback loop mechanism extensively used in industrial control systems and in a variety of applications that require a continuous modular control. The two controllers were implemented, to carry out a comparison under disturbances and set-point changes.

For the design of the PID and Fuzzy PID controllers, a transfer function of the plant was obtained using the System Identification Toolbox with a first order estimation. The obtained open-loop transfer function

$$G(s) = \frac{0.7996}{1 + 3.1498s} \tag{6}$$

Fig. 3. MPS PA compact workstation [16].

Had a fitting of 95%, according to the input-output curve method. Afterwards, a simulation with a PID controller was carried out utilizing the pidtool library, in which the desired output parameters were entered to obtain the Kp, Ki, Kd parameters. Figure 4 shows the model obtained, and the simulated PID controller.

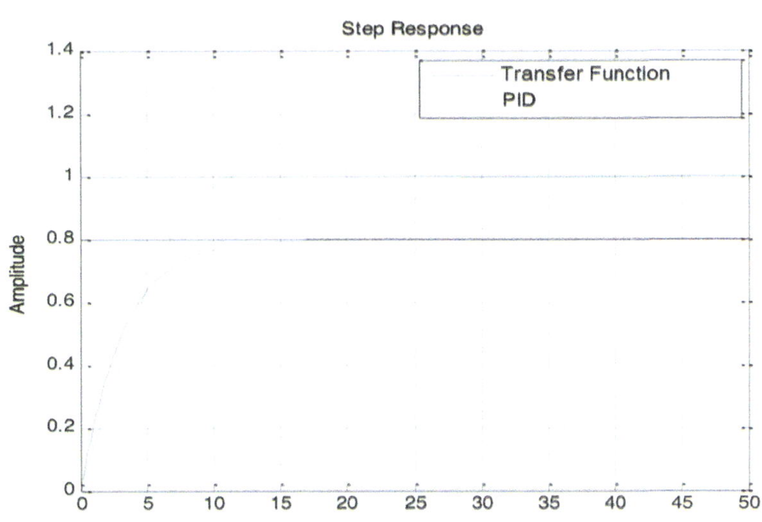

Fig. 4. Transfer function and PID controller.

The signal of the PID controller depends on three terms, which were obtained in the simulation are: Kp = 5, Ki = 0.05 and Kd = 6.5209e-05.

3.1 Design of the Fuzzy PID Controller

The structure of the proposed Fuzzy PID controller is illustrated in Fig. 5. It has two inputs, namely, the error 'e(t)' that depends on the set-point and on the current state, and the derivative of the error 'd(t)' that determines the time in which the change of the error occurs. The obtained outputs are u1, u2 y u3, given as the adjustment points of the control action, considering that such inputs are Δkp, Δki y Δkd, which in turn compensate the parameters Kp, Ki y Kd that are waiting for the value determined as a function of the state of the process.

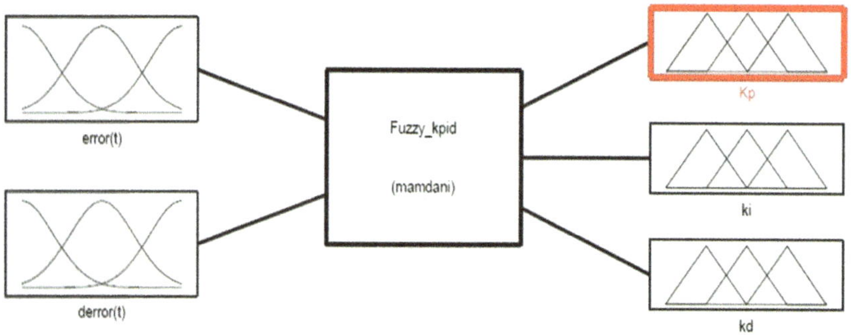

Fig. 5. Inputs and outputs of the fuzzy logic inference.

Utilizing an inference system of type Mamdani, Table 2 presents 25 fuzzy rules obtained for 5 input linguistic variables and 5 output linguistic variables. The input linguistic variables correspond to NB (negative big), NS (negative small), ZE (zero), PS (positive small) and PB (positive big), while the output linguistic variables are present in the membership functions S (Small), MS (Moderately Small), M (Medium), MB (Moderately Big) y B (Big).

Table 2. Fuzzy inference rules.

de(t)/e(t)	NB	NS	ZE	PS	PB
NB	S	S	MS	MS	M
NS	S	MS	MS	M	MB
ZE	MS	MS	M	MB	MB
PS	MS	M	MB	MB	B
PB	M	MB	MB	B	B

Fuzzification

The fuzzification is the transformation of numeric data to linguistic terms. In the present work utilizes two inputs, which should be defined in terms of linguistic variables. Both are ex-pressed in similar membership functions and have five linguistic variables in the input 'e(t)' and in its derivative.

In the process of filling the tank, the minimum and maximum volumes are considered as 1.2 L and 7 L, respectively. According to the parameters obtained for the conventional PID, five Gaussian membership functions are considered for each variable. The range of the error is from −1 to 5, while the range of the derivative of the error is from −0.05 to 0.05.

Rule Base and Fuzzy Inference

Fuzzy rules are expressed as IF-THEN sequences, that guide the control algorithm on the decisions that it should make. The law to construct the fuzzy rules is based uniquely on the expertise of the designer. In this case the fuzzy output of the system is denoted as Kp, Ki y Kd. These outputs are defined in terms of five linguistic variables. The rule base for this result in given in Table 3. The range for output Kp was established as [0, 10], for output Ki [0, 0.05] and for output Kd [0, 0.0025].

Table 3. Comparison of the control performance

Controller	Set point	Rise time (sec)	Overshoot (%)	Settling time (sec)
Fuzzy PID	2,5	43	2,4	70
	3,5	63	0,49	80
	5	113	2,16	120
Conventional PID	2,5	35	10,96	180
	3,5	69	5,42	228
	5	141	15,48	–

Defuzzification

Calculates the output employing different methods allocated in a search table, such that it interprets the degree of membership to the fuzzy sets in a specific decision, or the real value as a function of the current input. For this purpose, three outputs are used to estimate the Fuzzy PID controller, where the parameters ΔKp, ΔKi y ΔKd are compensators of the system. In this work, the Center of Gravity (COG) average defuzzification method is utilized, which calculates the precise value of the fuzzy quantity as the weighted average of the membership function.

3.2 Implementation of the PID and Fuzzy PID Controllers

The implementation of the PID controller is carried out in Simulink. Due to this, the Waijung libraries were installed, which enable the interaction with the peripherals of the STM32 card. Figure 6 shows the Fuzzy PID controller, which was designed using the Fuzzy Logic Controller Matlab library, to automatically adjust its parameters according to the changes and conditions of the process, reducing the intervention of the

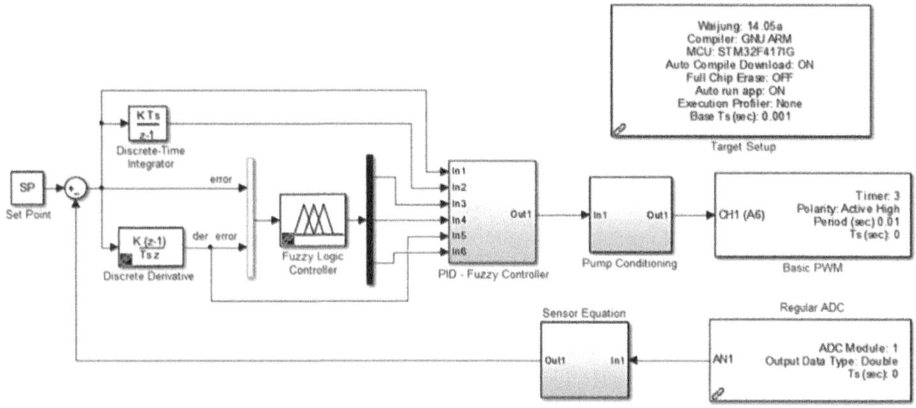

Fig. 6. Implementation of the self-tuning Fuzzy PID controller using MATLAB/Simulink.

operator; this can help to improve the total performance of the conventional PID controller.

The subsystem of the Fuzzy PID controller is shown in Fig. 7. This adopts a controller design that balances the two performance constraints, namely disturbance rejection and tracking of references, where it can be observed that the Fuzzy parameters are added to the Kp, Ki and Kd constants.

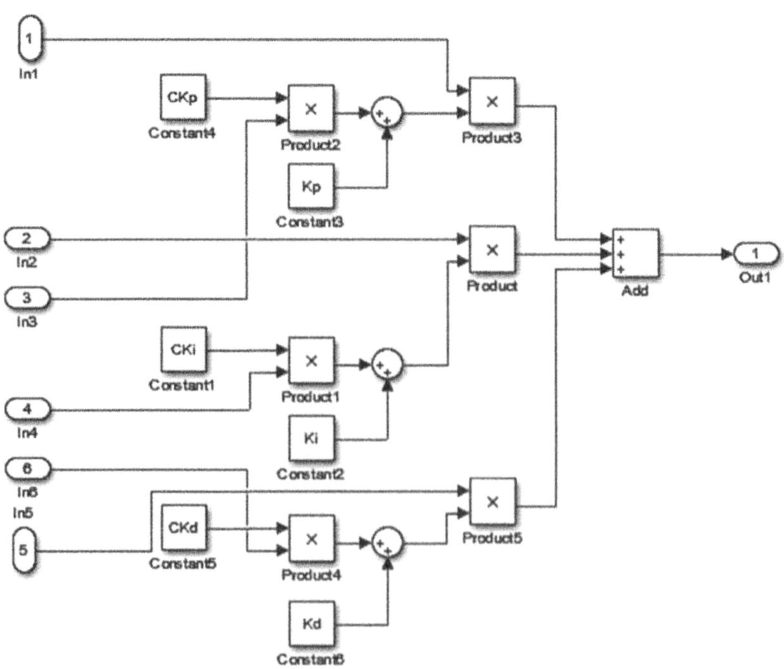

Fig. 7. Block diagram of the subsystem of units of the Fuzzy PID controller.

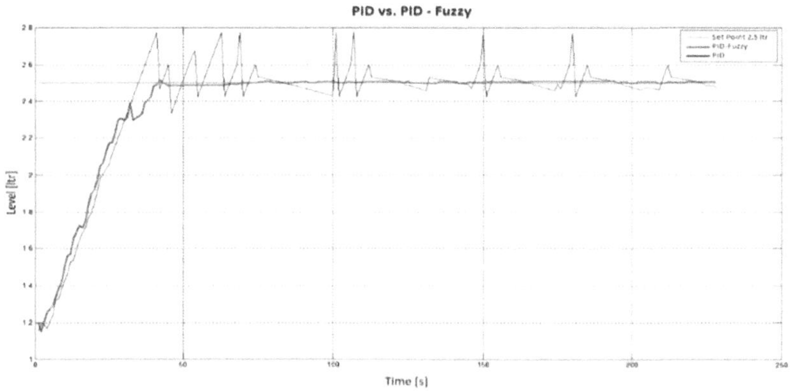

Fig. 8. Conventional PID vs. Fuzzy PID for a reference value of 2.5 L

3.3 Analysis of Results

This section describes the results obtained, where the experimental responses are compared for the conventional PID and Fuzzy PID. In order to compare the responses of the controllers, parameters such as the settling time, time of response and overshoot are analyzed for different reference values, as can be seen in Figs. 8, 9 and 10.

The algorithms of the Fuzzy PID and PID controllers were programmed for a 3.5 L set point, so it presents the best results as shown in Fig. 9.

Figure 10 shows that the pid controller does not have a good performance for different set point levels, that is, it is only optimal for the level that was set, while the fuzzy automatically calculates its parameters according to the set point.

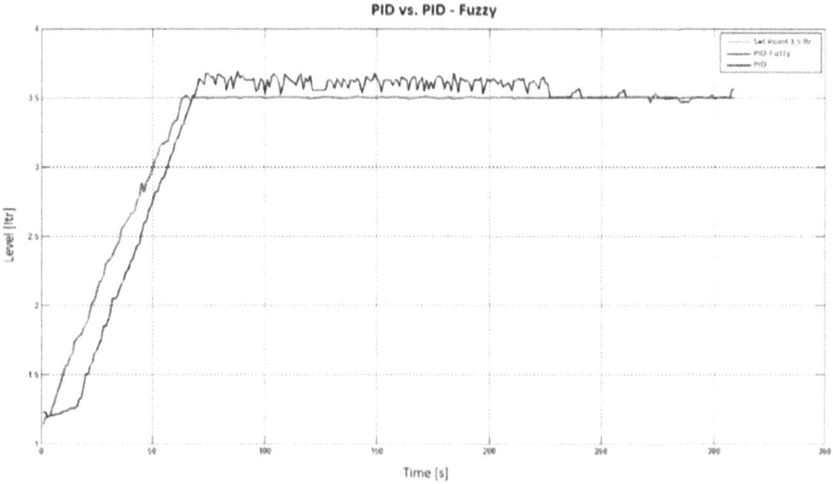

Fig. 9. Conventional PID vs. Fuzzy PID for a reference value of 3.5 L

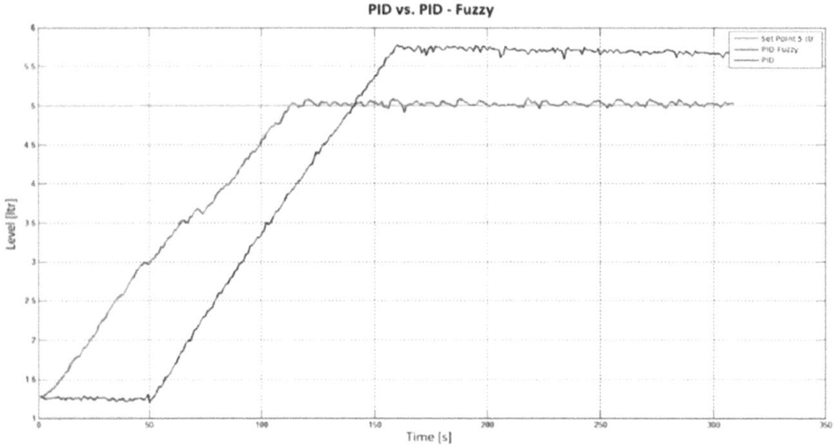

Fig. 10. Conventional PID vs. Fuzzy PID for a reference value of 5 L

Table 3 shows the performance of each controller, thus confirming the effectiveness and versatility of the Fuzzy PID controller, which indicates that a good tracking of the reference is achieved.

From the comparison of the controllers, the Fuzzy PID control algorithm yields better results for all reference values, since it reduces the rise time, settling time, overshoot, and it eliminates the steady-state error. This does not occur for the conventional PID controller, which is inefficient for large reference values in this system; on the contrary, the Fuzzy PID self-tunes its control parameters to avoid this situation. The 3.5 L reference value is analyzed, since the parameters of the optimum PID were obtained for this case; specifically, the rise time was reduced 8.6%, the settling time was reduced 64.9% and there is no overshoot with respect to the conventional PID.

4 Conclusions

By means of the implementation of the controllers, it can be observed that the Fuzzy PID control assigns the reference values for tuning based on the error input signal and error derivative. Improving the response of the system, since its parameters are automatically tuned according to the state of the process, thus raising the performance of the system. In addition, for optimal operation of the conventional PID it is required that the parameters be calculated for each change of variable, and it does not respond efficiently to disturbances.

The STM32F4 Discovery is a low cost, easy to programing card, with several options in programming languages such as C/C++ and block programming on the matlab platform with the help of the waijung library, which is applied for research development, offers various advantages for the development of intelligent control systems, since it can be programmed directly from matlab, including the use of the toolboxes designed in such platform, besides visualizing the state of the variables of the process in real time

Acknowledgment. To the Salesian Polytechnic University and the research Group in Electronics Control and Automation (GIECA) for the support given to the development of the project.

References

1. Sebastiao, A., Lucena, C., Palma, L., Cardoso, A., Gil, P.: Optimal tuning of scaling factors and membership functions for mamdani type PID fuzzy controllers. In: International Conference on Control, Automation and Robotics ICCAR, pp. 92–96. Universidade Nova de Lisboa, Singapore (2015)
2. Pruna, E., Andaluz, V.H., Proaño, L.E., Carvajal, C.P., Escobar, I., Pilatásig, M.: Construction and analysis of PID, fuzzy and predictive controllers in flow system. In: 2016 IEEE International Conference on Automatica (ICA-ACCA), pp. 1–7. IEEE, Curico (2016)
3. Sam, S.M., Angel, T.S.: Performance Optimization of PID controllers using fuzzy logic. In: 2017 IEEE International Conference on Smart Technologies and Management for Computing, Communication, Controls, Energy and Materials (ICSTM), pp. 438–442. IEEE, Chennai (2017)
4. Sharma, H., Palwalia, D.K.: A modified PID control with adaptive fuzzy controller applied to DC motor. In: 2017 International Conference on Information, Communication, Instrumentation and Control (ICICIC), pp. 1–6. IEEE, India (2017)
5. Xu, F., Wang, Z., Song, X., Wang, S.: The fuzzy PID controller design base on nonlinear U-model. In: 018 Chinese Control And Decision Conference (CCDC), pp. 5378–5383. IEEE, Shenyang (2018)
6. Ibrahim, I.N., Akkad, M.A.A.: Exploiting an intelligent fuzzy-PID system in nonlinear aircraft pitch control. In: 2016 International Siberian Conference on Control and Communications (SIBCON), pp. 1–7. IEEE, Russia (2016)
7. Fu, X.D., Cao, G.Z., Huang, S.D., Wu, C.: Fuzzy PID control of the planar switched reluctance motor for precision positioning. In: 2017 7th International Conference on Power Electronics Systems and Applications - Smart Mobility, Power Transfer & Security (PESA), pp. 1–4. IEEE, China (2017)
8. Chabni, F., Taleb, R., Benbouali, A., Bouthiba, M.A.: The application of fuzzy control in water tank level using Arduino. Int. J. Adv. Comput. Sci. Appl. **7**(4), 261–265 (2016)
9. Zhou, Y., Qi, B., Huang, S., Jia, Z.: Fuzzy PID controller for FOPDT system based on a hardware-in-the-loop simulation. In: 2 Proceedings of the 37th Chinese Control Conference, pp. 3382–3387. Researchgate, China (2018)
10. Prusty, S.B., Pati, U.C., Mahapatra, K.: Implementation of fuzzy-PID controller to liquid level system using LabVIEW. In: 2014 International Conference on Control, Instrumentation, Energy & Communication (CIEC), pp. 36–40. IEEE, India (2014)
11. Dehghani, A., Khodadadi, H.: Designing a neuro-fuzzy PID controller based on smith predictor for heating system. In: 2017 17th International Conference on Control, Automation and Systems (ICCAS 2017), pp. 15–20. IEEE, South Korea (2017)
12. Bai Y., Wang, D.: Fundamentals of fuzzy logic control-fuzzy sets, fuzzy rules and defuzzificatios. In: Advanced Fuzzy Logic Technologies in Industrial Applications, pp. 17–36. Springer, London (2006)
13. Thamma, M., Homchat, K.: Real-time implementation of self-tuning fuzzy PID controller for FOPDT system base on microcontroller STM32. In: 2017 2nd International Conference on Control and Robotics Engineering (ICCRE), pp. 130–134. IEEE, Thailand (2017)

14. Garcia, M.V., Perez, F., Calvo, I., Moran, G.: Developing CPPS within IEC-61499 based on low cost devices. In: IEEE International Workshop on Factory Communication Systems - Proceedings, WFCS (2015). https://doi.org/10.1109/WFCS.2015.7160574
15. Jacko, P., Kovac, D., Bucko, R., Vince, T., Kravets, O.: The parallel data processing by nucleo board with STM32 Microcontrollers. In: 2017 International Conference on Modern Electrical and Energy Systems (MEES), pp. 264–267. IEEE, Ukraine (2017)
16. Koszewnik, A., Nartowicz, T., Pawluszewicz, E.: Fractional order controller to control pump in FESTO MPS® PA compact workstation. In: 2016 17th International Carpathian Control Conference (ICCC), pp. 364–367. IEEE, Slovakia (2016)

Real-Time Hand Gesture Recognition: A Long Short-Term Memory Approach with Electromyography

Jonathan A. Zea(✉) and Marco E. Benalcázar(✉)

Departamento de Informática y Ciencias de la Computación (DICC), Escuela Politécnica Nacional, Quito, Ecuador
{jonathan.zea,marco.benalcazar}@epn.edu.ec

Abstract. Hand gestures are a non-verbal type of communication ideally suited for Human-Machine Interaction. Nevertheless, accuracy rates and response times still are a matter of research. One unattended problem has been the difficulty and vagueness of the evaluation of the models proposed in the literature. In this paper, a protocol for evaluating recognition is proposed. A Hand Gesture Recognition system using electromyography signals (EMG) is also presented. This model works in real time, is user dependent and is based in Long Short-Term Memory Networks. The model recognizes 5 different classes (wave in, wave out, fist, open, pinch) apart from the relax state. A data set with 120 people was collected using the commercial device Myo Armband. The data set was divided 50% for tuning and 50% for testing. Following the evaluation protocol proposed, the presented model achieves a 95.79% in classification and a 88.1% in recognition accuracy. An analysis of the characteristics of this model shows the advantage over similar models and its capability for being applied in all sort of fields.

Keywords: EMG · Hand Gesture Recognition · LSTM · Myo Armband

1 Introduction

Gestures are a type of non-verbal communication that is natural, meaningful and effective [34]. Hand gestures, specifically, may be well suited for Human-Machine interactions [14]. The Hand Gesture Recognition problem consists of identifying, from a certain movement of the hand, its corresponding class, and its time of occurrence [6]. In recent years, major research advances have been achieved with applications of Hand Gesture Recognition systems in different fields. For example, these systems have been applied in Sign Language Recognition [3,31,33], in prosthesis [35,38] and in robotics [23]. Hand Gesture Recognition models

The authors gratefully acknowledge the financial support provided by the Escuela Politécnica Nacional for the development of the research project PIJ-16-13.

J. Nummenmaa et al. (Eds.): CSEI 2019, AISC 1078, pp. 155–167, 2020.
https://doi.org/10.1007/978-3-030-33614-1_11

have also been applied in the medical field for data visualization [36] and image manipulation during medical procedures [10,17,40].

A Gesture Recognition model can be divided into 5 modules: data acquisition, preprocessing, feature extraction, classification and posprocessing [7].

Data acquisition consists of measuring, via some kind of sensors, signals generated when a person performs a gesture [7]. The type of sensor depends on the signal to be measured and, it is usually chosen based on the application. Hand Gesture Recognition models are often based on measures of movement (acquired from Inertial Measurement Units IMUs) [15,16,26], images or videos (acquired from cameras) [20,28]; measures of the fingers force or flexion (acquired through sensory gloves)[1,38]; or measures of the electrical muscle activity of the arm, known as electromyography EMG (acquired through electrodes) [8].

Many models proposed in the scientific literature for Hand Gesture Recognition are based on cameras. Some popular cameras are the Kinect [33], the Leap Motion [2] or dept cameras [37]. Using cameras has the advantage of being a non-invasive method for data acquisition, but it is affected by changes in light conditions, occlusion and, it is also susceptible to variations of the distance between the object and the camera. With respect to flexion sensors, those are placed in a sensory glove that may cause some annoyance to the user [31], and has to be designed for different hand sizes. Regarding electromyography EMG, there are two types: surface and intramuscular. Surface EMG uses superficial electrodes placed on top of the skin, and the intramuscular EMG uses invasive sensors like needles. Hence, surface EMG is non-invasive and performs well to capture features in static gestures [15]. Additionally, EMG is one of the few alternatives for data acquisition of movement intention in the case of upper limb amputees [13]. Because of these reasons, this work used surface EMG to implement a Hand Gesture Recognition model.

EMG is the measure of the electrical activity in the skeletal muscles of the human body. Motor units are responsible of the strength and movement of the muscles, both in reflex and voluntary contractions. A muscle is formed of several motor units composed of muscle fibers. The force that a muscle produces depends on the number of motor units that are activated by the brain. Each motor unit is innervated by a unique motor neuron. This motor neuron connects the spinal cord with the muscle fibers of a motor unit. When a motor neuron is activated, an electric pulse, known as action potential, is propagated trough the muscle fiber. In order to produce a movement, the brain constantly activates some motor units, generating a train of action potentials in each muscle fiber. The superposition of the action potential trains of all muscle fibers is known as Motor Unit Action Potential Train MUAPT. The number of active motor units (i.e. recruitment) and the time between consecutive pulses are considered random variables that depend on the muscle force required [11,25,30,39].

A Hand Gesture Recognition model may be based in the analysis of the activation of the motor units. However, decomposing the EMG into its MUAPTs is a very difficult task. Additionally, the EMG acquired depends on factors like temperature, fat, blood flow, skin thickness, and location of the sensors [39].

Thus, the activation of specific motor units is barely distinguished when using surface EMG. Instead, an approach based in supervised learning was carried out. For these reasons, a dataset with 120 people was collected for training the model proposed in this paper.

The second module in Hand Gesture Recognition models is preprocesing. It consists of preparing the acquired signals to match the feature extraction module. Common techniques are filtering for noise reduction [35], and normalization [4]. Normalization is very useful specially when signals of different nature are combined (i.e. sensor fusion) [15].

After data acquisition and preprocessing, the next module of a Hand Gesture Recognition model is feature extraction. Its goal is to extract distinctive and non-redundant information from the original signal [32]. Features are intended to share similar patterns between elements of the same class. Features can be extracted from time, frequency, and time-frequency domains [29]. Common features in the time domain are mean absolute value, nth-order autoregressive coefficients, zero crossing, modified mean absolute value, root-mean square value, sample mean and variance, log detector, average amplitude change, maximum fractal length, Willison amplitude, histogram, cepstral coefficients, and sample entropy. Regarding to the frequency domain, common techniques are the power spectrum, mean and median frequencies, frequency histogram, mean power, and spectral. In the time-frequency domain a common method is the discrete wavelet transform [22]. A common approach for feature extraction is to use a set of these variables selected heuristically. Finding the optimal set of features for a given classifier is not an easy task, since the problem of grouping features is combinatorial [12].

Classifiers are supervised learning algorithms that map a feature vector to a label. The label is chosen from a set that represents the different categories or classes. The most common classifiers used in Hand Gesture Recognition are Support Vector Machines SVM [2,38], feed-forward [2,38], Convolutional [24] and Recurrent Neural Networks [28], decision trees [19], Linear Discriminant Analysis LDA [18], k-nearest neighbors k-NN [21], and Hidden Markov models HMM [5].

The last module of a Hand Gesture Recognition system is postprocessing. Its objective is to adapt the responses of the classifier to the corresponding application (e.g. a drone, or robot). This operation, for example, filters spurious predictions, producing a smoother response [6].

Some applications require predictions of Hand Gesture Recognition models as soon as possible. It is clear that applications such as robotics will require fast responses of the system. This is usually referred as real time. A problem emerges when it is attempted to generalize this concept. In the context of human perception, a system is considered to work in real time when it returns a prediction in 300 ms or faster [9]. Achieving real-time systems usually restricts the complexity of the algorithms. Therefore, the challenge for research is developing a model with high recognition rate and low computational complexity. This paper presents a novel Real-Time Hand Gesture Recognition model using EMG. This model is based on Long Short-Term Memory LSTM networks because of its ability to learn from sequential information.

Following this Introduction, the remaining of this paper is organized as follows. Section 2 is about methodologies. It begins defining classification and recognition; later it describes each module of the proposed model. In Sect. 3, the experiments for testing the proposed model are described along with the analysis of the results obtained. Finally, Sect. 4 lists the findings of this research and outlines the future work.

2 Methodology

2.1 Classification and Recognition

In this paper, the notions of classification and recognition are differentiated. Classification will be understood as labeling feature vectors. This notion of classification leaves apart timing and order of labels. Recognition goes beyond this concept; it also includes identifying in time the begging and the end of a gesture. Following this approach, for a gesture (i.e. target), a classification will be correct when the output of the Hand Gesture Recognition model (i.e. prediction) matches the target's label. A real-time system should return predictions separated apart 300 ms or less (i.e. vector of predictions). Hence, the recognition evaluation involves comparing the vector of predictions with the actual instants of occurrence of the gesture in time (i.e. ground truth). This comparison, known as overlapping factor, is represented in Eq. 1 and Fig. 1.

$$\rho = 2 * \frac{|A \cap B|}{|A| + |B|} \tag{1}$$

Where A, B are the sets of instants of time with muscle activity, with A corresponding to the ground truth, and B corresponding to the vector of predictions. ρ is defined as the overlapping factor.

For the EMG signal shown in Fig. 1, a classification can be obtained by different means. One approach could be to use the whole signal passed to the feature extraction and then to the classifier, obtaining a unique output label. Another approach, would be to segment the muscle activity in the EMG signal and then follow the former approach. In this way, a unique output label is also obtained, but with the advantage of not considering the relax state of the EMG in the process of classification. Another approach could be to use sliding windows to generate not only a single prediction but a vector of predictions. This vector of predictions has to be analysed afterwards for obtaining an equivalent unique prediction (a technique that can be used for this reduction is the mode). If this reduced prediction matches the original label (i.e. target) the classification is correct.

The evaluation of recognition accuracy requires a valid vector of predictions with its corresponding appearances in time. This vector is valid only if it contains just one segment of equal consecutive labels (apart from the relax state). Otherwise, the overlapping factor ρ is not calculated and the recognition is stated as

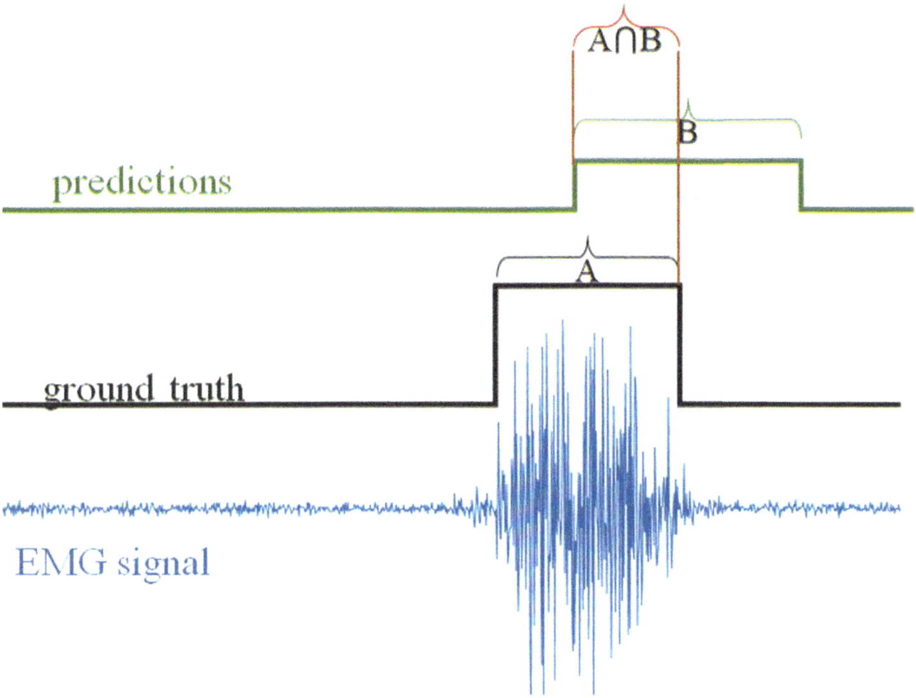

Fig. 1. In blue, an EMG signal from one channel recorded using the Myo Armband during 5 s. In black, the ground truth that delimits the muscle activity in the EMG signal. In green, the vector of predictions.

incorrect. A recognition is correct when the overlapping factor is greater than a threshold.

2.2 Data Acquisition

Hardware Characteristics: The Myo Armband is a Bluetooth device for Hand Gesture Recognition distributed by Thalmic Labs. It is a bracelet formed by 8 differential electrodes that measures the muscle electrical activity at 200 Hz with a resolution of 8 bits. It has an Inertial Measurement Unit of 9° (accelerometer, magnetometer and gyroscope) and haptic feedback. In Fig. 2(a) the Myo Armband is shown. This device also contains a proprietary recognition system, which is capable of identifying in real time 5 gestures of the hand: wave in, wave out, fist, pinch and open Fig. 2(c).

Dataset Description. For this work, a dataset of 120 volunteers was collected. Each volunteer performed 50 repetitions of every gesture. The dataset is formed by 6 classes; the same 5 gestures that the Myo Armband recognizes, and also it includes the relax pose. This is a gesture used to represent the remaining

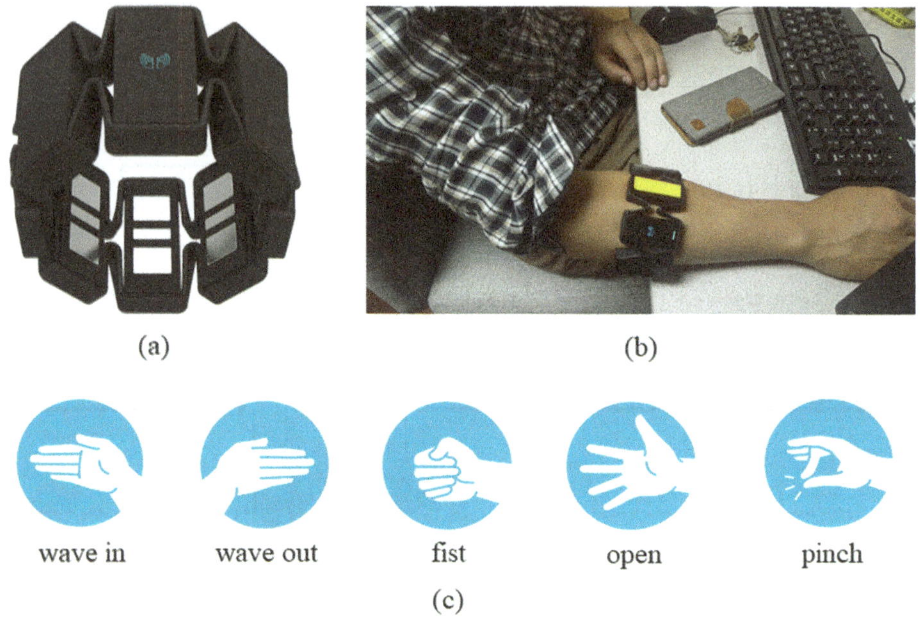

Fig. 2. (a) Myo Armband, (b) location of the Myo Armband, (c) set of hand gestures.

of all possible gestures. For the relax pose only 10 repetitions were acquired. The dataset is formed of 260 repetitions per volunteer, with a total of 31200 repetitions for the 120 volunteers. Each repetition was recorded during 5 s blindly (i.e. the person did not receive feedback from any Hand Gesture Recognition model). Volunteers were asked to wear the Myo Armband in their right forearm, as shown in Fig. 2(b). It was established that a recording starts from the relax pose, performs the gesture, and finishes coming back to the relax pose (i.e. transient gestures). The indices of beginning and ending of the gestures were manually registered (i.e. manual muscle activity segmentation). In addition to this information, gender and age was included. The dataset contains a 75% men and 25% of women between 17 a 29 years (all of them university student). The dataset was divided into 50% for training and 50% for testing. A sample of a recorded EMG is shown in Fig. 1.

2.3 Proposed Model

Preprocessing. The proposed model used a sliding window approach. For this matter, a window W will be understood as a partial observation of the EMG signal. A window is an $wx8$ matrix, where w is the number of samples, and it has 8 columns corresponding to the 8 channels of the Myo Armband. This proposed model uses $w = 150$ because it was a close value to the average duration of the muscle activity observed in the dataset. This number of points corresponds to 0.75 s of the EMG signal (since the Myo Armband transmits at 200 samples per second). The sliding window reloads its observation after a stride s. The stride is specified in terms of samples, $s = 20$; this value implies refreshing data in 0.1 s. In the case that the Recognition model returns a prediction faster than this time, will be considered in real time.

The sliding window is rectified and filtered using absolute value and a 4th order Butterworth filter with cut-off frequency of 5 Hz. This processed signal will be represented as $\Psi(W)$. The parameters for this procedure were selected to reduce the non-stationarity of the EMG by smoothing it out, as stated in [7].

Feature Extraction. The EMG features used for the classifier are the concatenation of specifically selected features $g(W)$ and the processed signal $\Psi(W)$, these features are described by [27]. The selected features $g(W)$ are the calculated from the following bag of functions: Mean Absolute Value, Slope Sign Changes, Waveform Length, Root Mean Square, and the Hjorth parameters. It is important to state that the bag of functions are only from the time domain because functions from the frequency and the time-frequency domains have higher computational complexities. The selected features $g(W)$ (obtained with the bag of functions in the window $wx8$) form a vector of length 56. The feature vector of a window $F(W)$, hence, is the concatenation of $\Psi(W)$ and $g(W)$, a vector with length 1256 ($150 * 8 + 56$, the points in the sliding window plus the selected features).

Classification. A Long Short-Term Memory (LSTM) network was chosen for classification because of its capability for learning from time series [4]. This architecture connects the LSTM with a softmax layer. A "last-state" configuration is used for training the LSTM layer that has 800 hidden units.

Data for Training and Generation of Sequences. As described in Subsect. 2.2, the dataset contains repetitions from 120 volunteers. The model was developed and tuned using 60 volunteers. Sequence examples are generated using the repetitions for training from each user. The model was trained with 10 repetitions per user. Although, it could have been trained with up to 25 repetitions, preliminary results did not show better performing with more repetitions. Moreover, the time spent in acquiring the EMG data is less than half when using 10 repetitions instead of 25.

A standardization process, that scales each feature to have a mean μ of 0 and standard deviation σ of 1, was applied to the training set of 10 repetitions per user. Each example (i.e. sequence) for training the LSTM was composed of 5 consecutive window observations.

The Stochastic Gradient Descend with Momentum SGDM was used to train the network, using up to 60 epochs with the following parameters: initial learning rate of 0.02, momentum of 0.35 and regularization factor of 0.001. Most of these parameters were set by heuristics. In Fig. 3 is shown a summary of the proposed Hand Gesture Recognition model.

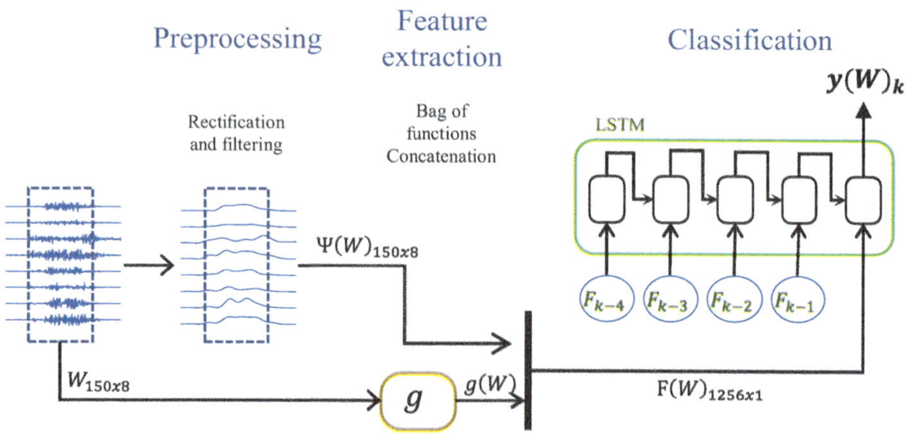

Fig. 3. Modules of the proposed Hand Gesture Recognition model.

3 Evaluation and Results

The proposed model was tested over the remaining 60 volunteers. Classification results are displayed in the confusion matrix (Table 1). For each class, 1500 targets were tested (25 repetitions per gesture times 60 volunteers). In the lower right corner of this table, can be seen that the classification accuracy in average for the proposed model is 95.79%. The highest and lowest sensitivities occur for the gestures fist 97.67% and pinch 93.87%, respectively. This means that fist had the largest number of repetitions correctly classified, whereas pinch had the lowest. Regarding precision, the highest and lowest values occur for the gestures fist 99.45% and wave out 97.72%, respectively. This means that when the gesture fist was predicted, it was a correct prediction 99.45%; whereas, when wave out was predicted, it was a correct prediction only the 97.72%. It is important to consider that a large number of predictions were relax. The classification error for each class is proportional to this value. For example, pinch was the gesture that was most times confused as relax (77 repetitions).

A possible explanation for this behavior is that the model, in cases of confusion, returns the relax as the default class. This error, although persistent throughout all the classes, is minimal compared to the total number of repetitions tested (216 out of 7500).

Following the evaluation procedure established in Subsect. 2.1, the alignment of the vector of predictions was considered for obtaining the recognition results. The recognition accuracy obtained is 88.1% with a standard deviation of 9.63% between users. A histogram is presented in Fig. 4 with the percentage of volunteers and the range of their recognition accuracy. The leftmost bar reveals that 1.67% of the people for testing had a recognition accuracy between 50%–52.5%. On the contrary, the rightmost bar reveals that around 15% of the people had an accuracy higher than 97.5%. An accumulative line (red) was included, it shows that less than 20% of the people had a recognition accuracy lower than 80%.

Table 1. Confusion matrix of the proposed models

	Targets					**Predictions count % Accuracy (Precision)**
	Wave in	Wave out	Fist	Open	Pinch	
Wave in	**1435**	9	1	3	0	**1448** **99.1%**
Wave out	10	**1460**	1	13	10	**1494** **97.72%**
Fist	3	0	**1465**	2	2	**1473** **99.45%**
Open	5	15	2	**1416**	3	**1441** **98.27%**
Pinch	7	4	2	8	**1408**	**1429** **98.53%**
Relax	40	12	29	58	77	**216** **0%**
Targets count % Accuracy (Sensitivity)	**1500** **95.67%**	**1500** **97.33%**	**1500** **97.67%**	**1500** **94.4%**	**1500** **93.87%**	**7500** **95.79%**

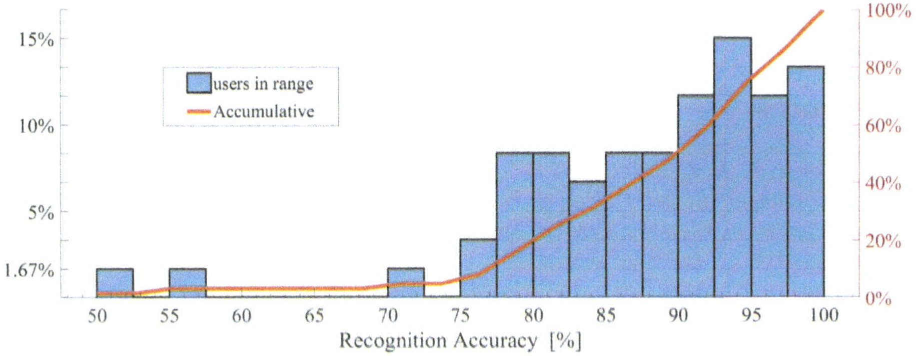

Fig. 4. In blue, a histogram of Recognition Accuracy by User; in red, the accumulative percentage of Recognition Accuracy.

Furthermore, it shows that around 50% of the people had an accuracy higher than 90%.

4 Concluding Remarks and Future Work

In this work, a Hand Gesture Recognition model for electromyography EMG was presented. This model used an LSTM network based on a filtered EMG signal and a set of time-domain selected features. A dataset was collected using the Myo Armband to evaluate this model. This dataset is formed by 120 volunteers that performed 260 repetitions of 5 gestures and the relax class. Real-time behavior was simulated using the sliding window approach. Training the system with 10 repetitions per class had a 95.79% of classification accuracy and 88.1% of recognition. The time responses of these tests allow us to claim that the model is suitable to work in real time.

Even though the accuracy rates obtained in this work are similar to other models proposed in the literature, an strict comparison can not be done. This is because datasets are formed by different type and number of gestures. Besides the procedure for testing is not always clearly specified. On this matter, this paper proposed a novel methodology for evaluating the recognition accuracy.

The gap between the classification and the recognition accuracy (95.79% and 88.1%, respectively) of this model is around 7.7%. It is hypothesized that the transitions between the relax state and the gesture may confuse the model, for it predicts a class before having enough information.

The Hand Gesture Recognition model proposed in this work achieves high classification rates, but it is user dependent. A user independent system has the advantage of not needing to be trained by the user, but variations in the armband's location as well as physiological differences turns it to be a much harder problem. Another issue with Hand Gesture Recognition models is the number of gestures. Even though, the architecture of the model presented in this work can include more gestures; it is a matter for future research. Larger datasets and deep learning models may solve these problems.

References

1. Abhishek, K.S., Qubeley, L.C.K., Ho, D.: Glove-based hand gesture recognition sign language translator using capacitive touch sensor. In: 2016 IEEE International Conference on Electron Devices and Solid-State Circuits (EDSSC), pp. 334–337. IEEE (2016)
2. Ameur, S., Khalifa, A.B., Bouhlel, M.S.: A comprehensive leap motion database for hand gesture recognition. In: 2016 7th International Conference on Sciences of Electronics, Technologies of Information and Telecommunications, SETIT 2016, pp. 514–519. IEEE (2017). https://doi.org/10.1109/SETIT.2016.7939924. http://ieeexplore.ieee.org/document/7939924/
3. Athira, P.K., Sruthi, C.J., Lijiya, A.: A signer independent sign language recognition with co-articulation elimination from live videos: an indian scenario. J. King Saud Univ. Comput. Inf. Sci. (2019). https://doi.org/10.1016/j.jksuci.2019.05.002. https://www.sciencedirect.com/science/article/pii/S131915781831228X
4. Barros, P., Maciel-Junior, N.T., Fernandes, B.J., Bezerra, B.L., Fernandes, S.M.: A dynamic gesture recognition and prediction system using the convexity approach. Comput. Vis. Image Underst. **155**, 139–149 (2017). https://doi.org/10.1016/j.cviu.2016.10.006. https://www.sciencedirect.com/science/article/pii/S107731421630159X
5. Belgacem, S., Chatelain, C., Paquet, T.: Gesture sequence recognition with one shot learned CRF/HMM hybrid model. Image Vis. Comput. **61**, 12–21 (2017). https://doi.org/10.1016/j.imavis.2017.02.003. https://www.sciencedirect.com/science/article/pii/S0262885617300471

6. Benalcázar, M.E., Anchundia, C.E., Zea, J.A., Zambrano, P., Jaramillo, A.G., Segura, M.: Real-time hand gesture recognition based on artificial feed-forward neural networks and EMG. In: European Signal Processing Conference, September 2018, pp. 1492–1496 (2018). https://doi.org/10.23919/EUSIPCO.2018.8553126
7. Benalcazar, M.E., Motoche, C., Zea, J.A., Jaramillo, A.G., Anchundia, C.E., Zambrano, P., Segura, M., Benalcazar Palacios, F., Perez, M.: Real-time hand gesture recognition using the Myo armband and muscle activity detection. In: 2017 IEEE 2nd Ecuador Technical Chapters Meeting, ETCM 2017, January 2017, pp. 1–6 (2018). https://doi.org/10.1109/ETCM.2017.8247458
8. Benatti, S., Rovere, G., Bosser, J., Montagna, F., Farella, E., Glaser, H., Schonle, P., Burger, T., Fateh, S., Huang, Q., Benini, L.: A sub-10 mW real-Time implementation for EMG hand gesture recognition based on a multi-core biomedical SoC. In: Proceedings - 2017 7th International Workshop on Advances in Sensors and Interfaces, IWASI 2017, pp. 139–144. IEEE (2017)
9. Bonato, P., D'Alessio, T., Knaflitz, M.: A statistical method for the measurement of muscle activation intervals from surface myoelectric signal during gait. IEEE Trans. Biomed. Eng. **45**(3), 287–299 (1998). https://doi.org/10.1109/10.661154. http://www.ncbi.nlm.nih.gov/pubmed/9509745, http://ieeexplore.ieee.org/document/661154/
10. Cartagena, P., Naranjo, J., Saltos, L., Garcia, C., Garcia, M.: Multifunctional exoskeletal orthosis for hand rehabilitation based on virtual reality. Adv. Intell. Syst. Comput. **884**, 209–221 (2019). https://doi.org/10.1007/978-3-030-02828-2_16
11. Dimitrova, N., Dimitrov, G.: Electromyography (EMG) Modeling. In: Wiley Encyclopedia of Biomedical Engineering. Wiley, Hoboken (2006). https://doi.org/10.1002/9780471740360.ebs0656. http://doi.wiley.com/10.1002/9780471740360.ebs0656
12. Duda, R.O., Hart, P.E.P.E., Stork, D.G.: Pattern Classification. Wiley (2001). https://www.wiley.com/en-ec/Pattern+Classification%2C+2nd+Edition-p-9780471056690
13. Farina, D., Jiang, N., Rehbaum, H., Holobar, A., Graimann, B., Dietl, H., Aszmann, O.C.: The extraction of neural information from the surface EMG for the control of upper-limb prostheses: emerging avenues and challenges. IEEE Trans. Neural Syst. Rehabil. Eng. **22**(4), 797–809 (2014)
14. Grif, H.S., Farcas, C.C.: Mouse cursor control system based on hand gesture. Procedia Technol. **22**, 657–661 (2016). https://doi.org/10.1016/J.PROTCY.2016.01.137. https://www.sciencedirect.com/science/article/pii/S2212017316001389
15. Huang, Y., Guo, W., Liu, J., He, J., Xia, H., Sheng, X., Wang, H., Feng, X., Shull, P.B.: Preliminary testing of a hand gesture recognition wristband based on EMG and inertial sensor fusion. In: Lecture Notes in Computer Science (Including Subseries Lecture Notes in Artificial Intelligence and Lecture Notes in Bioinformatics). vol. 9244, pp. 359–367. Springer, Cham (2015)
16. Iyer, D., Mohammad, F., Guo, Y., Al Safadi, E., Smiley, B.J., Liang, Z., Jain, N.K.: Generalized hand gesture recognition for wearable devices in IoT: application and implementation challenges. In: Lecture Notes in Computer Science (Including Subseries Lecture Notes in Artificial Intelligence and Lecture Notes in Bioinformatics), vol. 9729, pp. 346–355. Springer, Cham (2016). https://doi.org/10.1007/978-3-319-41920-6_26. http://link.springer.com/10.1007/978-3-319-41920-626

17. Jacob, M.G., Wachs, J.P., Packer, R.A.: Hand-gesture-based sterile interface for the operating room using contextual cues for the navigation of radiological images. J. Am. Med. Inform. Assoc. **20**(E1), e183–e186 (2013). https://doi.org/10.1136/amiajnl-2012-001212. http://www.ncbi.nlm.nih.gov/pubmed/23250787, http://www.pubmedcentral.nih.gov/articlerender.fcgi?artid=PMC3715344, https://academic.oup.com/jamia/article-lookup/doi/10.1136/amiajnl-2012-001212
18. Jiang, X., Merhi, L.K., Xiao, Z.G., Menon, C.: Exploration of force myography and surface electromyography in hand gesture classification. Med. Eng. Phys. **41**, 63–73 (2017). https://doi.org/10.1016/j.medengphy.2017.01.015. http://www.ncbi.nlm.nih.gov/pubmed/28161107, https://linkinghub.elsevier.com/retrieve/pii/S1350453317300176
19. Joshi, A., Monnier, C., Betke, M., Sclaroff, S.: Comparing random forest approaches to segmenting and classifying gestures. Image Vis. Comput. **58**, 86–95 (2017). https://doi.org/10.1016/j.imavis.2016.06.001. https://www.sciencedirect.com/science/article/pii/S0262885616300993
20. Kim, S.Y., Han, H.G., Kim, J.W., Lee, S., Kim, T.W.: A hand gesture recognition sensor using reflected impulses. IEEE Sensors J. **17**(10), 2975–2976 (2017). https://doi.org/10.1109/JSEN.2017.2679220. http://ieeexplore.ieee.org/document/7874149/
21. Li, G., Zhang, R., Ritchie, M., Griffiths, H.: Sparsity-based dynamic hand gesture recognition using micro-Doppler signatures. In: 2017 IEEE Radar Conference (RadarConf). pp. 0928–0931. IEEE (2017). https://doi.org/10.1109/RADAR.2017.7944336. http://ieeexplore.ieee.org/document/7944336/
22. Li Yang, Tian Yantao, Chen Wanzhong: Multi-pattern recognition of sEMG based on improved BP neural network algorithm. In: Proceedings of the 29th Chinese Control Conference. IEEE (2010). https://ieeexplore.ieee.org/abstract/document/5573567/similar#similar
23. Liu, H., Wang, L.: Gesture recognition for human-robot collaboration: a review. Int. J. Ind. Ergon. **68**, 355–367 (2018). https://doi.org/10.1016/j.ergon.2017.02.004. https://www.sciencedirect.com/science/article/pii/S0169814117300690
24. Mohanty, A., Rambhatla, S.S., Sahay, R.R.: Deep gesture: static hand gesture recognition using CNN. In: Advances in Intelligent Systems and Computing, AISC, vol. 460, pp. 449–461. Springer, Singapore (2017). https://doi.org/10.1007/978-981-10-2107-7_41. http://link.springer.com/10.1007/978-981-10-2107-741
25. Mordhorst, M., Heidlauf, T., Röhrle, O.: Mathematically modelling surface EMG signals. PAMM **14**(1), 123–124 (2014). https://doi.org/10.1002/pamm.201410049. http://doi.wiley.com/10.1002/pamm.201410049
26. Moschetti, A., Fiorini, L., Esposito, D., Dario, P., Cavallo, F.: Recognition of daily gestures with wearable inertial rings and bracelets. Sensors **16**(8), 1341 (2016). https://doi.org/10.3390/s16081341. http://www.ncbi.nlm.nih.gov/pubmed/27556473, http://www.pubmedcentral.nih.gov/articlerender.fcgi?artid=PMC5017504, http://www.mdpi.com/1424-8220/16/8/1341
27. Motoche, C., Benalcázar, M.E.: Real-time hand gesture recognition based on electromyographic signals and artificial neural networks. In: Lecture Notes in Computer Science (Including Subseries Lecture Notes in Artificial Intelligence and Lecture Notes in Bioinformatics), LNCS, vol. 11139, pp. 352–361. Springer, Cham (2018)
28. Palmeri, M., Vella, F., Infantino, I., Gaglio, S.: Sign languages recognition based on neural network architecture. In: Smart Innovation, Systems and Technologies, vol. 76, pp. 109–118. Springer, Cham (2018). https://doi.org/10.1007/978-3-319-59480-4_12. http://link.springer.com/10.1007/978-3-319-59480-412

29. Raez, M.B.I., Hussain, M.S., Mohd-Yasin, F.: Techniques of EMG signal analysis: detection, processing, classification and applications. Biol. Proced. Online **8**, 11–35 (2006). https://doi.org/10.1251/bpo115. http://www.ncbi.nlm.nih.gov/pubmed/16799694, http://www.pubmedcentral.nih.gov/articlerender.fcgi?artid=PMC1455479
30. Rodriguez-Falces, J., Navallas, J., Mal, A.: EMG modeling. In: Computational Intelligence in Electromyography Analysis - A Perspective on Current Applications and Future Challenges. InTech (2012). https://doi.org/10.5772/50304
31. Saggio, G., Orengo, G., Pallotti, A., Errico, V., Ricci, M.: Sensory systems for human body gesture recognition and motion capture. In: 2018 International Symposium on Networks, Computers and Communications, ISNCC 2018, pp. 1–6. IEEE (2018). https://doi.org/10.1109/ISNCC.2018.8531054. https://ieeexplore.ieee.org/document/8531054/
32. Scherer, R., Rao, R.: Non-manual control devices. In: Handbook of Research on Personal Autonomy Technologies and Disability Informatics, pp. 233–250. IGI Global (2011). https://doi.org/10.4018/978-1-60566-206-0.ch015. http://services.igi-global.com/resolvedoi/resolve.aspx?doi=10.4018/978-1-60566-206-0.ch015
33. Sidig, A.a.I., Luqman, H., Mahmoud, S.A.: Arabic sign language recognition using optical flow-based features and HMM. In: Lecture Notes on Data Engineering and Communications Technologies, pp. 297–305. Springer, Cham (2017). https://doi.org/10.1007/978-3-319-59427-9_32. http://link.springer.com/10.1007/978-3-319-59427-932
34. Sonkusare, J.S., Chopade, N.B., Sor, R., Tade, S.L.: A review on hand gesture recognition system. In: 2015 International Conference on Computing Communication Control and Automation, pp. 790–794. IEEE (2 2015). https://doi.org/10.1109/ICCUBEA.2015.158. http://ieeexplore.ieee.org/document/7155956/
35. Tavakoli, M., Benussi, C., Lourenco, J.L.: Single channel surface EMG control of advanced prosthetic hands: a simple, low cost and efficient approach. Expert Syst. Appl. **79**, 322–332 (2017). https://doi.org/10.1016/j.eswa.2017.03.012. https://www.sciencedirect.com/science/article/pii/S0957417417301574
36. Wachs, J., Stern, H., Edan, Y., Gillam, M., Feied, C., Smith, M., Handler, J.: A real-time hand gesture interface for medical visualization applications. In: Advances in Soft Computing, vol. 36, pp. 153–162. Springer, Heidelberg (2006). https://doi.org/10.1007/978-3-540-36266-1_15. http://link.springer.com/10.1007/978-3-540-36266-115
37. Wang, C., Liu, Z., Zhu, M., Zhao, J., Chan, S.C.: A hand gesture recognition system based on canonical superpixel-graph. Signal Process. Image Commun. **58**, 87–98 (2017). https://doi.org/10.1016/j.image.2017.06.015. https://www.sciencedirect.com/science/article/abs/pii/S0923596517301224
38. Wang, N., Lao, K., Zhang, X.: Design and myoelectric control of an anthropomorphic prosthetic hand. J. Bionic Eng. **14**(1), 47–59 (2017). https://doi.org/10.1016/S1672-6529(16)60377-3. https://www.sciencedirect.com/science/article/pii/S1672652916603773
39. Weiss, L.D., Weiss, J.M., Silver, J.K.: Easy EMG: A Guide to Performing Nerve Conduction Studies and Electromyography. Elsevier (2015)
40. Wipfli, R., Dubois-Ferrière, V., Budry, S., Hoffmeyer, P., Lovis, C.: Gesture-controlled image management for operating room: a randomized crossover study to compare interaction using gestures, mouse, and third person relaying. PLoS ONE **11**(4), e0153596 (2016)

Development of a Fuzzy Logic-Based Solar Charge Controller for Charging Lead–Acid Batteries

Fabricio Paredes Larroca[1(✉)], Erich Saettone Olschewski[1], Javier Quino Favero[1], Jimmy Rosales Huamaní[2], and José Luis Castillo Sequera[3]

[1] Universidad de Lima, 15023 Lima, Peru
{fparedes, esaetton, jquinof}@ulima.edu.pe
[2] Universidad Nacional de Ingeniería, 15333 Lima, Peru
jrosales@uni.edu.pe
[3] Universidad de Alcalá, 28871 Madrid, Spain
jluis.castillo@uah.es

Abstract. The design and implementation of a solar charge controller for lead–acid batteries is intended to supplement a component of the water purification module of the water treatment unit for natural disaster relief. This unit contains a solar panel system that supplies power to the module by charging batteries through a controller comprising an Atmega 328 processor. The solar panel feeds voltage to the batteries through fuzzy logic-based software, which allows up to 6 A DC to pass through the controller's power circuit. Consequently, the battery was charged in less time (an average of 7 h to reach maximum capacity), wherein battery lifespan is related to the charge wave frequency. Thus, our software may be adapted in different control algorithms without having to change hardware.

Keywords: Fuzzy logic · Controller · Battery charger

1 Introduction

The extensive use of microprocessor technology [1] yields simple solutions to complex problems through electronics, which allows both complex and moderately advanced algorithms to be quickly and easily programmed into new environments, such as MatLab. Thus, these algorithms may effectively contribute to solving problems from other fields. This hardware–software combination emerges as an embedded system used to address problems in multiple areas of technology, such as charging batteries through solar panels.

Presently, two types of solar energy charging systems are used: the maximum power point tracking that maximizes solar panel power output and the power width modulation (PWM), which slowly lowers the amount of power supplied to batteries when almost fully charged and is used herein. Feeding power to autonomous system components such as water purifications systems for disaster relief requires a system

J. Nummenmaa et al. (Eds.): CSEI 2019, AISC 1078, pp. 168–183, 2020.
https://doi.org/10.1007/978-3-030-33614-1_12

designed to supply large voltage amounts. Therefore, the battery charging system required must be arranged in two separate banks and powered using two different charging systems, in which one system actually charges the battery and the other supplies the necessary power to prevent halting of the water treatment system. Further, this arrangement of two individual intelligent exchange systems is intended to provide operating autonomy and safety to the autonomous water treatment unit. Moreover, the charging system requires a controller to efficiently manage the energy from the battery charging systems, which are usually solar panels. A solar charge controller is a device that quickly and efficiently charges batteries using solar panel, thereby extending battery lifespan and managing voltage consumption throughout the charging process. Fast charging of batteries is important for providing sufficient energy to the water treatment module regularly, thereby guaranteeing continuous water treatment operations 24 h a day. Conversely, the method used for charging lead–acid batteries such as pulse charging may reduce charging times and increase battery lifespan [2].

This proposal focuses on building a custom battery charger based on the features described above using electronic components available in the local market. This development method includes the following advantages: an ad hoc design suitable for outdoor operation in local weather conditions; easy access for the replacement of electronic components such as the control board and power board, which are different and independent herein; and the considerable reduction in importing costs for a finished product with these characteristics. In fact, this development may considerably contribute to the decrease in operating costs associated with off-grid rural electrification recorded in Perú since 1993 [3].

In this paper, Sect. 1 provides background information, Sect. 2 presents a conceptual framework for fuzzy logic, Sect. 3 describes the materials and methods used, Sect. 4 details the design and development of the proposed controller, Sect. 5 denotes our results, Sect. 6 discusses our conclusion, and finally, Sect. 7 suggests possible future works.

2 Fuzzy Logic Controller

Herein, the Mamdani model is used. This system features the following components (see Fig. 1).

2.1 The Fuzzifier

The value incoming from the voltage sensor [4] is converted to a language that can be processed by the Atmega 328 processor. Further, the data converted into fuzzy values will be processed by the Inference System. These fuzzy values correspond to the universe of input variables.

2.2 The Fuzzy Inference System

This structure contains the different membership levels originated in the fuzzifier. These are the values processed to create a fuzzy output.

2.3 The Fuzzy Rule-Based System

It is a set of rules that constitute the system's engine (see Fig. 2). These rules are based on the information provided via daily work procedures used by the system operator. Further, this procedure is interpreted using IF-THEN rules with a precedent and a result.

2.4 The Defuzzifier

The output generated through the Inference System is a fuzzy output that cannot be interpreted by the actuator of the system, indicating that the output must be converted to an analog value to achieve an adequate response. This analog value is obtained through the center of gravity for all possible answers, according to their membership degree [5].

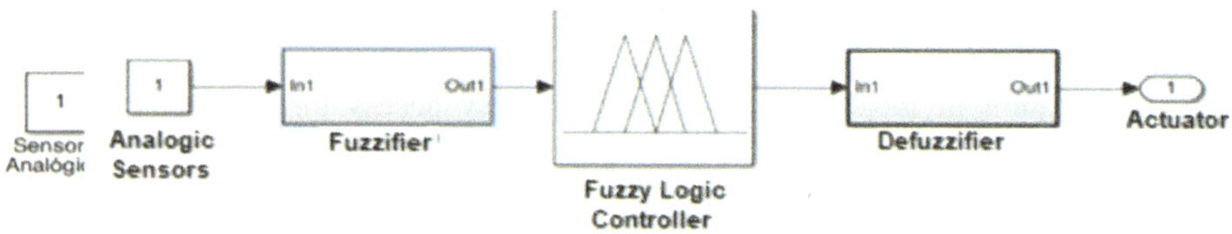

Fig. 1. Fuzzy logic system.

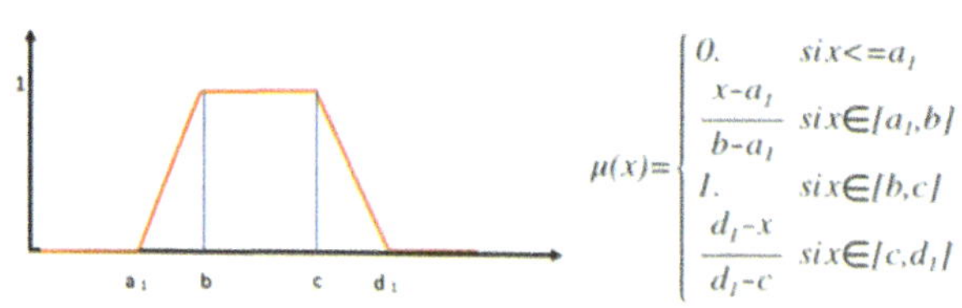

Fig. 2. Fuzzy control trapezoidal functions.

The program uses four membership functions [6] for the fuzzifier and defuzzifier, which are trapezoidal [7]. The membership functions define the degree to which physical values belong to terms in a set of linguistic variables. Membership functions indicate the degree to which each element of a given universe belongs to that set. For example, the function of belonging to the set S on a universe "Y" will be of the form: $\mu S: Y \rightarrow [0, 1]$, where $\mu S(x) = t$ if t is the degree to which "Y" belongs to S.

3 Materials and Methods

The circuit was designed using the Eagle® software from Autodesk. The electronic board was manufactured via a Roland CNC Milling Machine using a 0.3–mm-diameter drill bit to carve circuit tracks in the ceramic board.

Currently, the Atmega 328 processors support the creation of new electronic devices, in which the power of the processor may generate quite interesting solutions when coupled with an adequate software. The joint development of the power board and electronic board, which host the control software, creates unparalleled capacities for materializing the design of an electronic device.

Following the board development stage, our purpose is enclosing these electronics in an optimized model. Further, our initial sketches and designs are poured through electronic drawing tablets, which supplement the computers wherein the design software is framed, to create the new shapes that will contain the electronic boards designed in the previous stage. Eventually, these designs may be improved to craft something tangible for the first time through 3D printing, such that these prototypes closely resemble a final product. Hence, the Solidworks® software was used to design the packaging of the electronics, and the model was printed in polylactic acid.

3.1 Power System

To develop this solar charge controller for lead–acid batteries, the circuit features two protection systems: a fuse at the solar panel output and another at the battery input. Moreover, the circuit includes an MBR 2545 Schottky Barrier Rectifier diode as protection from reversal polarity. Additionally, the NTE 4941 diode is available as a protection measure against input overloads. The circuit developed has three IRF9540 Power MOSFET gates. The first gate is operated by a button, which disconnects the solar panels. The second gate drives the PWM signal [8], which charges the battery at the frequency that has been previously hardcoded into the microprocessor. The third gate can be used to disconnect the feeding load at the operator's discretion. The circuit also features a few free pins for connecting a set of additional sensors. For example, a voltage or a current sensor can be connected for future analysis through a datalogger (see Fig. 3–7) and 12–16 display the designed parts, along with their electronic components.

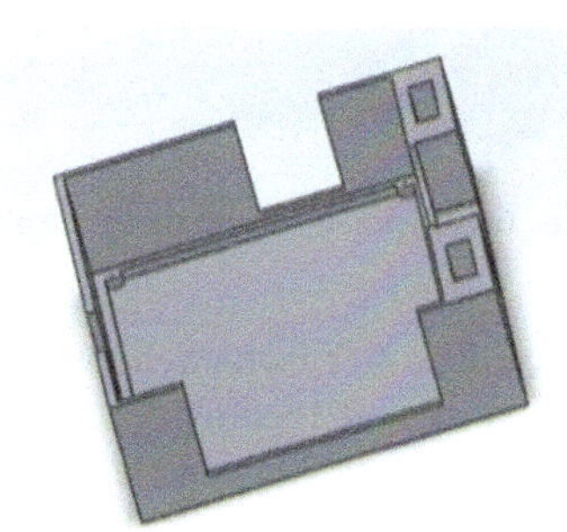

Fig. 3. Bottom view of the controller.

Figure 3 shows the controller box, which has been designed in two sections to facilitate assembly. In this box, we may observe the location where the power board will be placed, as the inside area must also allow cables to pass through to the other

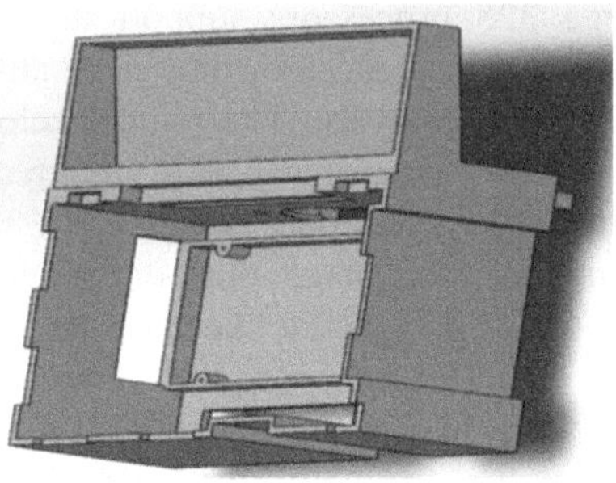

Fig. 4. Top view of the controller.

boards. Figure 4 denotes the box where the controller is located, and it will be bolted to the top section of the box.

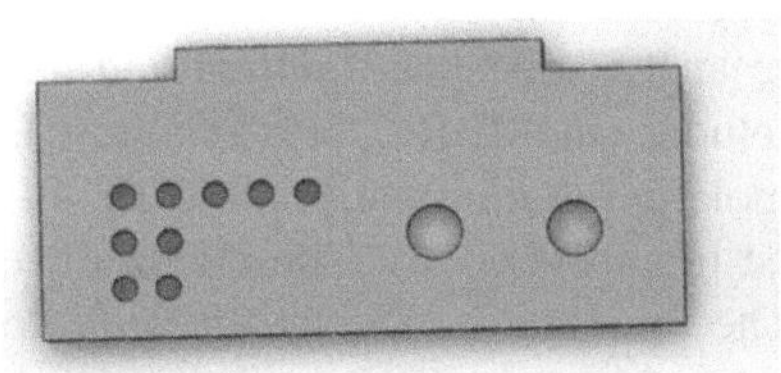

Fig. 5. LED indicator cover.

Fig. 6. Upper mask of the controller.

Figure 5 shows the indicator cover, which houses the LED assembly, and the buttons used to connect the panel and feeding load. Figure 6 shows the frame that contains the LCD screen, which features 20 characters and 4 lines. This screen provides information about solar panel voltage, whether the panel is currently connected or disconnected and whether charging is enabled. Figure 7 shows the assembly modeled in SolidWorks.

Before printing the assembly, the software may be used to verify whether the controller will fit perfectly, as the proposal herein was to design a system that suits a particular need. Therefore, calculations were made for a simple controller, which may properly charge the battery pack, uses components locally available in the market and

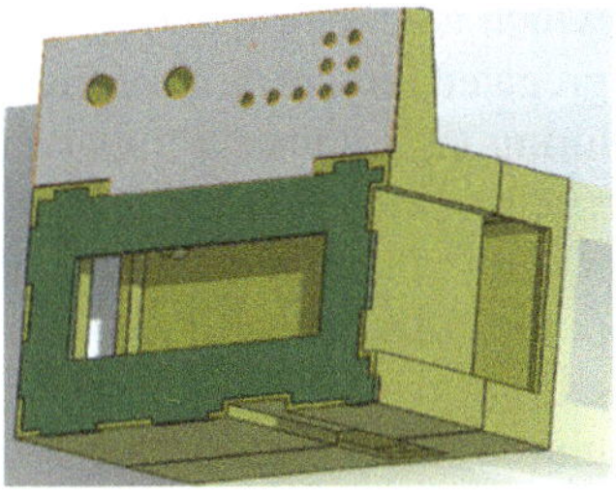

Fig. 7. Assembly modeled in SolidWorks.

that does not consume more than 0.6 A. The initial works were not as successful because the peripherals consumed considerable voltage; however, this problem was solved using the processor described herein at 5 V. Further work may include developing new electronic boards containing 3.3 V processors for lower energy consumption during operation. The power board was developed using the generic MOSFET gates to quickly and easily replace them using conventional welding tools.

The optimization of the fuzzifier rules was performed by adjusting the values of the fuzzy system coefficients in the fuzzicator rules, whose response is visualized through the simulation in the MatLab® with the Surface Viewer tool. The Surface Viewer generates the output surface of the diffuse inference, plotting the respective response ranges as reference values (see Fig. 8a); In this way, with the use of the adjusted coefficients, the graph showing the calculation of the error becomes more uniform (see Fig. 8b) which translates to a more uniform loading behavior.

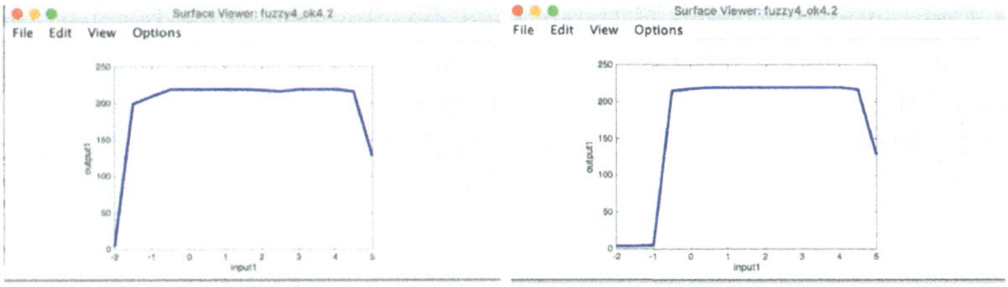

(a) (b)

Fig. 8. MatLab-surface viewer tool.

Subsequently, these new coefficients were transferred to the fuzzy logic library in the comprehensive IDE-Atmega328 development environment.

4 Controller Design and Development

The design foundations addressed are based on the robustness of the power board, its easy construction, the easy loading of the firmware and developed control program because the whole system must be replicable at a low cost. The control rule base comprises the error variable, which is given by the difference between the setpoint and voltage of the battery. The aggregated error was not considered for the solution because the processor slowed down during tests. Other points considered for the development were the use of the free access "Embedded Fuzzy Logic" library, a processor widely available in the local market, easy integration in the development board, and the algorithm supporting the proposed rules.

The software programmed for this process was initially developed using the MatLab® environment through the fuzzy logic library [9]. Once the coefficients were identified through the simulation, these values were used in the fuzzy logic library of the Atmega 328 processor (see Fig. 9) [10]. The calculated error is then transferred to the four trapezoidal input functions of the fuzzy system [11].

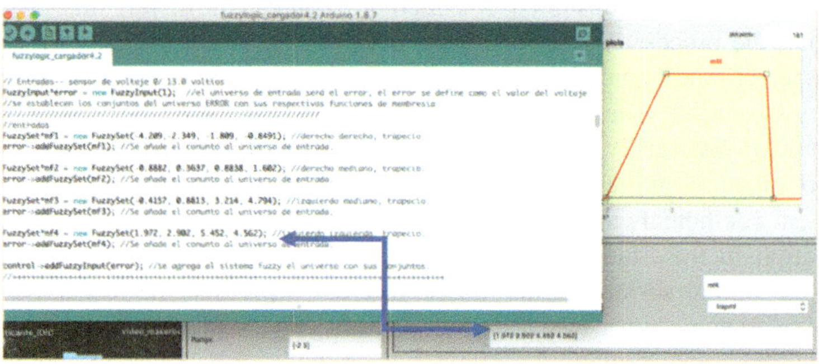

Fig. 9. MatLab values for the fuzzifier transferred to the fuzzy logic library of the processor.

Only four optimized rules [12] are used for this controller because the entire program that controls peripherals demands excessive processing time and memory consumption, thereby slowing down the operation [13]. The Table 1 shows the input rules in the fuzzifier, the fuzzy rules base system, and the Defuzziler outputs rules.

Table 1. Control rules transferred to the fuzzy logic library of the processor.

Fuzzifier rules (Input)	Fuzzy rule base system	Defuzzifier rules (Outputs)
Left input	If X then output Y, then ΔZ^1	Left output
Middle left input	If M then output N, then ΔZ^2	Middle left output
Middle right input	If L then output F, then ΔZ^3	Middle right output
Right input	If O then output P, then ΔZ^4	Right output

The output is the pulse-width modulation (PWM) signal, which is the response obtained through the membership rules [14] resulting from the calculation of the center of gravity given by the mathematical expression (1) [15].

$$Z^* = \frac{\sum_{i=1}^{M} \bar{Z} W_i}{\sum_{i=1}^{M} W_i} \tag{1}$$

$$\text{where } Wi = \mu(Mi) = \min\{\mu 1, \mu 2 \ldots \mu k\} \tag{2}$$

In Eq. (1), Z* represents the locus of the gravity center for all the possible rules involved in the solution, where M represents the number of fuzzy outputs. In Eq. (2), the term μk represents the fuzzy value involved in the antecedent of the i-th rule, where M represents the number of fuzzy outputs, with the corresponding W-weights for the maximum values of the results involved.

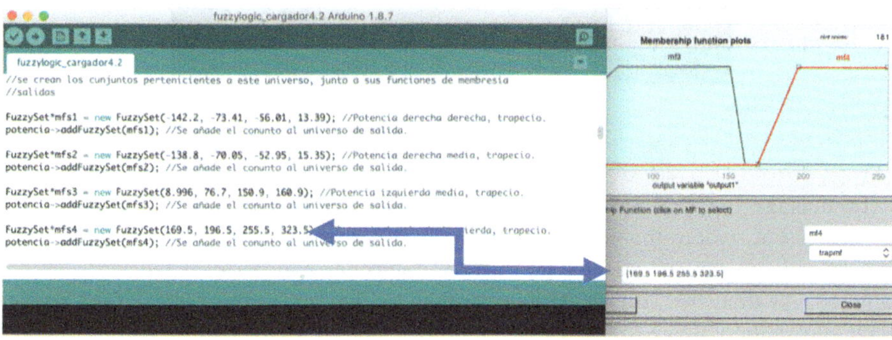

Fig. 10. MatLab values transferred to the fuzzy logic library of the processor.

In Fig. 10, the MatLab simulated and tested values [16] are transferred to the microprocessor's library, wherein the actual operation of the MatLab simulation is verified.

Through these development tools, the programming can take less time, which reduces the implementation hours required for a system developed in C++. The difference between the input voltage from the solar panel and battery voltage is the parameter used as a reference for the algorithm. Therefore, the board design incorporated voltage dividers to determine the voltage values for the battery and solar panel. Under low battery voltage conditions, the controller has supplied up to 6 A according to laboratory tests. With battery voltage at its peak, the work cycle decreases, and the system maintains battery voltage for the proposed setpoint.

4.1 Controller Implementation

The electronic boards are located inside the solar charger. One of them is the power board comprising three power transistors. The first transistor allows the PWM to charge the battery, the second transistor connects and disconnects the solar panel, and the third transistor activates or deactivates the load (consumption). This power board is prepared to support a workload of up to 8 A.

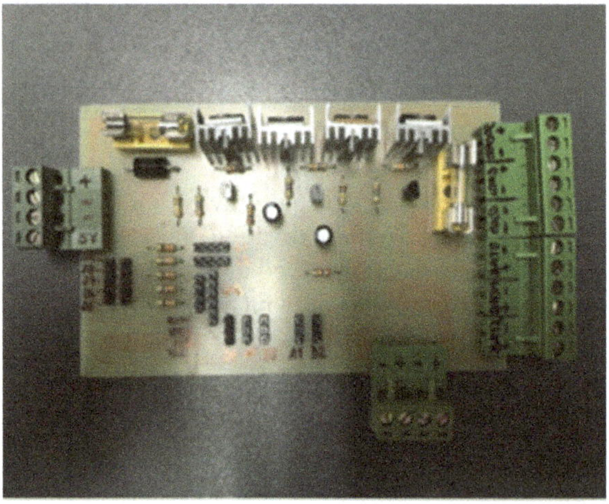

Fig. 11. Power board.

To manage the charging process for the lead–acid battery [17], an Atmega 328 microprocessor, where the control software is installed, is used. The voltage dividers serve as voltage sensors so that the battery controller can perform the work as programmed. At first, the voltage starts charging the battery through the PWM control, which allows the voltage to overcome the impedance of the battery. When the setpoint is reached, the controller disconnects the voltage from the solar panel.

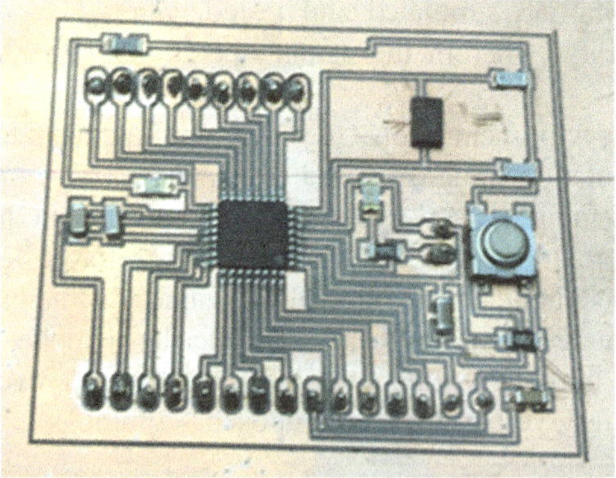

Fig. 12. Control board.

Figures 11 and 12 shows displays the control board using an Atmega328P-AU [18] with a 16 MHz crystal, where two 22 pF ceramic capacitors are used via the resonator. The software developed was installed on board with superficial mounted devices elements. Figure 13 denotes the prototype of a fully functional solar charge controller. The LEDs in Fig. 14 indicate whether the battery is being charged, fully charged, or

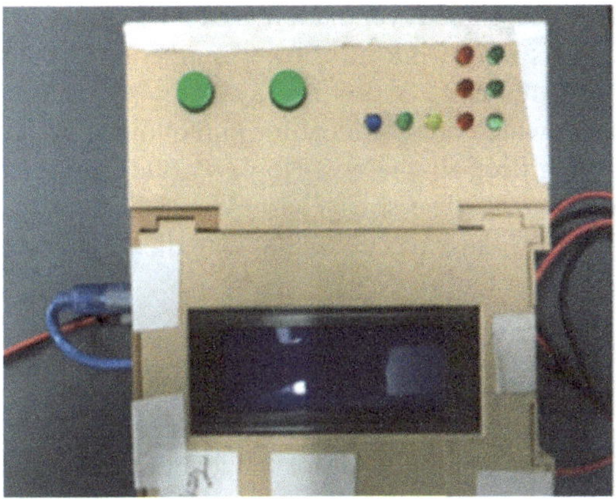

Fig. 13. Solar charge controller.

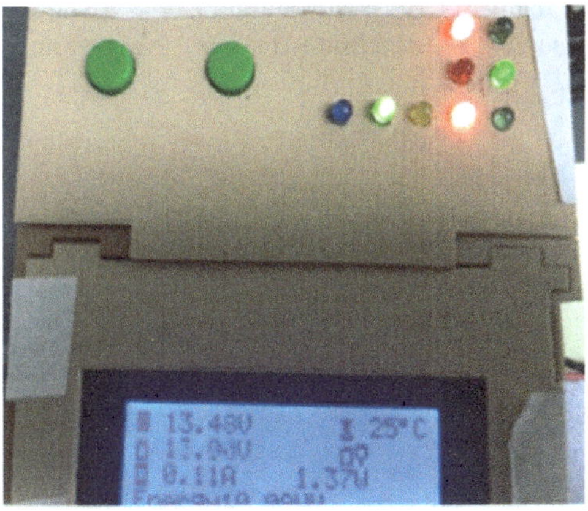

Fig. 14. Charging LED indicators.

completely drained. This figure also shows two buttons: the left button disconnects the solar panels and the right button deactivates the load that has been fed.

Fig. 15. Ventilation system.

Figure 15 displays the forced ventilation system, which is integrated into the system so that MOSFETs may work at a lower temperature, thereby extending their lifespan. The fan is powered at 12 V and is easy to replace and maintain. The system uses lead–acid batteries, which are not ideal for photovoltaic systems because they are not capable of supporting deep discharges. However, they are the most affordable and easily available in rural areas. The solar charge controller designed accounts for the inherent disadvantages of these batteries and adapts them based on their needs.

5 Results

Based on the tests performed during the two-month period in which the lead–acid battery was fully charged and drained for 60 cycles, the following results were obtained:

- The battery takes less time to fully charge than when using commercially available chargers. An approximate average of 7 h is obtained against the 12–14 h, which was recorded when using locally sold chargers.
- When the battery reaches its maximum charge, the system disconnects and stops charging the battery, thereby preventing voltage overloads.

The charts below show the working states for the controller in different charging regimes.

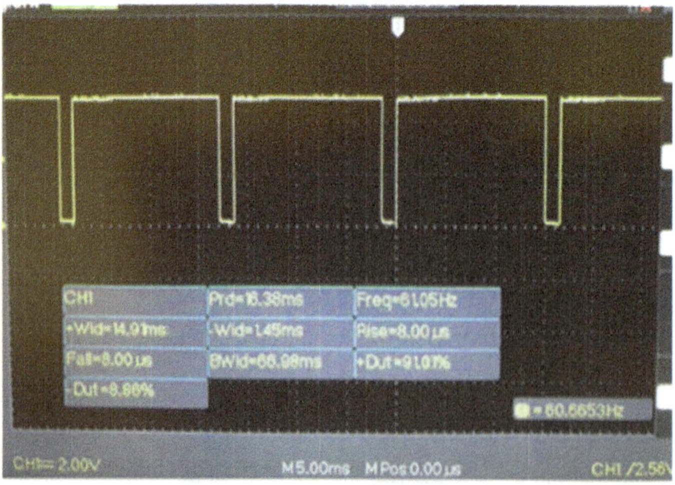

Fig. 16. PWM signal when the battery is drained.

When the MOSFET generates the PWM signal [19], its temperature increases and the forced air fan turns on.

The PWM is generated by the frequency that has been previously established in the microprocessor. The time percentage in which the pulse is at peak level with respect to the complete pulse period is known as the Duty Cycle, and it is expressed as a percentage. For example, a Duty Cycle of 91% at a frequency of 61.05 Hz, corresponds to a pulse of 16.40 ms, wherein maximum current is supplied for 14.91 ms before cutting supply off at 1.49 ms (see Fig. 16). Herein, the processor has applied a correspondence rule to operate at full load because the battery is below or equal to 10.5 V (see Fig. 17)

Fig. 17. Battery current.

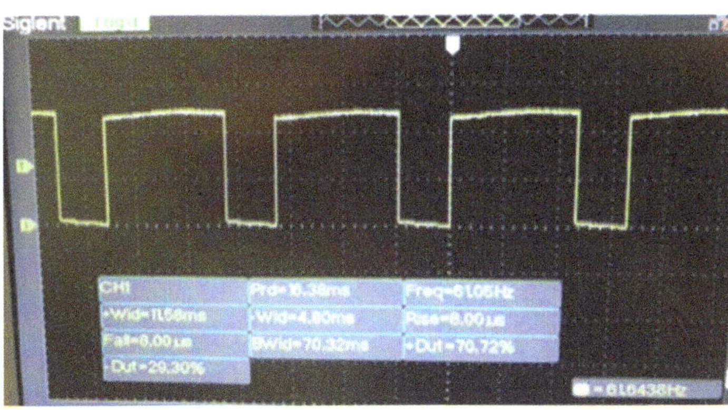

Fig. 18. PWM signal, the battery is reaching its maximum charge.

Conversely, when the battery charge reaches approximately 80%, the Duty Cycle changes to 70%, which means that the controller has applied another correspondence rule to start charging the accumulator when it is at approximately 12.8 V (see Fig. 18). The voltage supplied to the battery is controlled by the MOSFET, whose operation depends on the orders from the microprocessor exerting the fuzzy control. The voltage supplied continuously decreases as the battery is being charged. For example, if the battery is reaching maximum charge, the current decreases to 0.5 A, while the control algorithm disconnects the system and stops charging the battery. At this point, the

Table 2. Tests performed when charging a lead–acid battery. Controller behavior during a typical full charging cycle.

Hour	Duty cycle (%)	Voltage (V)	Current (A)
8:00	94.19	9.76	6.11
8:15	80.02	12.25	4.76
9:55	82.78	12.85	4.96
10:15	85.52	12.98	5.06
10:45	82.78	13.18	5.09
11:00	86.30	13.23	5.28
12:00	89.04	13.53	4.49
12:30	91.04	13.70	4.06
1:00	93.84	13.84	3.86
3:15	60.66	15.78	1.57

current and voltage from the panels is cut off, leaving the fan on so that the MOSFETs, which are still at high temperature, may cool down.

The Duty Cycle or period fraction decreases as the battery is charged up to the setpoint, where it cuts off the voltage supply, and the battery has been fully charged.

The data displayed in Table 2 correspond to a charging Duty Cycle, wherein the lead–acid battery with a rated voltage of 12 V and a capacity of 53 Ah took approximately 7 h to complete the charging process. About 60 charging and discharging cycles were tested. The Fig. 19 shows the charging time versus the charging voltage. The conventional charger takes up to 5 h longer to complete the same task. The top line corresponds to the fuzzy logic system.

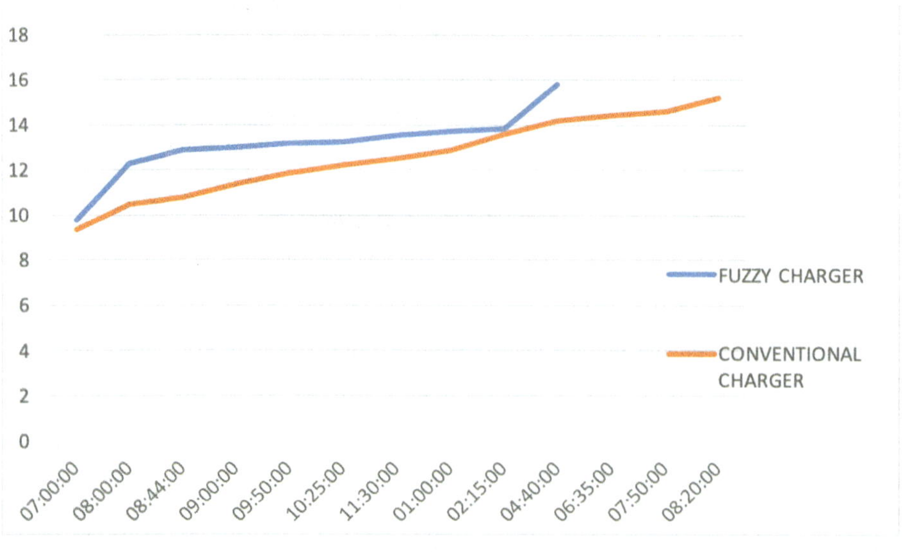

Fig. 19. Graphic comparison of Fuzzy charger (top line) and conventional charger (bottom line).

Fig. 20. Test bank using three solar panels and three controllers with different algorithms.

6 Conclusions

The proposed system has charged a 53 Ah lead-acid battery in 7 h, allowing great flexibility considering the hours of sunlight available in the area. This was done after realizing that small changes in the fuzzifier rules cause significant changes in the battery's duty cycle, resulting in longer or shorter charging times. Additionally, in rural areas with very low outdoor morning temperatures, changing the base of the fuzzifier rules could be used to increase the temperature of the battery assuring a better charge cycle. Due the fact that lead-acid batteries do not support deep discharges because it decreases dramatically its useful life, the controller also prevents discharges from falling below 10.5 V as a result of programming. It is worth mentioning that the controller was manufactured using components readily available in the local market to ensure that it can be replicated and repaired without trouble and with less expenses. The robustness of the system was demonstrated after performing laboratory tests for approximately four months without any difficulty.

7 Future Work

Using a single type of hardware for several types of algorithms allows the system developed herein to be deemed as a benchmark (see Fig. 20). This system currently uses three different control algorithms, external voltage sensors in the solar panel and battery, and current sensors to determine the amperes consumed by the load. Additionally, temperature sensors are present in the solar panel and battery, which collect data for future analysis.

Overall, the 16 sensors originate around 5500 data per day, which are transferred to a PC through a 16 analog channel National Instruments board for subsequent analysis of the algorithms contained in each load controller. This will help us to demonstrate the efficiency of our algorithms through Deep Learning techniques, which will allow us to find the best charging algorithm for the batteries. This test bank is valuable when determining the most effective algorithm contained in the three controllers for charging a lead–acid battery.

This test bank includes controllers with three different algorithms for future battery charging analyses. The first algorithm is the fuzzy logic algorithm, the second is an adaptive PID, and the third uses logical and comparison operators.

References

1. Jaisin, C., Intaniwet, A., Nilkhoa, T., Maneechukate, T., Mongkon, S., Kongkraphan, P., et al.: A prototype of a low-cost solar-grid utility hybrid load sharing system for agricultural DC loads. Int. J. Energy Environ. Eng. **10**(1), 137–145 (2019)
2. Lam, L.T., Ozgun, H., Lim, O.V., Hamilton, J.A., Vu, L.H., Vella, D.G., Rand, D.A.J.: Pulsed-current charging of lead/acid batteries—a possible means for overcoming premature capacity loss? J. Power Sources **53**(2), 215–228 (1995)
3. Javadi, F.S., Rismanchi, B., Sarraf, M., Afshar, O., Saidur, R., Ping, H.W., et al.: Global policy of rural electrification. Renew. Sustain. Energy Rev. **19**, 402–416 (2013)
4. Patil, A.R., Atar, K.D., Potdar, A.A., Mudholkar, R.R.: Embedded fuzzy module for battery charger control. Int. J. Adv. Res. Electr. Electron. Instrum. Eng. **2**(8), 4072–4078 (2013)
5. Paliwal, P., Patidar, N.P., Nema, R.K.: Fuzzy logic based determination of battery charging efficiency applied to hybrid power system. World Acad. Sci. Eng. Technol. **71**, 1164–1168 (2012)
6. Chevrie, F., Guely, F.: Fuzzy logic-Cahier technique no 191. Tekniske Dokumenter og Software, first issued (1998). http://www.studiecd.dk/cahiers_techniques/Fuzzy_logic.pdf
7. Sowah, R., Ampadu, K. O., Ofoli, A., Koumadi, K., Mills, G. A., Nortey, J.: Design and implementation of a fire detection and control system for automobiles using fuzzy logic. In: 2016 IEEE Industry Applications Society Annual Meeting, pp. 1–8. IEEE (2016)
8. Govidan, N., Rajasekaran Indra, M.: Smart fuzzy-based energy-saving photovoltaic burp charging system. Int. J. Ambient Energy **39**(7), 671–677 (2018)
9. Sivanandam, S.N., Sumathi, S., Deepa, S.N.: Introduction to Fuzzy Logic Using MATLAB, vol. 1. Springer, Heidelberg (2018)
10. Bawa, D., Patil, C.Y.: Fuzzy control based solar tracker using Arduino Uno. Int. J. Eng. Innov. Technol. **2**(12), 179–187 (2013)
11. Çelik, B., Birtane, S., Dikbıyık, E., Erdal, H.: Liquid level process control with fuzzy logic based embedded system. In: 2015 9th International Conference on Electrical and Electronics Engineering (ELECO), pp. 874–878. IEEE, Turkey (2015)
12. Lara, F., Sánchez, E.N., Zaldívar, D., Sur, P.L.M.: Real-time fuzzy microcontroller for a didactic level system. Electro **2001**, 153–158 (2001)
13. Malkhandi, S.: Fuzzy logic-based learning system and estimation of state-of-charge of lead-acid battery. Eng. Appl. Artif. Intell. **19**(5), 479–485 (2006)
14. Thao, N.G.M., Uchida, K.: A two-level control strategy with fuzzy logic for large-scale photovoltaic farms to support grid frequency regulation. Control Eng. Pract. **59**, 77–99 (2017)
15. Velázquez, R., Gómez, T., Rodríguez, J.: A pH process control embedded on a PLC using Fuzzy Logic. In: 2017 XIII International Engineering Congress (CONIIN), pp. 1–6. IEEE, Mexico (2017)
16. Driankov, D., Saffiotti, A. (eds.): Fuzzy Logic Techniques for Autonomous Vehicle Navigation, vol. 61. Physica, Sweden (2013)

17. Bandara, G.E.M.D.C., Ivanov, R., Gishin, S.: Intelligent fuzzy controller for a lead-acid battery charger. In: IEEE SMC 1999 Conference Proceedings, 1999 IEEE International Conference on Systems, Man, and Cybernetics (Cat. No. 99CH37028), pp. 185–189. IEEE, Japan (1999)
18. Atmel Mega 328p Datasheet (s.f.). En Atmel.com. Atmel 8-bit AVR Microcontrollers ATmega328/P Datasheet Summary. http://ww1.microchip.com/downloads/en/DeviceDoc/Atmel-7810-Automotive-Microcontrollers-ATmega328P_Datasheet.pdf. Accessed 27 Aug 2019
19. Silva, S., Soares, S., Valente, A., Marcelino, S.T.: Digital sound processing using Arduino and MATLAB. In: 2015 Science and Information Conference (SAI), pp. 1184–1191. IEEE, Portugal (2015)

Enabling Electronic System for Emergency Alerts

Santiago Manzano(✉), César Granizo, Homero Velasteguí, Jaime Guilcapi, Freddy Benalcazar, and Carlos Gordón

Facultad de Ingeniería en Sistemas, Electrónica e Industrial, Universidad Técnica de Ambato, Ambato 180150, Ecuador
{victorsmanzano, cesar_granizo, hvelastegui6534, jr.guilcapi, fg.benalcazar, cd.gordon}@uta.edu.ec

Abstract. Institutions of Higher Education of Ecuador for its intrinsic characteristics remain to face adverse trends anthropic events such as work accidents. Also, the location of the state in the Ring of Fire substantiates the existence of natural hazards such as earthquakes and volcanic eruptions. The management of these adverse events is done by precept, methods and resolutions previously planned; generating the tendency to confront disasters when managing identified risks is inadequate. The risks are directly influenced by the product of threats and vulnerabilities. The threat is the possibility of occurrence of an adverse event and vulnerability is a sensitive factor of a system exposed to a threat that has low resilience. Threats can only be detected, but not removed unlike the vulnerabilities, the threat detection and vulnerability reduction decrease risks in order to avoid disasters. The Huachi, Querochaca and Ingahurco campus, belong to the Technical University of Ambato, do not have an adequate system to inform the university community the need for evacuation to the occurrence of adverse events (fires, volcanic eruptions), turning this into a detected vulnerability. This project designed an electronic acoustic signaling system integrating the three campuses through a communications system. The purpose is to report immediately alert conditions for staff risk management protocols run evacuation drills and preventive actions. The system monitors and alerts locally or remotely, using a client-server architecture with Internet devices of things and Web Application Server using WebSockets.

Keywords: WebSocket · Server · Client · LAMP · Warning system

1 Introduction

The present research is a contribution to early warning systems, creating an activation scheme alerts through mobile devices. This system is an alternative action to an adverse event report that a protocol allowing emergency activates the internal community of an institution with high public turnout.

In this research, the current situation of emergency management at the Technical University of Ambato was analyzed, observing, as a consequence, the need to design an acoustic signaling system. An integration element of an alert system is carried out by

J. Nummenmaa et al. (Eds.): CSEI 2019, AISC 1078, pp. 184–198, 2020.
https://doi.org/10.1007/978-3-030-33614-1_13

a communication system enabling the remote monitoring and control of electronic signaling devices from any internal area of the institution.

Section 2 of this article details the methodology used. the problem and the current situation of the UTA is analyzed, the proposed solution is developed according to the needs of the Technical University of Ambato and guidelines established by the Ecuadorian Technical Standards NTE INE-ISO 7731: 2014 that determine the technical parameters of the warning system. The software required for system operation Activation Alerts is developed determining the sound pressure level specular buildings using the method of geometrical acoustics.

Section 3 of this article presents the results of performance testing of a prototype implemented at the Faculty of Engineering and Industrial Electronics Technical at the Technical University of Ambato.

Section 4 of the article contains the conclusions and recommendations of this research project.

2 Methodology

2.1 Current Situation and Requirements

Management risk activities relate emergencies and threats at the Technical University of Ambato are oriented to the recommendations of the Committee on Risk Management Manual of Ecuador. The roles of responsibility are: the Risk Management Committee and the Emergency Operations Committee CGR/COE united in one organization, institutionally the COE-I. The CGR focuses on risk reduction as a permanent and global function, and COE in attention and response during emergencies or disaster [1].

The COE-I of the UTA decides the priorities and actions to develop emergency and disaster in college. It is the responsibility of the authorities and of the standing committees support (LOOPS), prioritizing the required for handling adverse situations The highest authority of COE-I UTA is the principal, his function is being relieved to vice chancellor if he is absence.

Resolution 1590-CU-P-2015 is a document that formalizes the existence of the Unit for Risk Management, establishes that this Unit is responsible for the risks in the Technical Area Health and Safety at Work and also in the Technical area External Risk Management [2]. The structure of the Operations Committee Institutional Emergency of the University Technical of Ambato (COE-I-ALU) is comprised of:

- Principal and Vice Chancellors who are the highest authorities and with the direct assistance of the Risk Management Unit of the UTA (UGR-UTA), coordination and decision-making for risk management of the university is carried out.
- The Emergency Operations Committee is the entity that governs the decisions, priorities and actions to develop emergency and disaster in UTA.
- Loops are standing committees that will support risk management.
- Institutional Risk Management Units have technical delegates responsible for buildings on each campus [2].

Currently, the Risk Management Unit handles emergencies in an organized manner in the administrative. In each faculty, there is a focal point that is primarily responsible in making decisions, who should report the emergency situation to their superiors. Following the recommendations documented by the insurance company Axa/COPALTRIA, UGR-UTA is changing the structure of technical support to a scheme organized by brigades as shown in Fig. 1.

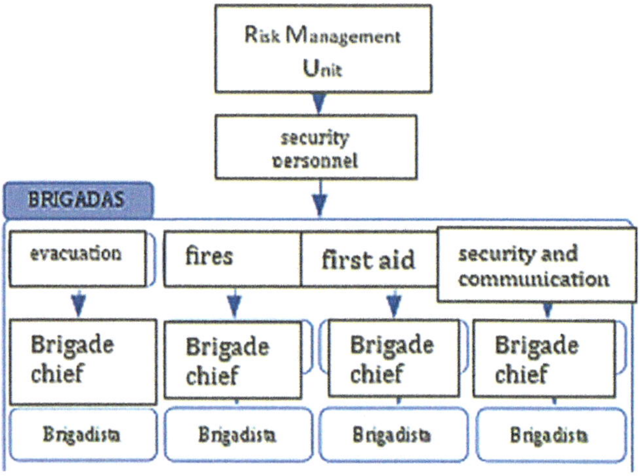

Fig. 1. Diagram of technical support in emergency situations. UGR-diagram provided by UTA.

The current system analysis alerts TAU performed in this investigation determined that 59% of the departments and academic units has not implemented a signaling system alerts, 18.2% has a hybrid signaling system and individual (acoustic light). The remaining 22.8% have a speaker system and the Faculty of Science and Food Engineering has implemented protocols involving the use of loudspeakers warning cases [3].

Currently, the information and alerts are carried through communication. To improve the current system and reduce the response time to the implementation of security protocols in the Unit Risk Management of UTA a monitoring station is implemented to inform their staff the location and type of activated alerts or detection hazardous situations [3].

Technical University of Ambato in their Huachi, Ingahurco and Querochaca campus requires the immediate implementation of a system of stimulating the senses signaling, in order to inform the internal community the occurrence of adverse events for the implementation of various emergency protocols.

The system requires a global design, to handle functions that enable the people of the university to identify trigger alerts, notifying the Risk Management Unit location and initial situation raised alert status. The main needs that the system must be designed are:

- Establish a communication medium that allows sending warning signals from the staff of the Risk Management Unit to the general population of the University.
- The general population requires immediate communication interface to notify the occurrence of events identified as hazardous.
- Unit Risk management requires a monitoring room status alerts to activate and review local situations alert, sectored or general.
- The signaling system should identify the status and origin of the alert using different types of signals for protocols fire, evacuations and simulations.

Among the requirements established by the risk management unit it is determined to be necessary to implement a signaling system alert status. Technical Standards for Protection NTP 889 provide that most concurrence public places must implement an alarm system. For this project, it is used a sound and light system according to Ecuadorian Technical Standard NTE INEN-ISO used: 7731: 2014 and UL 1638, respectively [4].

In the NTE INEN-ISO standard 7731: 2014 it states that the use of the light signals is required in environments where the ambient noise signal exceeds 100 dB. The literature investigated determines that the environments classrooms and offices the sound pressure level is about 50 dB. In addition, implementation of the system using sirens is cheaper and simpler compared to a light signaling system. Strobe lights signaling involves installing at least one device per classroom, office or laboratory; not so with acoustic signaling. The aforementioned background determines the best technology for the signaling system alerts, it is the acoustics.

NTE INEN Technical Standard ISO 7731: 2014, "Ergonomics. Warning signs in public places and workplaces. Acoustic Warning Signs" is an adaptation that sets the technical parameters of the ISO 7731: 2003 standard for acoustic signaling conditions. The signaling system of the UTA is based on the technical parameters of this policy, detailed below:

- The sound pressure level of an acoustic signal alerts anywhere in the reception area should be at least 65 dB and 118 dB not exceed.
- The acoustic warning signal must be overcome ambient noise sound by at least 15 dB. The ambient noise produced at the university is among the parameters weighted noise classrooms and conversations (according to the literature 50–60 dB) so, it is taken as reference 55 dB, having the minimum sound pressure level must be 70 dB.
- The acoustic signal must be maintained while the dangerous event and finalize the issue persists should ensure the feasibility of reuse.
- The frequency of the acoustic warning signal must be situated in the range of 500 to 2500 Hz.
- Preferred are pulsating signals danger continuous in time, the frequency of pulse repetition should be within 0.2 and 4 Hz.
- Wiring and installation of devices that are required in the activation system alerts should be within the protection settings recommended by NEMA or IP [4] equivalent.

2.2 Acoustic Signal Broadcasting

Coverage warning signal at the Technical University of Ambato depends on the propagation of acoustic emergency signal. The ideal location of the Sirens in each building are the isotropic points thereof, determined by the centers of gravity of their surfaces. The center of gravity of the buildings can be a place in space where there is no physical evidence to install sirens in the given case, the location of the sound source moves towards closer and with fewer obstacles surface.

The coverage analysis of the acoustic signal at the university is done by a study of the sound pressure level specular the signal emitted by the siren in buildings, by determining the amount required to meet the NTE INEN-ISO 7731 standards: 2014. In requirements analysis determined that the minimum sound pressure level of the alert signal in buildings, for purposes of equipment sizing, is 70 dB; establishing a slightly higher margin and taking recommended as an ideal values exceeding this threshold.

The maximum coverage distance reaches a siren used is calculated by Eq. 1 [5, 6]; wherein the sound pressure level reference to a meter from the source is 112 dB and the minimum required level is 70 dB. Solving equation r2 it is determined that the siren delivers an adequate level of sound pressure at distances less than 125 m. SLP_0

$$SLP_f = SLP_0 - 20\log(\frac{r2}{1}) \tag{1}$$

Where:

SLP_f: It is the sound pressure level is calculated at a certain distance from the source (r2).
SLP_0: It is the sound pressure level to known reference distance measured from the source (1 m).
$r2$: The distance from the source to the point of analysis. Substituting the data has:

$$70 = 112 - 20\log(\frac{r2}{1})$$
$$r_2 = 10^{2.1} = 125.89\,\text{m}$$

The analysis of the sound pressure emitted by the sirens in buildings is performed using the ray tracing method of geometrical acoustics, according to the following procedure:

- The isotropic point for the location of the siren the building calculating the approximate center of gravity of their surfaces and shifting the location to the nearest wall with fewer obstacles.
- Elemental radiation pattern is located in the planes to display the partial coverage of the siren.
- The ray tracing method assumed that the energy is transported by a series of particles follow a straight line from the source. The hard smooth surfaces reflect the signal almost entirely [7, 8].
- A source image is placed in the symmetric plane to the sound source to determine coverage of the reflected signal [9, 10].

- Specular sound pressure of the reflected waves value is determined by Eq. 1, assuming the distance r2 as the sum of the distance wave incident on the reflection point with distance from the reflection point to the point of analysis (place within the further from the siren) building.
- By the method of geometrical acoustics, an omnidirectional siren propagation is assumed and the source image, determining whether the acoustic propagation gives the building coverage or if required an additional siren [8, 10].

$$SLP_f = SLP_0 - 20\,log(\frac{r2}{r1})$$

The radiation pattern elemental analyzes the horizontal and vertical spread, considering the directivity characteristics of the siren. The model selected siren has a coverage angle of 110° on the horizontal axis and 60° on the vertical axis. The radiation pattern is shown in Fig. 2, wherein by Eq. 1, limits sound pressure has to the ends of each color is set [11, 12].

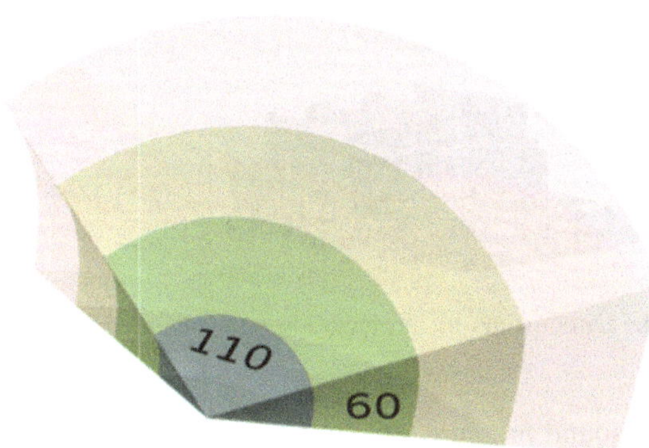

$$SLP_A \geq 112 - 20log(40) \geq 80dBSLP_V \geq 112 - 20log(80) \geq 74dBSLP_N \geq 112 - 20log(120) \geq 70dBSLP_R \geq 112 - 20log(170) \geq 67dB$$

Fig. 2. Radiation pattern of the sound source. Blue 40 m, 80 m green, orange-brown 120 m, dark pink 170.

Autocad software in visual scheme is performed using elemental radiation pattern to have a general idea of the spread and sound pressure level buildings UTA. The proposed location scheme for sirens below, considering it as an ideal sound system area within the blue and green areas and spherical projections, taking as an example the building 24 of the Huachi campus where the Faculty of Engineering is located in Electronics and Industrial Systems (FISEI).

In Fig. 3 the block 24 decomposed based on interleaved plants, the right side is observed can be seen the entire building in central plants 2, 4, 6 are shown, and 8 while the left plants 1 3, 5 and 7.

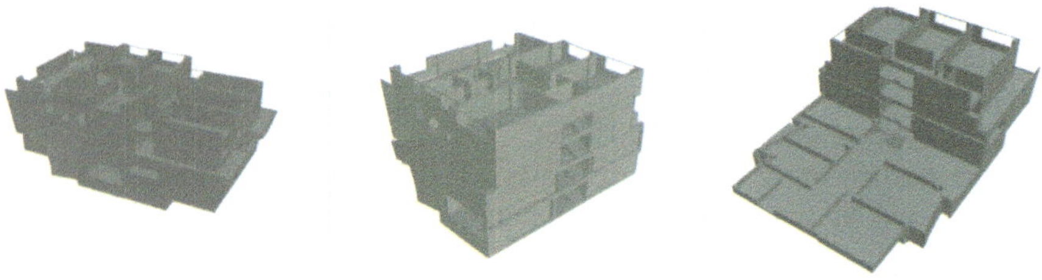

Fig. 3. Architectural structure decomposed block 24 of the UTA.

The proposal for the siren location is in the center of the building, which is the isotropic point for the propagation of the acoustic signal. The building has a height of 12.6 m 6.3 m being its center to the ground, because at this point is not possible to install the device location 50 cm moves upwards, the roof of the plant # 5 of 6.8 m ground level in the center of the bleachers according to Fig. 4. From the spot recommended, the sound signal spreads to the classrooms and laboratories through reflections on the walls of the corridors [13].

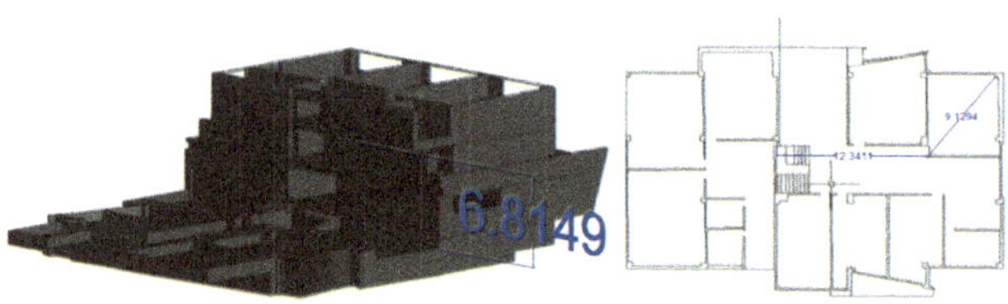

Fig. 4. Dimensions for propagation analysis siren sound of FISEI.

The analysis of sound pressure level is performed at the farthest point to the source inside the building, using the method of acoustic geometric ray tracing. The linear distance is taken from the source to the point of analysis, summing the distances between the reflection points [9].

$$r_t = r_1 + r_2 + r_3 \tag{2}$$

$$r_T = 6,81\,\text{m} + 12,34\,\text{m} + 9,13\,\text{m} = 28,28\,\text{m}$$

The sound pressure level at the point of analysis is given by Eq. 1 [5, 9]. Knowing the siren used has a SLP of 112 dB at 1 m source and the maximum distance of 28.28 m, the sound pressure level at the point of analysis is 82,97 dB.

$$\begin{gathered} SLP_f = SLP_0 - 20\log\left(\frac{r2}{r1}\right) \\ SLP_f = 112 - 20\log\left(\frac{28{,}28}{1}\right) = 82,97\,\text{dB}. \end{gathered} \tag{3}$$

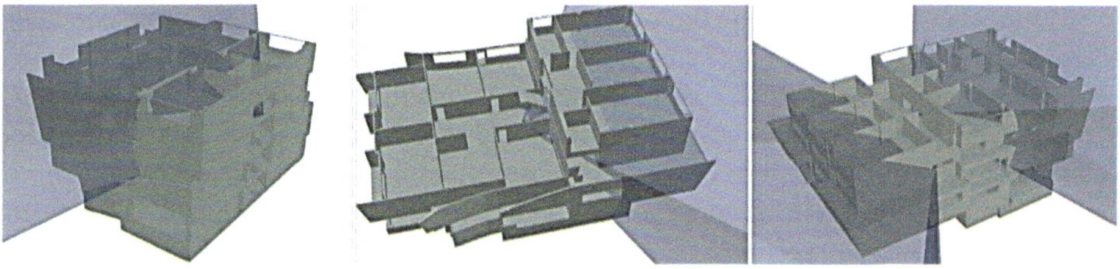

Fig. 5. PA area C block 24.

Radiation from the sound source and the source image block 23 shown in Fig. 5 wherein a side building coverage is observed. Areas without coverage, shown on the right of the image would be voiced by the propagation of waves reflected by the aisles [14].

2.3 Communications Network

The communications system implemented in the UTA Ethernet is very stable for sending information alert system activation, this due to the heterogeneous characteristics of backup power and data traffic produced by high amount of network connected devices. However, local university network is high performance, so it is used as a support network activation system alerts and for sending system data communications between the three campus (main system), they are made VLANs on links existing fiber.

Wireless technology allows devices the system can move within the coverage area and facilitates the addition of new devices to the network. Installing a system using wireless technology is simpler compared to wired technologies. Because of these reasons, the main communications system uses a communication based on wireless technology, adapting better to the needs of risk management unit, where the terminal devices of the system are mobilized without difficulty; giving the opportunity to the university community to manipulate the alert mechanisms remotely.

WiFi supports a link for mobile devices (laptops, smartphones, tablets) and allows access to the network suitable for application activation system alerts distances, also via settings repeaters using configurations distributed wireless systems (WDS) is gives full coverage to a particular area. This technology allows handling communication protocols in the network layer (IP) transport (TCP) and application (HTTP). By using these protocols it has a connection-oriented transmission, guaranteeing the sending of information [15–20].

2.4 Activation System Alerts

The proposed system consists of four blocks: Control Unit Monitoring System, Application Server and Communication System as shown in Fig. 6.

The Control Unit is a group of items and electronic devices, responsible for the activation of the warning signs that are performed remotely. The unit is designed so that installation can be performed at any accessible location of university buildings, considering the NEMA 4 protection standards.

Monitoring Unit is a terminal equipment via web clients allows the user to manage access to perform actions on, off and monitoring of electronic signaling devices Control Units, using as an intermediary communications system.

The web application server is a software that controls access and information Activation System Alerts. The exchange of information between devices with Control Units Monitoring System requires a communications system. (Designed Wi-Fi technology in the previous section).

Requests originating from users via web clients are identified by the application server. If the request made is correct and authenticated, the needed information is returned to the customers and they can require to Web servers, changing state of the signaling devices.

The Control Unit continuously monitors the status of the signaling devices, when a change occurs, the client of this unit updates the information on the Web Application Server, the same informing the Monitoring Unit.

Activating devices and users Control Units exchange information according to the procedures shown in Fig. 7 by means of requests and services. The processes are divided into two groups: those performed by the server and executed in the Control Unit.

The server used is the set of software LAMP (Linux, Apache, MySql and PHP) installed on Centos 7. The software programmed into the server is the interface between the user, monitoring devices (cell phones, computers) and Control Units. Users monitor the status of the sirens via a website linked to a server Websockets.

Websockets server establishes a link with customers while the website remains open. The link established allows the server to send information to customers when events are detected scheduled to be heard on port 9000. Communication between Control Units and LAMP-WebSocket server is performed by the client server architecture with HTTP POST requests using port 80.

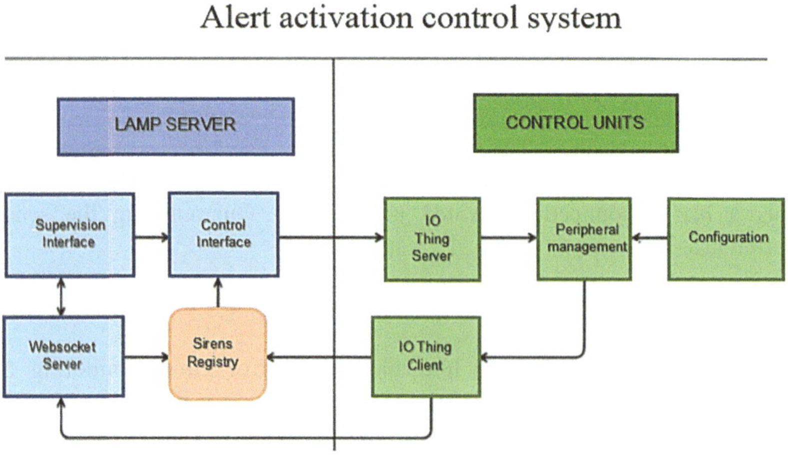

Fig. 6. Diagram exchange of information between elements of Activation System Alerts. Designed in Calligra Flow

As shown in the illustration Alerts are triggered from the website control, the system prompts the user to identify and through consultations in the database sending requests to authorizing Control Units. Control Units act as clients and servers. In client mode, the status information is updated in units database when a change is detected in the sensors. Posteriorly an event is created on the server to send information to the website of the connected clients.

Server mode of Control Units permanently serving external requests, analyzing the information generated warning signals or turn off electronic signaling devices. The outline of the overall system design Activation Alert UTA is shown in Fig. 8.

In the Huachi campus, Faculty of Accounting and Auditing is the main server of the university, where a sub-domain is created to install the web Application Server. In each area Huachi campus (ZA to ZD) and the Ingahurco and Querochaca campus you have a backup server, which is operated to control the devices in their area, when communication is lost with the main server.

Backup servers, SR-ZX, 6 are installed on desktops, 4 Huachi (one area), one and one Ingahurco Querochaca. The SR-ZA server is located on the Risk Management Unit; SR-ZB at the Faculty of Accounting and Auditing; SR-ZC in the Faculty of Management; Monitoring the room where General is located, so installing a smart TV for control and monitoring Control Units recommended; and SR-ZD in the Department of Languages. Ingahurco campus in the SR-ZI server is located in block 03 and the SR-ZQ Querochaca is located in the building 01 Agricultural Sciences.

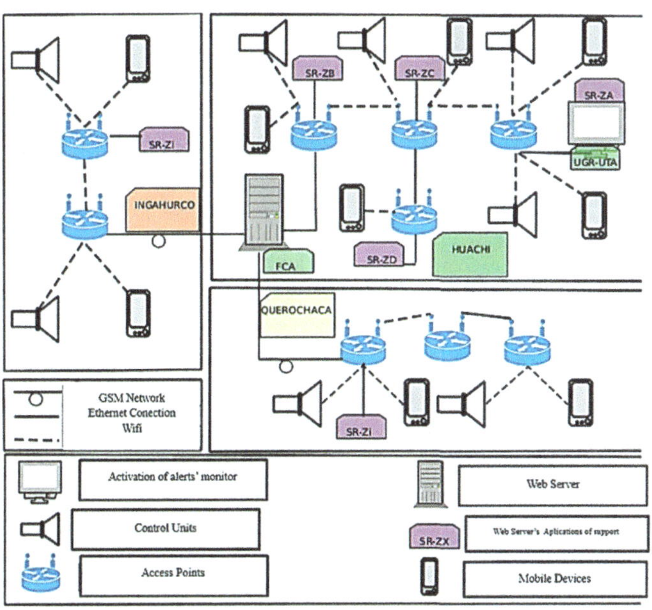

Fig. 7. Diagram activation system alerts the UTA.

Computers backup servers are installed in the offices of the focal points. The brigade can monitor and activate or deactivate acoustic signaling devices Control Units across the University from these computers.

Generic university community connects wirelessly to the activation system alerts via laptops and cell phones, taking the feasibility of enabling or disabling alerts.

In each building campuses must be installed Control Units according to the number of required sirens to sound to form sectorized the university buildings, ensuring complete signal coverage in academic units.

3 Results

Validation of the data was performed by measurements with Cesva sonometer model SC102 under different conditions within the laboratory as shown in Fig. 9. With the open door has a pressure level average sound of 83 dB, a maximum of 85 dB and a minimum of 81 dB warning signal. The sound pressure levels attenuate approximately 10 dB when closing the door of the laboratory, however, the acoustic signal is more than 10 dB the average level of ambient noise (50 dB), considered as suitable to the design conditions.

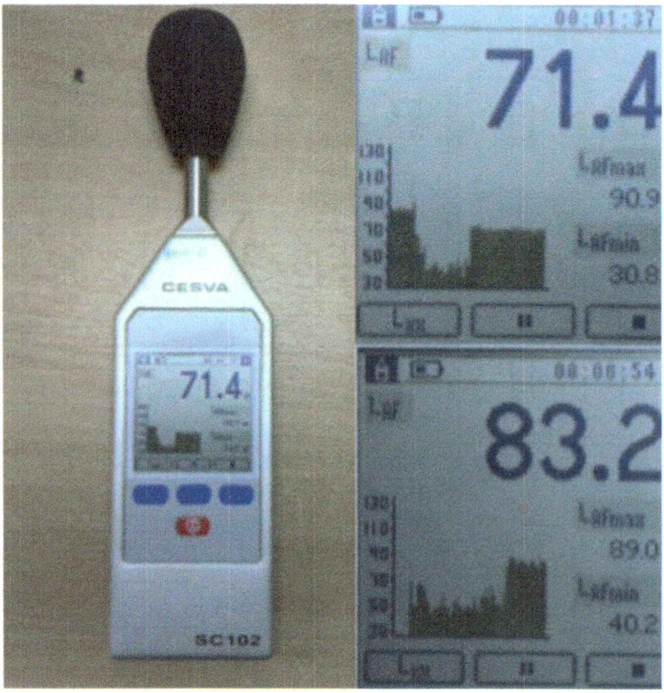

Fig. 8. Measurements of sound pressure level realized in the FISEI-UTA.

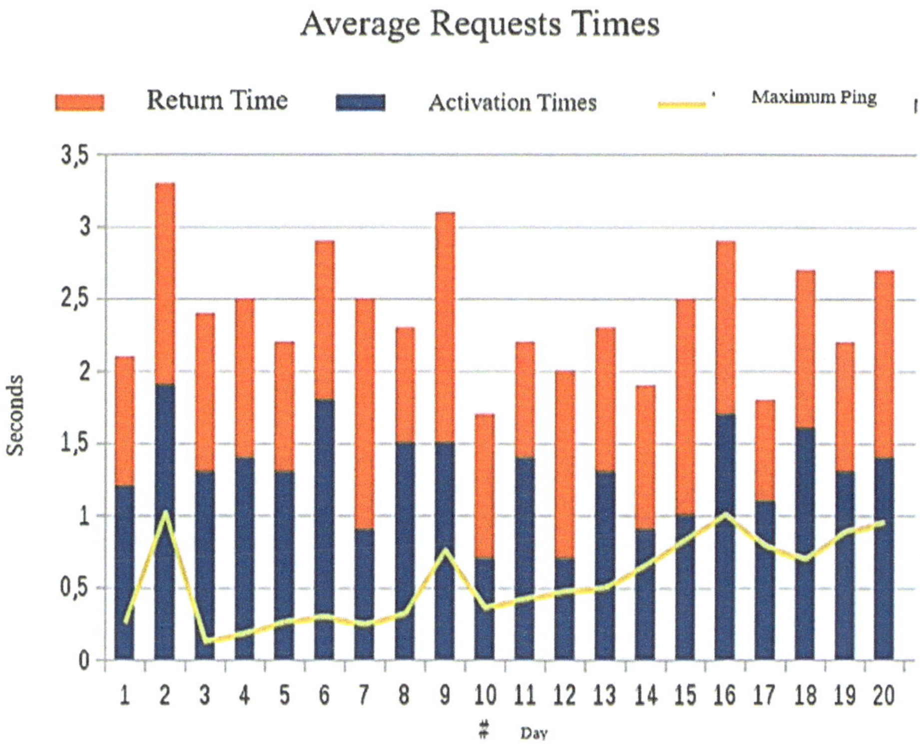

Fig. 9. Statistics peak times requests and ICMP packets.

Performance tests activation devices were performed for 20 consecutive days with 200 random daily testing over time and sends packets using the ICMP protocol. In Fig. 8, it is shown how the activation time, the return time, the maximum ping time and packet loss statistics, are recorded.

In Fig. 9 a graph showing the average time of 200 requests daily shows. The average activation time of an alert signal is 1.3 s and 2.4 s return.

The probability of failure of a request is shown in Fig. 10, 4000 of the requests made 0.05% do not respond to the first request, but has immediate response to the subsequent request without changing system parameters.

The percentage loss of ICMP packets from mobile devices made up a level greater than −75 dBm signal is not more than 7%. The use connection-oriented causes the percentage of failed requests is less than 0.1%, even under conditions of significant losses ICMP packet protocols.

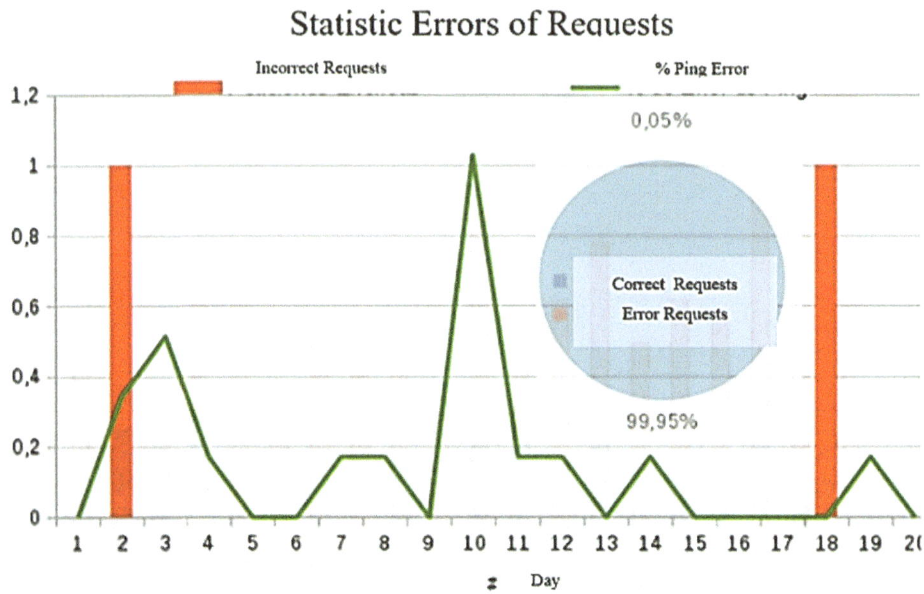

Fig. 10. Statistical Alert activation errors.

4 Conclusions

Understanding the coverage of the acoustic signal in university buildings it is developed by using acoustic methods geometric ray tracing and images. The ideal location of mermaids are architectural spaces that form cavities and passageways that connect classrooms and laboratories offices. The method of ray tracing used determines the siren (112 dB at 1 m) must deliver a signal level appropriate to the spaces below without obstacles 125 m away, considering the sum of lengths of reflection points.

Client server architecture Websockets software technology delivery system alert enable the ability to reduce response time in sending alert requests. Supports activation signals from mobile devices with an event-driven reducing packet traffic control network programming. Compared to continuous traffic required in traditional architecture, it is more efficient.

Performance tests determine that the sound pressure difference signal between the calculated values and (82,9 dB) measured (81 to 85 dB) are in ranges of variation dB, maintaining the signal within the acceptable threshold in acoustic signaling. The average activation time of an alert sent from a mobile device to the siren is 1.3 s and the return of the control unit to the monitoring interface 2.4 s with a margin of error in the request customer less than 0.1% ±2.

Acknowledgements. The authors thank the Technical University of Ambato and the "Dirección de Investigación y Desarrollo" (DIDE) for their support in carrying out this research, in the execution of the project "Plataforma Móvil Omnidireccional KUKA dotada de Inteligencia

Artificial utilizando estrategias de Machine Learnig para Navegación Segura en Espacios no Controlados", project code: PFISEI27.

References

1. Secretaría Nacional de Gestión de Riesgos: Manual de Comité de Gestión de Riesgos. http://reliefweb.int/sites/reliefweb.int/files/resources/Informe_completo_20.pdf. Accessed 21 Apr 2019
2. Universidad Técnica De Ambato, Consejo Universitario: Resolución: 1590-CU-P-2015. http://redi.uta.edu.ec/jspui/handle/123456789/13711. Accessed 15 July 2019
3. Unidad De Gestión De Riesgos, Universidad Técnica De Ambato.: Manual de Normas de Seguridad, Salud y Reducción de Riesgos en los Laboratorios y Áreas de Simulación. ISO 7731:2003 IDT, pp. 10–20 (2007)
4. Malaysian Standard MS ISO 7731:2007.: Ergonomics-Danger Signals for Public and Work Areas-Auditory Danger Signals. ISO 7731:2003 IDT, pp. 4–7 (2007)
5. Martínez Franco, V.D., Ramírez Gómez, I., Rabadán Malda, M., Sánchez Sánchez, M.: Diseño de Alerta Sonora para Alarma Sísmica de la Esime Zacatenco, Tesis de Ingeniería, Instituto Politécnico Nacional, Esc. Sup. de Ingeniería Mecánica y Eléctrica, México D.F. (2015). http://tesis.bnct.ipn.mx/handle/123456789/14076. Accessed 21 Apr 2019
6. Carrión Isbert, A.: Diseño Acústico de Espacios Arquitectónicos, 1st edn. Ediciones UPC, Barcelona (1998)
7. Tippens, P.E.: Física Conceptos y Aplicaciones, 7th edn. Mc Graw Hill, Peru (2011)
8. Miyara, F.: Acústica y sistemas de sonido. Editorial de la Universidad Nacional de Rosario, Argentina (1999)
9. Arau-Puchades, H.: La arquitectura del sonido en la geometría de los espacios. https://www.arauacustica.com/files/publicaciones/pdf_esp_52.pdf. Accessed 21 Apr 2019
10. Ruiz, A.: Competencia digital y TICs en interpretación: «renovarse o morir». EDMETIC **8**(1), 55–71 (2019)
11. Disaster Alert Device Systemand Method. https://patents.google.com/patent/US20070296575A1/en. Accessed 23 Apr 2019
12. Aranda, J.E., Jimenez, A., Ibarrola, G., Alcantar, F., Aguilar, A., Inostroza, M., et al.: Anális de la percepción del riesgo de terremotos en la zona metropolitana del valle de México. http://tesis.bnct.ipn.mx/handle/123456789/14076. Accessed 23 Apr 2019
13. Yao, Y., Glisic, B.: Detection of steel fatigue cracks with strain sensing sheets based on large area electronics. Sensors **15**(4), 8088–8108 (2015)
14. Kumagai, H., Sakurauchi, H., Koitabashi, S., Uchiyama, T., Sasaki, S., Noda, K., et al.: Development of resilient information and communications technology for relief against natural disasters. J. Disaster Res. **14**(2), 348–362 (2019)
15. Nemoto, Y., Hamaguchi, K.: Resilient ICT research based on lessons learned from the Great East Japan Earthquake. IEEE Commun. Mag. **52**(3), 38–43 (2014)
16. Kotabe, S., Komukai, T., Shimizu, Y., Tohjo, H.: Disaster resilience using movable and deployable ICT resource unit (MDRU), IEICE, J100-C(3), pp. 141–148 (2017)
17. Nishiyama, H., Suto, K., Kuribayashi, H.: Cyber Physical systems for intelligent disaster response networks: conceptual proposal and field experiment. IEEE Netw. **31**(4), 120–128 (2017)
18. Peng, L., Toshiaki, M., Kun, W., Song, G., Weihua, Z.: Vehicle-assist resilient information and network system for disaster management. IEEE Trans. Emerging Top. Comput. **5**(3), 438–448 (2017)

19. Shimizu, Y., Suzuki, Y., Sasazawa, R., Kawamoto, Y., Nishiyama, H., Kato, A., et al.: Development of movable and deployable ICT resource unit (MDRU) and its overseas activities. J. Disaster Res. **14**(2), 363–374 (2019)
20. Barti, R.: Acústica Medioambiental, vol. I. Editorial Club Universitario, Alicante (2013)

Evaluation of Internet of Things Protocols for Shopfloor Communication Integration

Carlos S. Leon[1], David I. Ilvis[1], Edison G. Remache[1], Williams R. Villalba[2], Carlos A. Garcia[1], and Marcelo V. Garcia[1,3](✉)

[1] Universidad Tecnica de Ambato, UTA, 180103 Ambato, Ecuador
{cleon9397,dilvis9234,eremache9607,ca.garcia,mv.garcia}@uta.edu.ec
[2] Escuela Politecnica del Chimborazo, ESPOCH, 60155 Riobamba, Ecuador
wvillalba5@hotmail.com
[3] University of Basque Country, UPV/EHU, 48013 Bilbao, Spain
mgarcia294@ehu.eus

Abstract. The productivity and revenue of a company depend clearly on the correct corporate decision making and the management of productive data. Since most industries want speed and impact in their systems, the implementation of an IoT network manages to connect both field-mounted devices and sensors, as well as automated devices such as robots, which provides a considerable advantage within the planning. In this paper, two conventional IoT communication protocols are compared, such as the AMQP and CoAP, both within an automated system, making use of low-cost devices such as RaspBerry cards. The results obtained show that the CoAP protocol is designed to be so small that it fits inside a microcontroller, however, it does not provide satisfactory communication speed, but it can be fully applied in cyber-physical environments, in another aspect the AMQP protocol is more complex, there is no official support and you need bigger installation packages; but it provides a higher communication speed. In the present paper, analyze these behaviors and other questions with respect to both IoT protocols, how and why they should be implemented in an automated system.

Keywords: AMQP · CoAP · Industrial Internet of Things (IIoT) · Low-cost automation

1 Introduction

The current industrial business focuses on issues such as decision-making processes and the management of process data vital characteristics for a company to improve productivity and financial income. Factories today decide to invest in automated systems to perform operations autonomously to achieve ideal production quality levels [9,12].

The relationship between the IoT and computer technologies in the cloud allow effective decision making to improve the productive capacity of a factory.

J. Nummenmaa et al. (Eds.): CSEI 2019, AISC 1078, pp. 199–213, 2020.
https://doi.org/10.1007/978-3-030-33614-1_14

This improvement in the industry is called the fourth industrial revolution or Industry 4.0, which brings new capabilities in the environment [2–5].

In this paper, It is evaluated and compared the use of two conventional communication protocols of IoT, AMQP, and CoAP, within an automated system. The main characteristics of this system are the requirement of real-time data transmission between the devices, as well as a diverse number of messages, due to the fact that the system consists of a sensor that partially sends the signal to activate the operation of the robotic manipulator arm (Scorbot). It should be noted that there is no intention of establishing which protocol is better than the other because the ideal protocol depends on the type of automation application being carried out. The aim is to capture the behavior of both protocols in the same automation environment.

This document is organized as follows: In Sect. 2 is analyzed related works where the performance of the AMQP and CoAP protocols is evaluated and their contributions to research are highlighted. It describes the environment and the different work devices that are used in the implementation of both protocols. Section 3 shows the state of the art in the work, as well as the automation environment in which the protocols will be used. In Sect. 4, the case study and the automation environment in which the work is developed are presented. In Sect. 5 the results obtained from the research are discussed. Finally, in Sect. 6 conclusions of the project are established.

2 Related Work

In the following investigations and works carried out with the communication protocol CoAP and AMQP, the usefulness and versatility provided by these protocols are analyzed, as well as the applications and uses that have been developed by them.

The research developed by Alvear [1] relates the IoT technology with artificial vision concepts in which it is intended to obtain data in real-time and at the same time remotely, which will be stored in databases. The authors denote that the collected data will be used to carry out statistical and probability studies in certain areas based on the activities or characteristics of people in environments or environments where the application of electronic systems is possible, either to improve processes or identify weak points them. This whole process started with the analysis of image processing and the capture of videos to be used in an algorithm focused on IOT, for this reason the authors address the application of the CoAP protocol, because it is a specialized protocol for data transfer to be able to directly relate to HTTP and at the same time integrate into the Web.

In the study conducted by Naik [17] a comparison of IoT protocols is made, with communication standardization as a priority and, as an important factor, real-time transmission and transfer of data that are important aspects in IoT applications. The author mentions that the choice of a standard communication protocol that is also effective is a work that deserves considerable study because of the nature of the system to be implemented as well as the communication requirements that must be generated. The scientific article mentions an evaluation of communication protocols such as CoAP and AMQP used in IoT systems, to identify their strengths and limitations.

On the one hand, the author mentions that the communication systems that use the AMQP protocol are binary systems that generally use 8-byte headers with small or small messages, in addition that this protocol uses TCP as the default transport protocol and TLS/SSL and SASL for security. Similarly, the author mentions characteristics of the CoAP protocol that, unlike AMQP, uses fixed 4-byte headers with small messages and uses UDP as the transport protocol and DTLS for its security. An important point that the author considers is that these messaging protocols with the passage of time have been evolving according to the processes or needs that must cover, so it could be considered, devices, resources and the specific applications of IoT in which they will be employees [17].

A comparison made by the author refers to M2M/IoT compared to standardization, speaking of AMQP this study is successful worldwide and adopted the international standard ISO/IEC 19464: 2014 is currently used in projects of great importance as Nebula Cloud Computing from NASA and Indias Aadhar Project, however CoAP has not been left behind and in recent years has gained momentum and has been employed by large companies such as Cisco, Contiki, Erika and IoTivity in addition to having a specialized IETF standard to integrate IoT and the Web thanks to Eclipse Foundation. And when talking about security, AMQP presents a high level of security while CoAP uses DTLS and Ipsec useful tools for integrity, authentication and encryption [5].

Fernades [16] mentions that there are certain problems when talking about services or communications because a lot of data is commonly sent to databases and the agglomeration of data can affect the performance of the systems significantly; AMQP is a protocol that appears to address this problem and solve it. The study that the authors propose is the analysis of message exchange in a certain time, observing that when there is a high volume of message exchange the most favorable results are generated by AMQP because it can be connected with different applications and different platforms. In this analysis we use the AMQP protocol and the storage of data in a relational database created in MySQL, the exchange of data between clients and servers determined that when exchanging data in bulk the number of messages that can be sent per second it reduces with what causes a high consumption of resources and as a solution to this problem the AMQP is used.

In the scientific article carried out by Urkia [14] reference is made to the fact that in the last few years on the internet of things it has been developing with the incorporation of communication protocols. The authors mention that the CoAP protocol stands out from the rest of the existing protocols. The study was implemented with the use of Raspberry Pi as an industry platform 4.0. In this work we find characteristics and behaviors that use CoAP, the analysis focuses on libCoAP, smcp, microCoAP, FreeCoAP, Californium, node-CoAP and CoAPthon and determined that they are interoperable, when talking about performance they detail that when implemented in C coding you have good results including the SMCP and libCoAP libraries are the fastest what does not happen with FreeCoAP and microCoAP who are at a low level. Regarding the speed of the clients, the authors recommend the use of the CoAPthon (Python), h5.CoAP and node-CoAP (Node.js), Califor-nium (Java), node-CoAP (Node.js) languages for the transmission of data after having carried out the study.

Talaminos [19] mentioned that there are very few studies focused on IoT and communication protocols, however, their study shows the benefits of CoAP talking about factors such as response time compared to other protocols and energy consumption as well as latency and performance. To mention this benefits the authors experiment in M2M test banks comparing CoAP and HTTP estimating the bandwidth consumed and the response time.

As mentioned previously there are studies that to compared protocols IoT and certain characteristics presented by these protocols, varying according to the application or analysis that have been focused, is why the aim of this work is a comparison between CoAP and AMQP protocol as a way to help when making IoT systems and adopt the most appropriate protocol according to the requirements they must fulfill.

3 State of Art

3.1 Internet of Things

The IoT is based on three main foundations related to the capacity of objects that must have communication capabilities, computational capacity and may have interaction capacity [13,20]. It is called communication capacity because the IoT objects must have a minimum set of communication capacity. What we mean by this is not only a channel of communication, but also everything related to it, in order to make an efficient communication, such as an address, identifier and name. The objects can have all these characteristics or some of them [20]. The objects must have a basic or complex computational capacity to process data and network configurations. For example, receiving commands on the communications channel, administering network tasks, saving the status of a sensor, activating an effect, receiving signals and managing and controlling data.

IoT technology can have an interaction capacity in terms of detection and activation. This can be done either by sensors and/or actuators. Sensors are those that detect the real world environment (for example, light, humidity, temperature, movement, voice, etc.). The effectors are those that change or affect the real world as, switches that allow you to activate or activate/deactivate anything that can change the real word like engines, beepers, cameras, etc [3].

3.2 CoAP (Restricted Application Protocol)

The CoAP communication protocol is used to communicate simple and inexpensive electronic devices such as PLC'S, RaspBerry and low power sensors. This

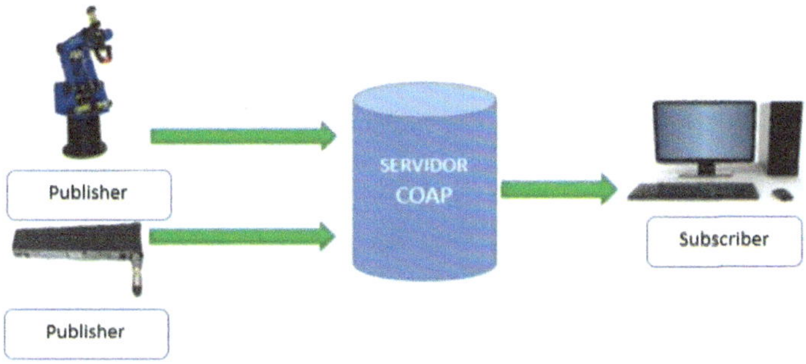

Fig. 1. CoAP architecture

protocol is a derivation of the HTTP protocol, but it is added several requirements such as multicast, overhead and simplicity, which are very important for the Internet of Things (IoT), reason why the protocol is applicable to develop the connectivity of intelligent objects using the Internet [4,12]. See Fig. 1.

It is a specialized protocol for the use of limited and limited low power wireless nodes that can communicate interactively through the internet, its client/server interaction model is similar to that of HTTP with the difference that CoAP performs these interactions (exchanges of messages) asynchronously by means of the UDP transport protocol [5].

3.3 AMQP (Advanced Message Queue Protocol)

The Advanced Message Queuing Protocol is also a publication/subscription protocol based on a reliable message queue. It has been commonly used in the financial sector. This community uses services such as commerce and banking systems that often require extremely high levels of performance, scalability, reliability and manageability [18]. AMQP uses TCP as its main transport protocol for the exchange of messages. Application level messages have a header to route them to the respective queue (See Fig. 2). The AMQP architecture is composed of two main components: Queues and Exchanges [19].

Queues represent the main concept of AMQP. All messages end in a queue that stores them before forwarding them to recipients. These queues can be organized by service levels with respect to implementation performance characteristics such as latency and availability [20].

3.4 RabbitMQ Broker

Both AMQP and MQTT are communication protocols based on intermediaries. As discussed in Sect. 2, they contain a central entity, called an agent, in charge of managing peer-to-peer communication in the network. In this work, we use RabbitMQ, a popular open source message agent. RabbitMQ is an Erlang based

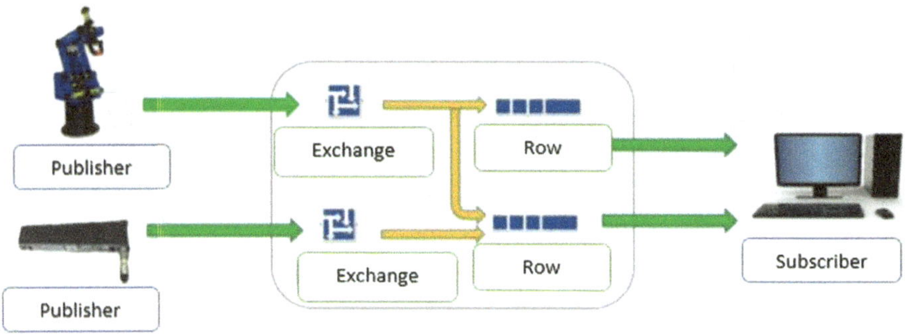

Fig. 2. Architecture AMQP

technology that allows asynchronous communication between devices. Initially, it was developed to implement AMQP and then to support MQTT [12].

The exchanges distribute the messages to the respective queues according to the predefined rules. As a new message arrives at the intermediary, the exchange evaluates the message and stores it in a queue, ready to be forwarded. Figure 2 represents the main messaging process in an environment based on AMQP. First, editors who want to send messages to potential subscribers, send them to a broker. The intermediary has exchanges and queues. As mentioned above, the exchanges receive the messages and forward them to the respective queues. In turn, the queues send these messages to the clients that previously subscribed to the given queue [11].

3.5 Raspberry Pi 3

It is a low cost hardware platform that includes all the elements offered by a computer. Nowadays it has acquired great importance in the market due to its diversity of options for projects in computer networks, electronic circuits, robotics, domotics, security, programming, among other technological areas. Even some authors like Saari et al. [10] have used the Raspberry Pi as a solution for the Internet of Things (IoT).

The most current model of Raspberry Pi 3 is B, which has a storage unit with MicroSD Card Slot and is equipped with 35000 packages and pre-compiled programs in a format that facilitates installation. In addition, despite being adapted to the perfection of the board, it is not an operating system affiliated with the Raspberry Pi foundation, since it was created by a small team dedicated to developers, so it allows the installation of a great variety of operating systems, including Noobs, Ubuntu MATE and Windows, although the most commonly used is Raspbian based on Debian, with online support to update it [7,8,15].

4 Case Study

This paper develops a communication of two PLCs Siemens S7-1200 used for the mobility of a Robotic Arm and a Conveyor Belt for its manipulation. And carry out the palletizing process. This communication allows the interaction between the PLCs through the use of IoT and visualizes in a web address the changes of the states of the inputs and outputs of the PLC's to monitor the operation of the system from a remote area without the need of programming.

The proposed process is integrated by actuator elements such as Robotic Manipulator Arm and a Conveyor Belt managed by S7-1200 PLC communicated through the use of conventional IoT, like CoAP and AMQP protocols. The communication of all the components is given with a data transmission speed of 10 Mbps to have a minimum error rate in the communication, managing the processes in real time and monitoring the active or inactive state of devices connected to the environment. Figure 3 presents the structure of the communication made. Arm Manipulator performs palletizing long as it starts to move the conveyor belt that is when there are objects on it, this communication machine machine (M2M) is achieved by means of IoT that links the two Raspberry Pi 3 and exchange data with each employee protocols CoAP or AMQP for its communication achieving the synchronization of the movements of the actuator devices.

The development of the proposed system covers a simple structure. In order for the CoAP library to communicate with other devices, it is necessary for its information to be centralized in one device, in this case in one of the RaspBerry Pi 3 that acts as a server accepting registration requests. This microcomputer also acts as a client exchanging data from the PLC1, to the other RaspBerry Pi 3 that acts only as a client to publish the PLC2 states, as well as to modify them if there are changes in the PLC1. Through the implementation and application of conventional IoT protocols such as CoAP and AMQP.

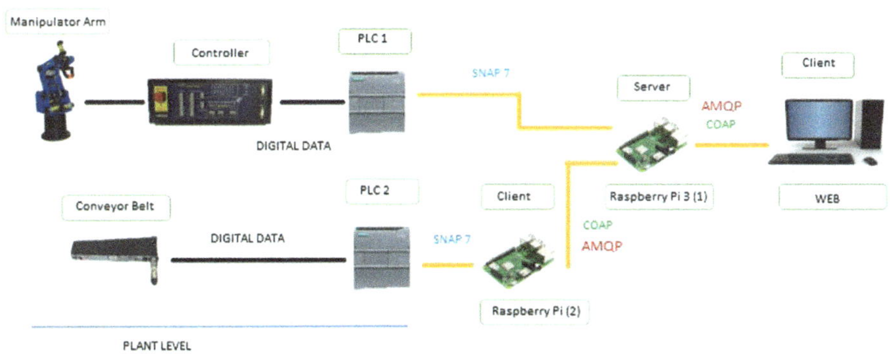

Fig. 3. Case study Hardware Architecture Platform

4.1 SNAP 7 Implementation

Applying the Snap 7 (C++) library, it is intended to obtain data from the S7-1200 PLCs through the reading of the internal databases of the PLC using the Raspberry Pi 3. This library uses the Communication interface. S7 Siemens Ethernet for reading and writing PLC data (inputs, outputs, memories, timers, counters) and has three independent components: client, server and partner. Next, the definition of the Snap 7 client is represented for the CoAP client for data transmission. See Listing 1.1.

```
Client= new TS7Client();
   Client->SetAsCallback(CliCompletion,NULL);
   signal(SIGINT, intHandler);
   if (CliConnect())
   {
    byte DI[1] = {0b00000000};
    int datodb = Client->EBWrite(0, 1, &DI);
    int datowa = Client->WriteArea(0x82,1,0,1,0x02,&DI)
   }
```

Listing 1.1. Snap 7 library execution for data transfer

The advantages of using the Snap7 library are many, because it is written in C++, reading data from Ethernet compatible PLCs, as long as the requests to Ethernet are not restricted. In the work done, the data is written and read through bytes, but you can use variables of type: Word, Double Word and Real used extensively in the programming of the Siemens language [6].

The reading in the program is done through a byte vector, within the MultiRead function, which is obtained with the ReadMultiVars method.

The use of this library is possible when it is identified that the PLC is acting, ie PLC1 or PLC2, and the IP address of each of the PLCs is specified. When configuring and setting these parameters, the program monitors the I0.x and Q0.x outputs of each of the S7-1200 PLCs and, upon detecting changes in the conformation of the received byte, sends it through the Client. The data writing in the output is achieved through the WriteArea command with the byte formed to activate the output as shows in Listing 1.2.

```
  if (findtrue!=std::string::npos || findfalse!=std::
   string::npos){
        if (findtrue!=std::string::npos ){
            byte DI = 0b10000;
            Client->WriteArea(0x81,1,0,1,0x02,&DI);
            Client->WriteArea(0x82,1,0,1,0x02,&DI);
        }
    }
```

Listing 1.2. Writing data in PLC

4.2 CoAP Protocol Implementation

The libCoAP library was used to communicate through the IoT to the Raspberry Server, while Snap7 is used to obtain the PLC data and change the states of the output through the manipulation of the databases.

When programming within the free Linux software, the library libCoAP (C) Librarian used to communicate by Iot to the Raspberry Pi Server is used, but since it is too extensive, it is used of the programming algorithm described below. When the client is initialized, to call CoAP-client within the program, GET or PUT actions can be performed, in this case a get is used to obtain information of states of the connected PLCs, later the IP address of the server is entered, then port 5683 is entered and finally the address to PLC1 or PLC 2. To execute the CoAP program that has already been carried out, the string code output = exec (cstr) is typed. With the line of code executed the data transmission is initialized. See Fig. 4.

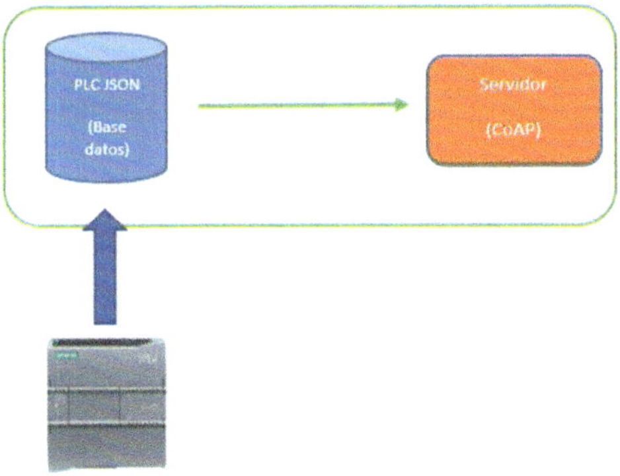

Fig. 4. Server data transfer

When a request arrives at a resource from the client, the module of the server Server.on takes care of it according to what type of information requirement is, for example, the GET method receives information of the I0.5 input of the selected device, while the PUT method writes the value of each input and output of the PLC in which it is being executed. See Fig. 5.

4.3 AMQP Implementation

For this communication, rabbitmq was used, which is a message broker for AMQP, simple amqpclient to define the Client for AMQP and the C++ library ampqlib which is a Javascript library for subscribing AMQP messages. The Listing 1.3 shows the import of simple AMQP libraries.

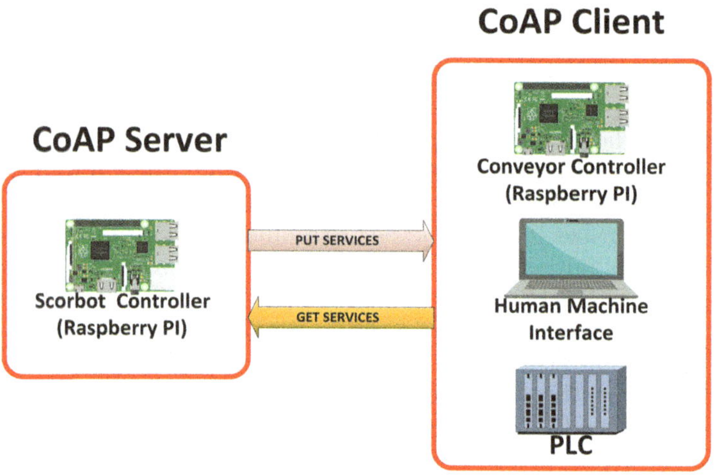

Fig. 5. Sending data from the server to different clients

```
#include <SimpleAmqpClient/SimpleAmqpClient.h>
#include <chrono>

AmqpClient::Channel::ptr_t connection;

Address=argv[1];
Rack=atoi(argv[2]);
Slot=atoi(argv[3]);
ipservidor = argv[4];
plc = argv[5];
samples = atoi(argv[6]);
connection = AmqpClient::Channel::Create(ipserver,5672, "uta",
    "industrial");
connection->DeclareQueue(plc, false, false, false, false);
```

Listing 1.3. Creation AMQP

5 Discussion of Results

In the present study, different tests have been performed to compare both protocols and define in which situations they behave better. The first test is about the execution time introducing different bandwidths (5, 2.5, 1 Mbit), latencies (0, 5, 10 ms) and lost packet rates (0, 2.5 Y5)%, it should be noted that the tests were measured in seconds. Fig. 6(a) shows a uniform behavior over time, with a variation between tests no greater than 7% the time.

As the main reason to be implemented the CoAP protocol is to run on hardware and minimal infrastructure, the execution time is clearly greater than AMQP due to the implementation problems required by this library. In the case

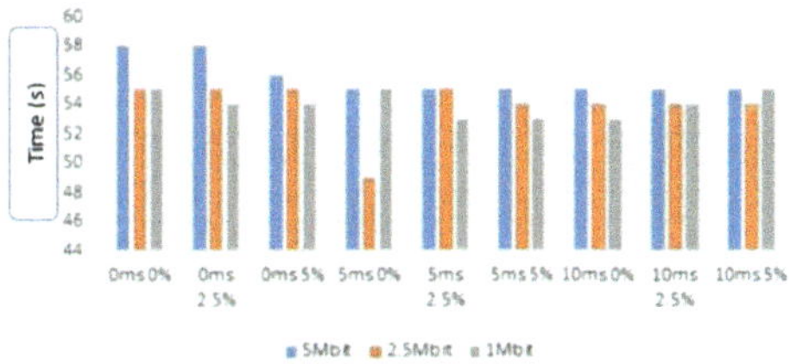

(a)

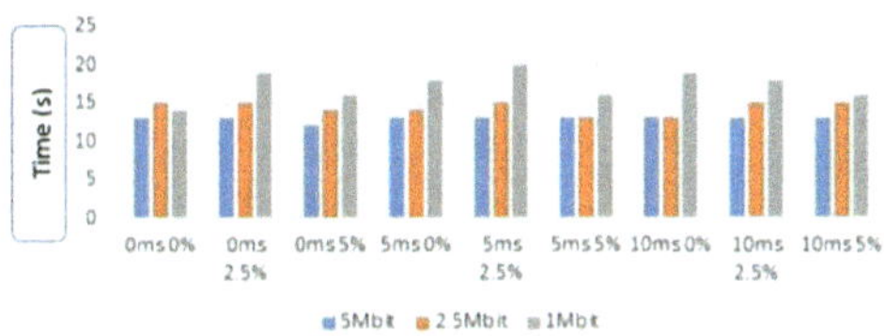

(b)

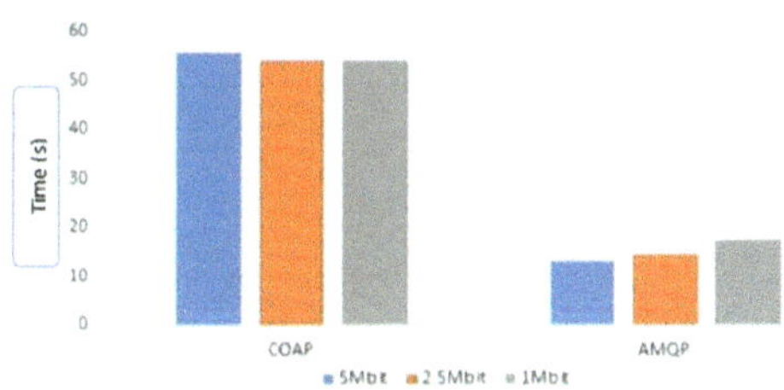

(c)

(d)

Fig. 6. Comparison of AMQ and CoAP protocols. (a) CoAP Protocol execution time. (b) Execution time AMQP protocol. (c) Execution time per bandwidth. (d) Average execution times

of the AMQP protocol, the speed of execution is unparalleled, with an average of 0.09s being much faster than CoAP, under normal conditions the protocol is optimal. The problem arises when the network conditions are less favorable, it shows a 17% variability with respect to the average. See Fig. 6(b).

Regarding the execution time, it can be noted that the AMQP protocol significantly exceeds the CoAP protocol with an average of 0.09 s under normal conditions, given the implementation problems required by the CoAP library. See Fig. 6(c).

Regarding the variability presented by the protocols, it can be noted that the CoAP protocol being restricted remains constant over time (variability of 1%), while AMQP suffers when the network conditions are not optimal (variability up to 30%). Figure 6(d) shows that AMQP is superior to CoAP.

In the 200 samples that were taken from the Scorbot Robot, it demonstrates a speed of 380% with respect to the process by CoAP. Throughput performance graphs, are done in a controlled and noise-free environment, have similar characteristics, so if an industrial user wants to get a successful transfer of packets, both protocols fulfill their purpose. As can see in Fig. 7, AMQP-based architecture produced lower latency. MQTT messages had experienced lower delays

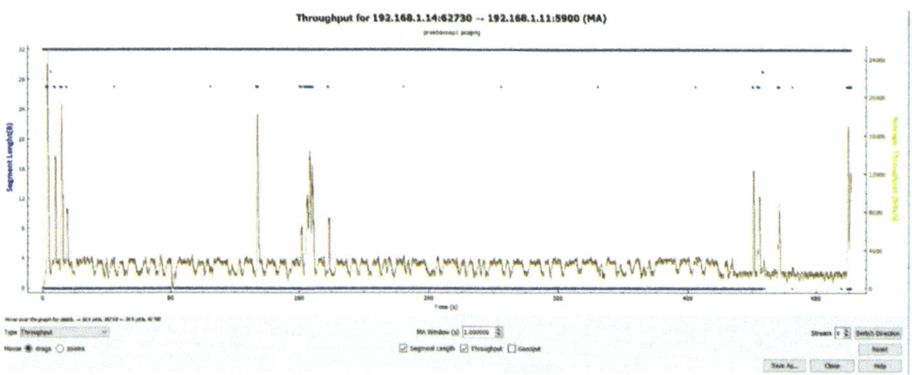

(a)

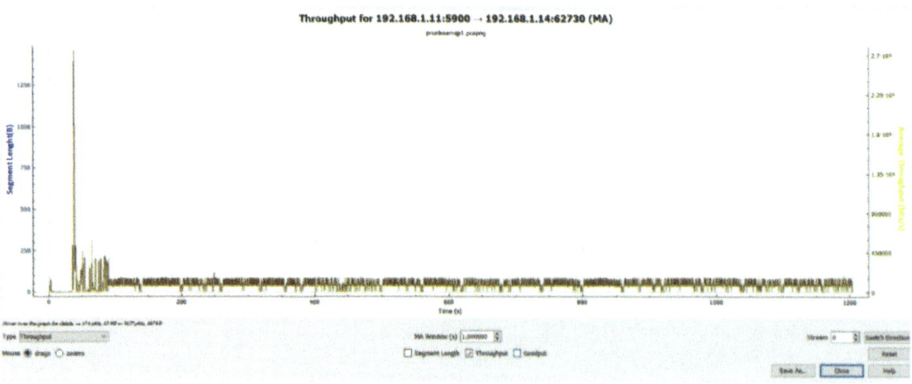

(b)

Fig. 7. AMQ and CoAP protocols performance. (a) CoAP protocol performance. (b) AMQP protocol performance.

than CoAP for lower packet loss and higher delays than CoAP for higher packet loss. The comparison of these two protocols shown that the average CoAP was more than 20% shorter than AMQP. Regarding the latency, AMQP latencies were measured in the order of milliseconds, and CoAP latencies as low as hundreds of microseconds. However, it is important to notice that in the cases of less reliable networks, AMQP's underlying TCP protocol will be an important advantage and the results would be different.

AMQP and CoAP have been analyzed using the common instruction operations of the Scorbot Robot in terms of bandwidth consumption that was measured as total data transferred per message. In the cases where message size is small as in this study case, and independently of the increase of packet loss rate, CoAP consumed less bandwidth than AMQP. CoAP based in Fig. 7 showed a comparably lower bandwidth consumption that did not increase with increased network packet loss or increased network latency, unlike AMQP where bandwidth consumption increased.

The power/energy consumption is essential in every IoT based system, and the choice of protocols affects the same. Analyzing average energy consumed by AMQP and CoAP for a Scorbot Robot with experimental results showing that CoAP is more efficient in terms of energy, though both of them proved to be efficient. Due to the lack of compatibility, AMQP would not be compatible with small devices. In contrast, AMQP has reception feedback, so if it is implemented in the code, the rate of lost packets is zero in critical processes.

6 Conclusion and Future Work

In the present work presents a study in order to make a comparison between two IoT communication protocols; AMQP and CoAP, due to the constant revolution in the industry and the implementation of Process Automation, where the Industrial Internet of Things (IoT) has been inserted as part of Industry 4.0, its development is on track.

In this way, the authors have developed research with the purpose of comparing both protocols and to give a verdict of which one is the most convenient in different types of scenarios, measuring the total time of the process (TPT) in the same conditions and for each protocol, for the analysis of AMQP the data is stored in MySQL, we have found that AMQP was faster than CoAP with an average of 0.09s in normal conditions, however CoAP does not stay behind, since it does not require a broker and in the analysis of the variation in bandwidth, being a restricted protocol, it remains constant over time, for the case of AMQP, it presents variation is when the network conditions are not optimal.

When implementing the two protocols for the application of the palletizing of the Scorbot Er-4u Manipulator Arm, it was noted that AMQP is superior to the CoAP protocol in the 200 samples taken, showing a 380% higher benefit. However, each IoT protocol has its own characteristics, so each one of them in different scenarios can be of vital importance in terms of communication and in different cases, the CoAP protocol would be the best option.

As a future work, we have proposed to implement as an application the analysis of the FESTO parts sorting station with IoT protocols, since it makes use of sensors and actuators allowing us to change the environment and open ourselves to the field of pneumatic together with the PLCs and the mechanics.

Acknowledgment. This work was financed by Universidad Tecnica de Ambato (UTA) and their Research and Development Department (DIDE) under project CONIN-P-0167-2017.

References

1. Alvear, V.: Internet de las Cosas y Visión Artificial, Funcionamiento y Aplicaciones: Revisión de Literatura (Internet of Things and Artificial Vision, Performance and Applications: Literature Review). Enfoque UTE **8**(1), 244–256 (2017). http://ingenieria.ute.edu.ec/enfoqueute/
2. Ancillotti, E., Bruno, R., Vallati, C., Mingozzi, E.: Design and evaluation of a rate-based congestion control mechanism in CoAP for IoT applications. In: 19th IEEE International Symposium on a World of Wireless, Mobile and Multimedia Networks, WoWMoM 2018, pp. 14–15 (2018). https://doi.org/10.1109/WoWMoM.2018.8449736
3. Bahashwan, A.A.O., Manickam, S.: A brief review of messaging protocol standards for internet of things (IoT). J. Cyber Secur. Mobility **8**(1), 1–14 (2018). https://doi.org/10.13052/jcsm2245-1439.811
4. Bhatia, R., Gupta, B., Benno, S., Esteban, J., Samardzija, D., Tavares, M., Lakshman, T.V.: Massive machine type communications over 5G using lean protocols and edge proxies. In: IEEE 5G World Forum, 5GWF 2018 - Conference Proceedings, pp. 462–467 (2018). https://doi.org/10.1109/5GWF.2018.8517086
5. Bolettieri, S., Tanganelli, G., Vallati, C., Mingozzi, E.: pCoCoA: a precise congestion control algorithm for CoAP. Ad Hoc Netw. **80**, 116–129 (2018). https://doi.org/10.1016/j.adhoc.2018.06.015
6. Din, S., Paul, A., Hong, W.H., Seo, H.: Constrained application for mobility management using embedded devices in the Internet of Things based urban planning in smart cities. Sustain. Cities Soc. **44**, 144–151 (2019). https://doi.org/10.1016/j.scs.2018.07.017
7. Garcia, M.V., Irisarri, E., Perez, F., Estevez, E., Orive, D., Marcos, M.: Plant floor communications integration using a low cost cpps architecture. In: 2016 IEEE 21st International Conference on Emerging Technologies and Factory Automation (ETFA), pp. 1–4, September 2016. https://doi.org/10.1109/ETFA.2016.7733631
8. Garcia, M.V., Perez, F., Calvo, I., Moran, G.: Developing CPPS within IEC-61499 based on low cost devices. In: 2015 IEEE World Conference on Factory Communication Systems (WFCS), pp. 1–4, May 2015. https://doi.org/10.1109/WFCS.2015.7160574
9. Gohar, M., Choi, J.G., Koh, S.J.: CoAP-based group mobility management protocol for the Internet-of-Things in WBAN environment. Future Gener. Comput. Syst. **88**, 309–318 (2018). https://doi.org/10.1016/j.future.2018.06.003
10. Granjal, J., Silva, J.M., Lourenço, N.: Intrusion detection and prevention in CoAP wireless sensor networks using anomaly detection. Sensors (Switzerland) **18**(8) (2018). https://doi.org/10.3390/s18082445

11. Halabi, D., Hamdan, S., Almajali, S.: Enhance the security in smart home applications based on IOT-CoAP protocol. In: 6th International Conference on Digital Information, Networking, and Wireless Communications, DINWC 2018, pp. 81–85 (2018). https://doi.org/10.1109/DINWC.2018.8357000
12. Harish, M., Karthick, R., Rajan, R.M., Vetriselvi, V.: ICCCE 2018. Proceedings of the International Conference on Communications and Cyber Physical Engineering 2018, vol. 500, pp. 497–511 (2019). https://doi.org/10.1007/978-981-13-0212-1
13. Herrero, R.: Dynamic CoAP mode control in real time wireless IoT networks. IEEE Internet Things J. **6**(1), 801–807 (2019). https://doi.org/10.1109/JIOT.2018.2857701
14. Iglesias-Urkia, M., Casado-Mansilla, D., Mayer, S., Urbieta, A.: Validation of a CoAP to IEC 61850 mapping and benchmarking vs HTTP-REST and WS-SOAP. In: IEEE International Conference on Emerging Technologies and Factory Automation, ETFA 2018-September, pp. 1015–1022 (2018). https://doi.org/10.1109/ETFA.2018.8502624
15. Iglesias-Urkia, M., Orive, A., Urbieta, A., Casado-Mansilla, D.: Analysis of CoAP implementations for industrial Internet of Things: a survey. J. Ambient Intell. Humaniz. Comput. **00**(2016), 1–14 (2018). https://doi.org/10.1007/s12652-018-0729-z
16. Fernandes, J.L., Lopes, I.C., Rodrigues, J.J., Ullah, S.: Performance evaluation of RESTful web services. Comput. Eng. **1**(3), 72–78 (2010). http://www.scopus.com/inward/record.url?eid=2-s2.0-77953530128&partnerID=40&md5=925a7830b01475e316b66bc34593c0a2
17. Naik, N.: Choice of effective messaging protocols for IoT systems: MQTT, CoAP, AMQP and HTTP. In: 2017 IEEE International Symposium on Systems Engineering, ISSE 2017 - Proceedings (2017). https://doi.org/10.1109/SysEng.2017.8088251
18. Rathod, D., Patil, S.: Security analysis of constrained application protocol (CoAP): IoT protocol. Int. J. Adv. Stud. Comput. Sci. Eng. **6**(8), 37–41 (2017). https://search.proquest.com/docview/1947842952?accountid=17242
19. Talaminos-Barroso, A., Estudillo-Valderrama, M.A., Roa, L.M., Reina-Tosina, J., Ortega-Ruiz, F.: A machine-to-Machine protocol benchmark for eHealth applications - use case: respiratory rehabilitation. Comput. Methods Programs Biomed. **129**, 1–11 (2016). https://doi.org/10.1016/j.cmpb.2016.03.004
20. Vallati, C., Righetti, F., Tanganelli, G., Mingozzi, E., Anastasi, G.: ECOAP: experimental assessment of congestion control strategies for CoAP using the wishful platform. In: Proceedings - 2018 IEEE International Conference on Smart Computing, SMARTCOMP 2018, pp. 423–428 (2018). https://doi.org/10.1109/SMARTCOMP.2018.00040

Embedded System for Hand Gesture Recognition Using EMG Signals: Effect of Size in the Analysis Windows

Juan Mantilla-Brito[1], David Pozo-Espín[1](✉), Santiago Solórzano[1], and Luis Morales[2]

[1] Facultad de Ingeniería y Ciencias Aplicadas, Universidad de las Américas, Quito, Ecuador
{juan.mantilla.brito,david.pozo, santiago.solorzano}@udla.edu.ec

[2] Departamento de Automatización y Control, Escuela Politécnica Nacional, Quito, Ecuador
luis.moralesec@epn.edu.ec

Abstract. The electromyography signals (EMG) analysis in the field of robotics has had a great impact due to its application in prosthesis and system control using electrical signals resulting from the action of muscles associated with different parts of the human body. In this article, an embedded system of hand gesture recognition based on EMG signals measured in the forearm is implemented. The EMG signals are acquired through the Myo Armband sensor and processed with a 32-bit STM microcontroller. The signals are classified with the Naïve Bayes algorithm (NB) and by recognizing an established pattern (open or closed), the system sends control signals to a robotic hand to replicate the movement. As main contribution, a comparative analysis of the performance of the embedded system is presented based on: number of samples (analysis windows), acquisition and processing times, as well as the macro and micro-evaluation of multiclass metrics (Accuracy, Precision, Recall and F-Score) for the recognition of movements.

Keywords: Myo Armband · Naïve Bayes · Embedded system · 32-bits microcontroller

1 Introduction

Electromyography signals (EMG) are the electrical activities generated by the muscles when they relax or contract. Through these signals, it is possible to develop applications in fields like robotics, medicine, home automation, among others [1]. The real time analysis of the different EMG signals in order to identify hand gesture recognition becomes a challenge depending of the number of signals to be processed and the gestures to be identified. Several works have been developed in order to study the hand gesture recognition, however, in most of the cases the data processing and recognition algorithms are implemented on computers. The develop of embedded systems with reduced size, low energy consumption, low cost and portable capability becomes an

J. Nummenmaa et al. (Eds.): CSEI 2019, AISC 1078, pp. 214–225, 2020.
https://doi.org/10.1007/978-3-030-33614-1_15

important field study because of its hardware limitations compared with traditional computers systems or even the use of single board computer devices.

The present study is focused on the development of an embedded system, using a 32-bit microcontroller, that executes the classification algorithm Naïve Bayes for the process of identifying two hand movements (open and closed) based on EMG signals provided by a Myo Armband sensor located on the forearm. The main objective is to analyze the behavior of the system in terms of processing time and performance through macro and micro evaluation metrics (Accuracy, Precision, Recall and F-Score). The features used for the analysis are the Mean Absolute Value (MAV), Root Mean Square (RMS), Standard Deviation (STD) and Variance (VAR) based on the previous study carried out in [2]. In addition, an analysis of the system behavior is performed based on the number of samples for the identification process. For this, tests are run with different time windows which have 30, 50, 70, 90, 100 and 125 samples in real time.

2 Background Literature

At the beginning, the EMG signals were used in neuroscience to diagnose neuromuscular diseases or disorders by electromyographic exams [3]. In psychiatry, through neural networks and EMG signals it is possible to establish the state of mind of a patient by the expression in his face [4]. On the other hand, in fields like obstetrics, electrohysterography (EHG) signals that are signals EMG produced by the uterine walls, allow through the application of classifiers to identify normal pulses of pregnancy or labor contractions [5]. In works such as [6], these signals are used in the muscles of the leg in order to study the evolution and to establish quantitative biomarkers for Parkinson's disease.

The analysis of EMG signals has been used in people who have suffered cerebrovascular accidents, developing games that require precise movements of the hand in order to rehabilitate the mobility [7]. In the same way, rehabilitation systems based on EMG measurement have been designed for upper limbs though the application of serious games [8].

On the other hand, the interaction of systems and people through EMG signals allows developing applications in the field of robotics [9, 10], teleoperation [11], musical interaction [12], or even prototypes development that include EMG and IMU signals for interaction with visualization devices [13].

Myo Gesture Control Armband is a sensor to perform the EMG signals acquisition, which communicates with the computer and control functionalities such as mouse pointer, presentations, games, among others [14]. Thalmic Labs is constantly responsible for the development of applications that exploit the potential of the sensor. The most popular is "MYO Diagnostics", that shows the muscular activity of the forearm in a graphical interface, being a great help in physiotherapy [15]. The performance of the Myo Armband has been evaluated in works such as [16], where a comparison was made between sensors for the control of a robotic arm simulated in Unity 3D. In this development it was established that Myo Armband presents the best results. Due to its versatility and functionality, Myo Armband has been used in applications related to the identification of gestures using SVM [17] and e-learning for manual tasks using neural networks and models of Markov [18].

The application of classification algorithms for EMG signals is a field study where several works have been developed. In 2017, several classifiers of EMG signals were compared, among which were applied the algorithm of Feed-Forward Neural Networks (FFN), Support Vector Machines (SVM), the Naïve Bayes Classifier (NBC) and Linear Discriminant Analysis (LDA). In this experiment, the Myo Armband was located on the forearm area of eight participants to identify some hand gestures [2]. In 2019, an application for the recognition of hand gestures through EMG signals was created, which offered an improvement for the optimization of the General Regression Neural Networks (GRNN) with the aim of reducing redundant information, improve efficiency and accuracy of recognition in real time, achieving an accuracy of 95.1% and with a recognition time of 0.19 s [19]. Likewise, a project was developed to recognize three gestures of the hand (open, closed and extension of the wrist) using a QNET electromyographic module trained and tested by four people. From 83% to 98% was the range of accuracy achieved by applying neural networks [20].

The analysis of the behavior of the EMG signals incorporated in portable devices takes force especially in fields such as sports, where it is possible to monitor the behavior and the muscular state of a person; this is the case of [21], where a multilayer application was designed for Android devices for EMG activity monitoring. In addition, in works like [22] they present the implementation of machine learning algorithms in microcontrolled systems for the identification of different movements of the hand based on the acquisition of EMG signals.

3 Methodology

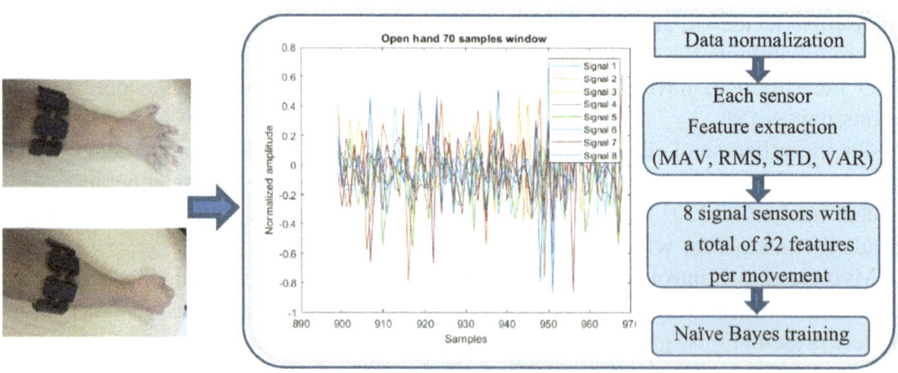

Fig. 1. Data acquisition of the eight sensors of the Myo Armband.

The system consists of two stages: off-line training (see Fig. 1) and the data processing into the embedded recognition system (shown in Fig. 2). The signals used for the study are obtained from the Myo Armband which has 8 dry sensors with a sampling rate of

200 Hz. The sensor has been placed in the same position on the forearm for each user. The embedded recognition system is built with a microcontroller ATmega328P, which receives the information from the Myo Armband and a microcontroller STM32f103c8t6 used to process the EMG signals and perform the gesture recognition for the control of a robotic hand.

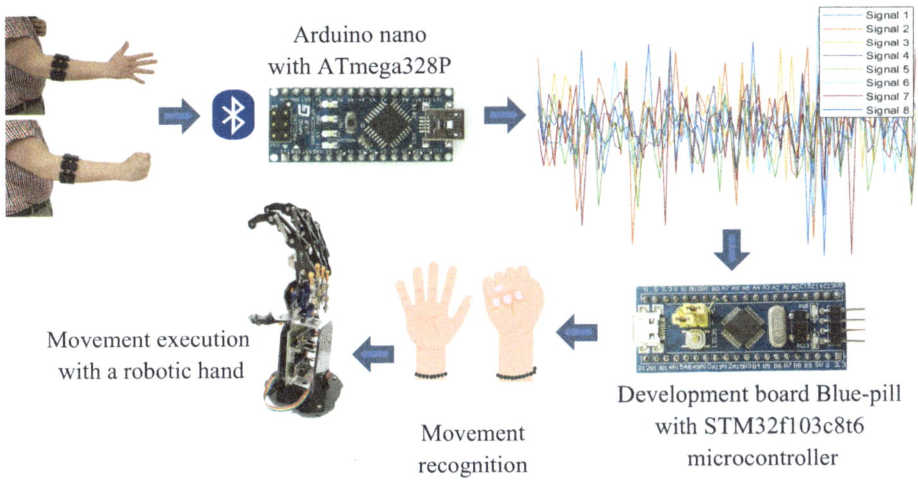

Fig. 2. Embedded system scheme.

3.1 Off-Line Training

The training stage uses the information of seven people of different physical context and gender, each of whom made two hand movements: open and closed. The data acquisition for each movement lasted 10[s] (Fig. 1). The data acquisition process for the training is developed in MATLAB, where the normalization of each of the 8 signals is performed based on their maximum absolute values. Then, we proceeded to obtain the features x_i (MAV, RMS, STD and VAR) associated to each signal, obtaining $n = 32$ features, where $i \in \{1, 2, 3.\ldots\ldots n\}$ for each movement X_j, where $j \in \{1, 2\}$, are open and closed respectively. The expressions for the calculation of the features are shown below:

$$MAV = \frac{1}{W}\sum_{k=1}^{W}|S_k| \tag{1}$$

$$RMS = \sqrt{\frac{1}{W}\sum_{k=1}^{W}S_k^2} \tag{2}$$

$$VAR = \frac{1}{W-1}\sum_{k=1}^{W}|S_k - \bar{S}| \tag{3}$$

$$STD = \sqrt{\frac{1}{W-1}\sum_{k=1}^{W}|S_k - \bar{S}|^2} \quad (4)$$

Where: S_k is the k-th sample, W are the number of samples, and $\bar{S}$ is the mean of data in a time window.

The features are obtained from a defined time window, where the time windows analyzed in this work have $W = \{30, 50, 70, 90, 110, 125\}$ samples. Taking care that the sampling frequency of the sensor is 200 Hz, the total samples acquired in 10[s] for each movement will be 2000. The number of sample windows to be analyzed in each case is defined by: V = 2000/W. Figure 4 shows an example of a time window based on 70 samples obtained from Fig. 3.

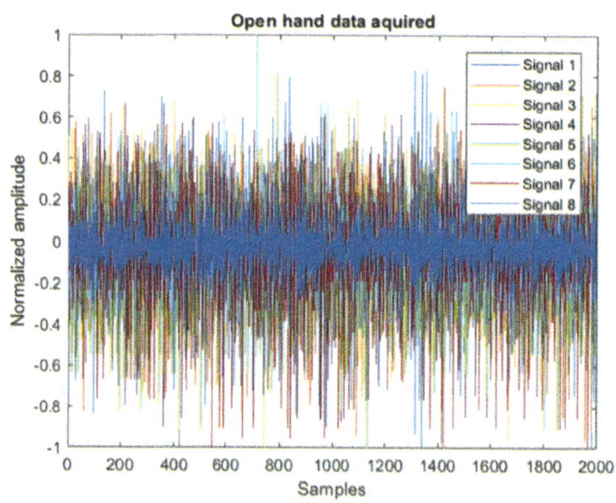

Fig. 3. Data acquisition of the eight sensors of the Myo Armband.

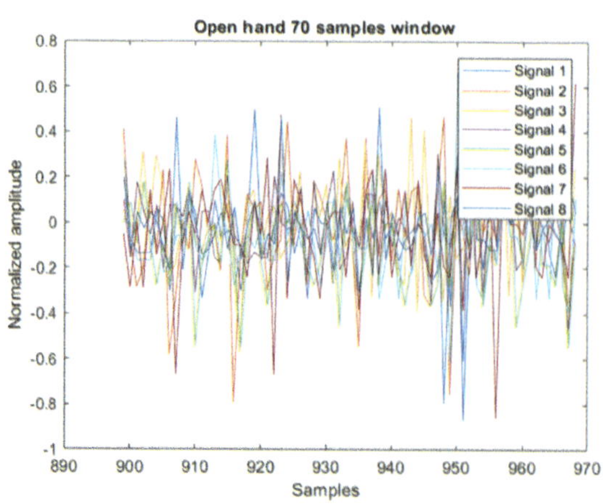

Fig. 4. Selection of analysis window.

Once the features spaces for a defined analysis time window have been obtained, we proceed to obtain the constants for the probability density function (5) used for the Bayes theorem.

$$P(x_i|C_j) = \frac{1}{\sqrt{2\pi\delta_{ic}^2}} * e^{-\frac{(x_i - u_{ic})^2}{2\delta_{ic}^2}} \tag{5}$$

Where: δ_{ic} is the standard deviation of i-th training feature of a class j, and u_{ic} is the mean of i-th training feature of a class j.

3.2 Embedded Recognition System

The embedded system receives the data from the Myo Armband in a defined time windows, this data is acquired through an ATmega328P microcontroller, which manage the information requests to the sensor. Once the eight EMG signals have been received in the ATmega328P microcontroller, this sends them to the STM32f103c8t6 microcontroller, which process the signals, performs de gesture recognition and control of the robotic hand.

The samples of each of the eight sensors that are received through the time windows have range of values between −127 to 127, which will be normalized according to the absolute maximum value of each signal. Let W the number of samples for each sensor in a defined time window and W_u the matrix containing the W for each sensor, where $u \in \{1, 2, 3 \ldots .8\}$. The normalization for each signal is performed by the expression (6). These standardized values will be used for the feature extraction and classification through Naïve Bayes algorithm.

$$N_u = \frac{W_u}{\max(abs(W_u))} \tag{6}$$

Once the off-line training is done obtaining the model parameters (δ_{ic}, u_{ic}) for the distribution function (5) of each class (open and closed), the microcontroller STM32f103c8t6 implements the Bayes theorem (7).

$$P(C_j|X) = \frac{P(X_j|C_j) * P(C_j)}{P(X)} \tag{7}$$

$$P(X|C_j) = \prod_{i=1}^{n} P(x_i|C_j) \tag{8}$$

$$P(X) = \sum_{j=1}^{m} \left(\prod_{i=1}^{n} (P(x_i|C_j)) * P(C_j) \right) \tag{9}$$

Where: $P(C_j|X)$ is the posterior probability, $P(X_j|C_j)$ is the conditional probability, $P(C_j)$ is the prior probability of class j, X is the feature space of all classes, X_j is the feature space of a class j, $P(X)$ is the evidence, C_j is the class to be analyzed where j $\in \{1,2,3......m\}$, and x_i are the features of X_j where $i \in \{1,2,3......n\}$.

Let $P(x_i|C_j)$ being the probability density expressed as (5) and $P(C_j) = 0.5$, because there are two classes whose probabilities of concurrence are considered equal. The expression (5) uses constants calculated in the offline training for each feature and classes. As in the off-line training, Eqs. (1), (2), (3) and (4) have to be programmed in the microcontroller STM32f103c8t6 in order to obtain the x_i features of each analysis window in time and for each class C_j.

Then, the probability of each class is expressed from 0% to 100% where 0% means that there is no probability that the processed signals belong to a certain class and 100% represents a high probability that the treated data belong to the class. According to the identified class, the microcontroller will send five control signals to the robotic hand in order to make the identified movement. In addition, a minimum acceptance threshold for a class has been stablished, in which, if none of the classes exceeds this threshold, the system will interpret as an undefined class, in which case, the robotic hand keeps the last movement performed. The labels to make the final decision in the identification of classes can be observed in the expression (10).

$$\begin{gathered} P(C_1|X) < P(C_2|X) > \textit{threshold} \rightarrow \textit{closed movement class} \\ P(C_2|X) < P(C_1|X) > \textit{threshold} \rightarrow \textit{open movement class} \\ (P(C_2|X)\&\&P(C_1|X)) < \textit{threshold} \rightarrow \textit{undefined movement class} \end{gathered} \quad (10)$$

4 Tests and Results

Once the off-line training is done and the algorithm is implemented in the embedded system, the necessary tests are carried out to analyze the performance of the classifier. Because data discrimination is multiclass, micro and macro evaluation metrics are used [23, 24], represented by the μ and M indexes respectively. The performance metrics computed are detailed in Table 1.

Table 1. Evaluation of the classifier in the embedded system for the test group. tp – true positive, fp – false positive, fn – false negative, l – number of classes.

Metric	Equation	Evaluation objective
$Accuracy_\mu$	$\frac{tp+tn}{tp+fp+tn+fn}$	Measure the relationship of the correct predictions of a positive class with respect to the classes evaluated
$Precision_\mu$	$\frac{tp}{tp+fp}$	Measure positive patterns that are correctly predicted from the total projected patterns for a positive class
$Recall_\mu$	$\frac{tp}{tp+fn}$	Measure the correct classification of a positive class
$F-Score_\mu$	$2 * \frac{Precision_\mu * Recall_\mu}{Precision_\mu + Recall_\mu}$	Represent the average harmonic relationship between Recall and Precision

(continued)

Table 1. (*continued*)

Metric	Equation	Evaluation objective
$Accuracy_M$	$\frac{\sum_{i=1}^{l} \frac{tp_i + tn_i}{tp_i + tn_i + fp_i + fn_i}}{l}$	Represents the effectiveness average of all classes
$Precision_M$	$\frac{\sum_{i=1}^{l} \frac{tp_i}{tp_i + fp_i}}{l}$	Represents an average of the Precision by class
$Recall_M$	$\frac{\sum_{i=1}^{l} \frac{tp_i}{tp_i + fn_i}}{l}$	Represents an average of the classifier's effectiveness to identify the classes
$F1 - Score_M$	$2 * \frac{Precision_M * Recall_M}{Precision_M + Recall_M}$	Represents an average of the F-Score per class

In addition, a group of four test users will be included for the study of the system performance. In all cases, we implement the acquisition of 25 windows for several samples $W = \{30, 50, 70, 90, 110, 125\}$ for each class. The threshold was chosen heuristically, setting a value of 60%. In order to check its applicability, the users made

Table 2. Evaluation of the classifier in the embedded system for the test group.

Analysis windows (number of samples)	CLASSES	Accuracy	Precision	Recall	F1-Score
30	Open	0,87	0,79	0,92	0,85
	Closed	0,83	0,71	0,88	0,79
	Undefined	0,80	0,85	0,50	0,63
Macro – average of the metric		**0,83**	**0,78**	**0,77**	**0,77**
50	*Open*	0,85	0,73	0,88	0,80
	Closed	0,90	0,87	0,86	0,86
	Undefined	0,86	0,86	0,70	0,77
Macro – average of the metric		**0,87**	**0,82**	**0,81**	**0,82**
70	*Open*	0,86	0,75	0,90	0,82
	Closed	0,87	0,89	0,73	0,80
	Undefined	0,91	0,88	0,86	0,87
Macro – average of the metric		**0,88**	**0,84**	**0,83**	**0,83**
90	*Open*	0,81	0,79	0,64	0,71
	Closed	0,80	0,67	0,82	0,74
	Undefined	0,88	0,86	0,83	0,85
Macro – average of the metric		**0,83**	**0,77**	**0,76**	**0,77**
110	*Open*	0,82	0,86	0,63	0,73
	Closed	0,80	0,73	0,71	0,72
	Undefined	0,79	0,66	0,86	0,74
Macro – average of the metric		**0,80**	**0,75**	**0,73**	**0,74**
125	*Open*	0,66	0,58	0,36	0,44
	Closed	0,67	0,56	0,53	0,55
	Undefined	0,71	0,60	0,86	0,70
Macro – average of the metric		**0,68**	**0,58**	**0,58**	**0,58**

a third movement (resting hand) that has not been considered in the training stage and that will be interpreted as an undefined class.

In Table 2, a comparison of the different time windows is presented based on the statistics: Accuracy, Precision, Recall and F-Score.

In Fig. 5 is presented the comparison of the micro and macro evaluation metrics based on the different time windows analyzed in the embedded system.

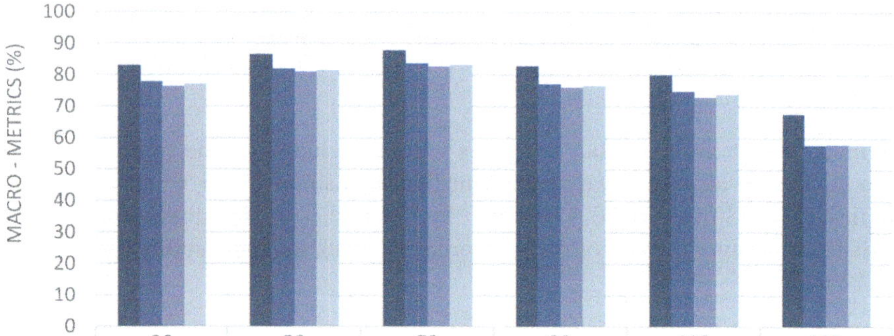

	30	50	70	90	110	125
Accuracy	83.26	86.79	88.02	83.05	80.37	67.81
Precision	78.12	82.21	83.93	77.38	75.04	58.06
Recall	76.67	81.33	83.00	76.33	73.24	58.33
F-Score	77.39	81.77	83.46	76.86	74.14	58.19

SAMPLE WINDOWS

Fig. 5. Micro and Macro – evaluations metrics for each time window analyzed.

The results obtained in each time window show to be consistent in each case, that is, by increasing accuracy, it also increases the F-score, which is adequate due to the balancing of classes (same number of individuals per class). When performing experiments with different time windows, it can be noticed that the best performance of the classifier is obtained handling time windows of 70 samples (Accuracy: 0.88, F-score: 0.83), which, we consider one of the most significant contributions of this work, evidencing that not necessarily the fact of working with time windows of more samples (larger), improves the performance of the classification algorithm (model) e.g. time window of 90 samples (Accuracy: 0.83, F-score: 0.77). It is evident that by increasing the number of samples the performance decreases considerably (see the results for time windows of 125 samples).

Figure 6 shows the acquisition and processing time of the embedded system related with the different time windows used. The acquisition and processing time increases when working with time windows with a greater number of samples, this is evident since it increases the number of data and operations to be calculated by the micro-controller system, however, it can be noticed that working with a time window of 70 samples are handled times that are in the average of all the experiments carried out and that could be improved when working with electronic devices with better processing speed.

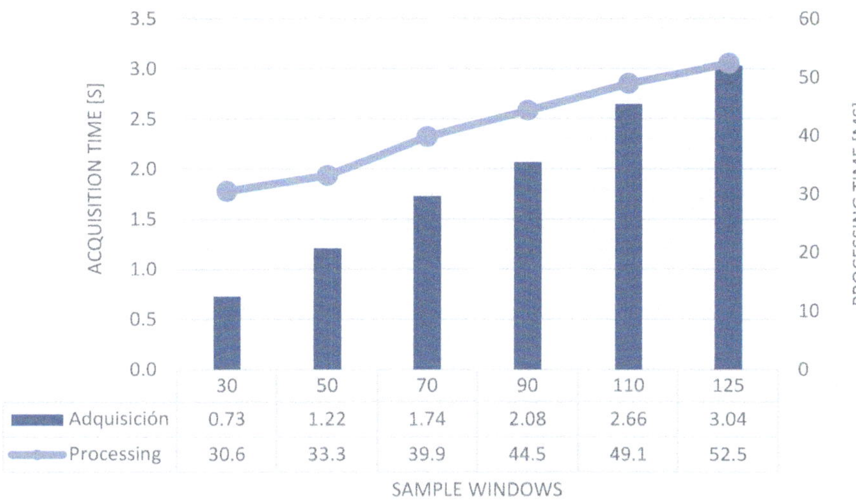

Fig. 6. Average times of acquisition and processing of the embedded system for defined time windows.

5 Conclusions

The developed experimentation allows to conclude that the best electromyography range to obtain good results are with samples from 30 × 8 to 70 × 8. The configuration of windows that are outside this range can produce long response times, too low precision percentages and greater consumption of the microcontroller system resources.

Although no classification algorithm is 100% accurate, the implementation of these in the machine learning area is essential to achieve objectives and results close to success. The use of Naïve Bayes in this study allowed to obtain accuracies from 68% to 83% with processing times of less than 53 ms.

One of the primary properties that is considered in a machine learning algorithm is its behavior against cases or users who are not registered within its configuration. In the experiments carried out, it was proposed to evaluate this, achieving a good adaptability of the system and reaching acceptable prediction percentages that allowed to conclude that the system can be used by any user.

32-bit microcontrollers can be used to develop embedded machine learning systems. This is explained because the algorithms work with a high amount of resources and processing because they work with large blocks or information that arrive constantly and that need to be processed quickly.

References

1. Shroffe, E.H., Manimegalai, P.: Hand gesture recognition based on EMG signals using ANN. In: International Conference on Consumer Electronics, ICCE, Berlin, vol. 2013, no. 3, pp. 174–178. IEEE, Berlin, April, 2013

2. Morales, L., Pozo, D.: An experimental comparative analysis among different classifiers applied to identify hand movements based on sEMG. In: 2nd Ecuador Technical Chapters Meeting 2017, ETCM, vol. 2017, pp. 1–6, IEEE, Ecuador, January 2018
3. Mishra, V.K., Bajaj, V., Kumar, A., Sharma, D., Singh, G.K.: An efficient method for analysis of EMG signals using improved empirical mode decomposition. AEU - Int. J. Electron. Commun., 200–209 (2017)
4. Gruebler, A., Suzuki, K.: Design of a wearable device for reading positive expressions from facial EMG signals. Trans. Affect. Comput. **5**(3), 227–237 (2014)
5. Tsipouras, M.G.: Uterine EMG signals spectral analysis for pre-term birth prediction **8**(5), 3310–3315 (2018)
6. Flood, M.W., Jensen, B.R., Malling, A.S., Lowery, M.M.: Increased EMG intermuscular coherence and reduced signal complexity in Parkinson's disease. Clin. Neurophysiol. **130**(2), 259–269 (2018)
7. Ghassemi, M., et al.: Development of an EMG-controlled serious game for rehabilitation. Trans. Neural Syst. Rehabil. Eng. **PP**(c), 1 (2019)
8. Bastos-Filho, T., Longo, B., Sime, M.: Serious game based on Myo Armband for upper-limb rehabilitation exercises, vol. 70/2, pp. 701–704 (2019)
9. Correa-Figueroa, J.L., Morales-Sánchez, E., Huerta-Ruelas, J.A., González-Barbosa, J.J., Cárdenas-Pérez, C.R.: Sistema de adquisición de señales SEMG para la detección de fatiga muscular. Rev. Mex. Ing. Biomed. **37**(1), 17–27 (2016)
10. Morais, G.D., Neves, L.C., Masiero, A.A., Castro, M.C.F.: Application of Myo Armband system to control a robot interface. In: BIOSTEC, vol. 4, pp. 227–231 (2016)
11. Xu, Y., Yang, C., Liang, P., Zhao, L., Li, Z.: Development of a hybrid motion capture method using MYO Armband with application to teleoperation. In: Conference Mechatronics and Automation, IEEE ICMA 2016, pp. 1179–1184. IEEE (2016)
12. Nymoen, K., Romarheim, M., Alexander, H., Jensenius, R.: MuMYO—evaluating and exploring the MYO Armband for musical interaction. In: New Interfaces Musical Expression, vol. 179 (2015)
13. Haque, F., Nancel, M., Vogel, D.: Myopoint: pointing and clicking using forearm mounted electromyography and inertial motion sensors. In: Proceedings 33rd Annual ACM Conference on Human Factors in Computing Systems - CHI 2015, vol. 1, pp. 3653–3656 (2015)
14. Rawat, S., Vats, S., Kumar, P.: Evaluating and exploring the MYO ARMBAND. In: Proceedings 5th International Conference System Modeling and Advancement in Research Trends, SMART 2016, pp. 115–120 (2016)
15. Sathiyanarayanan, M., Rajan, S.: MYO Armband for physiotherapy healthcare: a case study using gesture recognition application. In: 8th International Conference on Communication Systems Networks, COMSNETS 2016, pp. 1–6 (2016)
16. Shin, H.-S., Ganiev, A., Lee, K.-H.: Design of a virtual robotic arm based on the EMG variation. Adv. Sci. Technol. Lett. **113**, 38–43 (2015)
17. Akhmadeev, K., Rampone, E., Yu, T., Aoustin, Y., Le Carpentier, E.: A testing system for a real-time gesture classification using surface EMG. IFAC-PapersOnLine **50**(1), 11498–11503 (2017)
18. Kutafina, E., Laukamp, D., Bettermann, R., Schroeder, U., Jonas, S.M.: Wearable sensors for eLearning of manual tasks: using forearm EMG in hand hygiene training. Sensors (Switzerland) **16**(8), 1–10 (2016)
19. Lobov, S.A., Krylova, N.P., Anisimova, A.P., Mironov, V.I., Kazantsev, V.B.: Optimizing the speed and accuracy of an EMG interface in practical applications. Hum. Physiol. **45**(2), 145–151 (2019)

20. Mane, S.M., Kambli, R.A., Kazi, F.S., Singh, N.M.: Hand motion recognition from single channel surface EMG using wavelet & artificial neural network. Procedia Comput. Sci. **49** (1), 58–65 (2015)
21. Karimpour, M., Parsaei, H., Sharifian, R., Rojhani, Z., Yazdani, F.: An android application for estimating muscle onset latency using surface EMG signal: In: J. Biomed. Phys. Eng. (2019)
22. Benatti, S., Casamassima, F., Milosevic, B.: A versatile embedded platform for EMG acquisition and gesture recognition, pp. 1–11 (2015)
23. Sokolova, M., Lapalme, G.: A systematic analysis of performance measures for classification tasks. Inf. Process. Manag. **45**(4), 427–437 (2009)
24. Hossin, M., Sulaiman, M.: A review on evaluation metrics for data classification evaluations. Int. J. Data Min. Knowl. Manag. Process **5**(2), 1–11 (2015)

A Novel Technique for Improving the Robustness to Sensor Rotation in Hand Gesture Recognition Using sEMG

Victor H. Vimos[1(✉)], Marco Benalcázar[2], Alex F. Oña[3], and Patricio J. Cruz[4]

[1] Escuela Politécnica Nacional, Quito, Ecuador
victor.vimos@epn.edu.ec
[2] Departamento de Informática y Ciencias de la Computación, Escuela Politécnica Nacional, Quito, Ecuador
marco.benalcazar@epn.edu.ec
[3] Escuela de Formación de Tecnólogos, Escuela Politécnica Nacional, Quito, Ecuador
alex.ona@epn.edu.ec
[4] Departamento de Automatización y Control Industrial, Escuela Politécnica Nacional, Quito, Ecuador
patricio.cruz@epn.edu.ec

Abstract. Hand gesture recognition consists of identifying the class among a set of classes of a hand movement given. Surface electromyography (sEMG) measures the electrical activity generated by voluntary contractions of skeletal muscles. The performance of a recognition system is affected significantly by the orientation of the armband. This orientation could change every time that the user wears the armband. In this paper, a novel technique to improve the robustness in a recognition system with variation in the orientation of the armband is proposed. To test the performance of the proposed model, 4 experiments at recognizing 6 hand gestures are executed. In these experiments the proposed method shows a recognition accuracy of 92.4% versus 59.5%, which corresponds to the accuracy of a traditional recognition model without the correction of orientation.

Keywords: Hand gesture recognition · sEMG · SVM

1 Introduction

Surface electromyography (sEMG) measures the electrical activity generated by voluntary contractions of skeletal muscles. This technique has several uses in biomechanic, robotic and mechatronic systems [1,2]. Gesture recognition consists of identifying the class among a set of classes of a hand movement given [11]. Hand gestures recognition system using sEMG can be used to control mechatronic devices [3–5]. The amplitude and frequency content of sEMGs are affected by different factors including: skin thickness, muscle strength, muscle volume, physiological interference, external noise and electrodes placed incorrectly.

J. Nummenmaa et al. (Eds.): CSEI 2019, AISC 1078, pp. 226–243, 2020.
https://doi.org/10.1007/978-3-030-33614-1_16

Hand gestures recognition systems work well when the sensors are placed exactly in the same position that were used to acquire signal for training [6,7]. However, placing a sensor in exactly the same position is difficult because the physical characteristics of people's arms are different.

There are different devices for sensing sEMG such as Myo armband [12], gForce [10], DTing [13] and eCon [14]. These sensors allow to easily obtain electromyography data, in addition to the advantage of their portability. These sensors are easy to place making it practical to implement several application in any user. The Myo armband manufactured by Thalmic Labs is an EMG sensor that allows the registration of 8 bipolar signals on a person's forearm. The manufacturer of the Myo armband suggests to place the armband in a certain position on the forearm to have good performance (see Fig. 1). However, placing the armband in the same position implies to know exactly the coordinates of the sensor with respect to a given point of reference in the forearm and at the test time placing the sensor in exactly the same coordinates. This requires measuring distances accurately for every single time that the sensor is used which is very difficult for practical applications.

There are two types of gesture recognition systems: general and user specific. General recognition systems are trained with a finite dataset acquired from a group of people and tested by any user. On the other hand, user specific systems require to be trained and tested with the data from the same person each time that the system is used. The use of general systems implies that the armband must be placed in the same position both for training and testing which is difficult to achieve in practice [8]. User specific models do not require placing the sensor in the same position because these systems are trained for each user and for each time they are used. However, training a model for each user and for each time it is used is time consuming, making thus difficult their use for practical applications. In practice, a user simply wants to wear the sensor and start using the recognition system right away. Therefore the best option for practical applications is the use of general recognition models, which should have a system that compensates the variation in the orientation of the sensor for the recognition systems to work well.

In [9] an algorithm is proposed to compensate the variation in sensor rotation. The sensor rotation decrements the performance of the recognition model and sometimes even makes inapplicable the use of the recognition model built in one position. In this work, the armband is rotated every 45° and the data is recorded with that rotation. A remapping is performed according to the predicted angle and the distribution is marked on the user's arm prior to the signal recording. In addition to the high complexity of the proposed algorithm, the correction of the orientation can be done in steps of 45°.

In [11] a general model is proposed to classify 40 gestures in real time. The proposed model works in both the right and the left hand and use the Myo armband for data acquisition and a support vector machine for the classification. The paper shows the real time classification of the gestures made. To compare the results obtained with the Myo armband's own recognition system, users wear strictly the armband in the position recommended by the manufacturer.

Unfortunately, the authors do not give any further details about the results when the armband is placed in different positions. However, authors show the possibility to classify the gestures independent of the arm on which the armband is worn.

In another hand gesture recognition system such as the one proposed by Weissmann and Salomon [15] provide good recognition results up to 100% recognition rate, but the need to wear a glove can restrict the user's freedom of movements and it is not for practical applications.

To solve the problem of the variation in the orientation of the armband, a novel method is proposed in this paper. This method is based on the maximum amplitude detection (MAD) which identifies the sensor with the maximum activity in the sEMG and based on this detection related to a sensor, the data is rearranged by creating a new matrix with the reordered data. The maximum amplitude sensor is calculated using the movement *wave out* in a calibration process that is executed for every time that a person wants to use the recognition system. The recognition model used to test the algorithm for correcting the orientation is based on common features (mean value, windowing, energy, curve envelope, standard deviation) and a SVM classifier.

Following this introduction, the remaining of this paper is organized as follows. The proposed material and model section (Sect. 2) describes the materials used for data collection, how each data matrix is handled and how the new matrix is organized. The experiment section (Sect. 3) describes the 4 experiments with training and testing data. The results and analysis section (Sect. 4) shows a comparison between the traditional method and the proposed method.

2 Materials and Proposed Model

2.1 MYO Armband

The Myo armband is an electronic device that measures sEMGs. This armband consists of 8 bipolar channels which work with a sampling frequency of 200 Hz. Data are transmitted via Bluetooth to a personal computer. The measured data matrix consists of *8* columns and n rows, the rows depend on the recording time of the sEMG. For 1 s the number of rows is 200. Each column of the data matrix represents the measurements of each sensor.

The manufacturer of the armband suggests to place the armband on a specific position (see Fig. 1) on the forearm for achieving good recognition accuracy. The Myo armband brings a proprietary recognition system whose performance is sensitive to the variations of the recommended position. The Myo armband was tested by rotating it and the recognition data showed that the system has difficult to recognize gestures when the armband is rotated from its suggested position.

2.2 Datasets

The dataset is organized as follows:

1) *Training data*

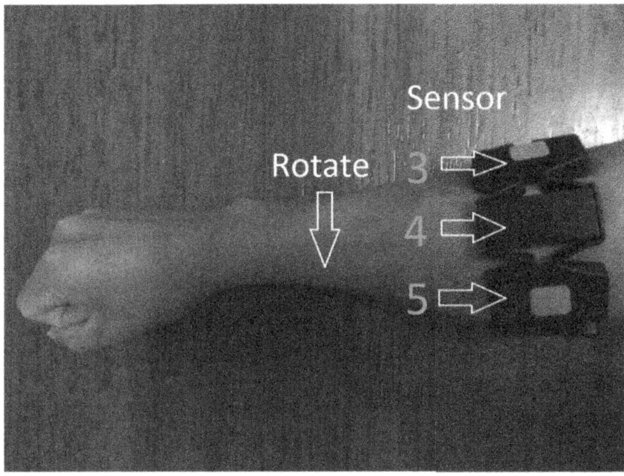

Fig. 1. Myo armband base position suggested by manufacturer.

2) *Testing* $data_1$ (for experiment 1 and 3)
3) *Testing* $data_2$ (for experiment 2 and 4)

The *training data* was recorded using the same armband position for all users (position suggested by manufacturer). In order to compare the performance of the traditional system with the proposed one, the same training dataset for both systems are used. The Eq. (1) shows in general how is composed the matrix of the training data.

Training data from 40 people is used, 25 men and 15 women. The training data consists of 15 repetitions per gesture for each category and for each user. Each training data matrix has eight columns and their values are normalized. Each column has the sEMG data measured by each sensor.

$$\begin{gathered} Dtr_{general} = [(V_1, D_1), \ldots, (V_i, D_i)] \\ Dtr_{general} \in R^{1000x8}, V_i \in [-1, 1]_{1000x1}, D_i \in [1, 8] \end{gathered} \tag{1}$$

The categorical variable is represented by $Y \in \{out, in, close, thumb, relax, tap\}$ and denotes the label for the gesture signal. The total training data per user consists of 90 rows.

The *testing data* consists of two datasets, test $data_1$ and test $data_2$. The test $data_1$ was recorded using the position suggested by Myo armband manufacturer (see Fig. 1) and test $data_2$ was recorded placing the armband in different positions (see Fig. 2). For test $data_2$, people took the armband off and they put the armband back on the forearm in any position they wanted and no specific angle was rotated. Each recording was made during 5 s per gesture and user. The Eq. (2) shows in general how is composed the testing data matrix.

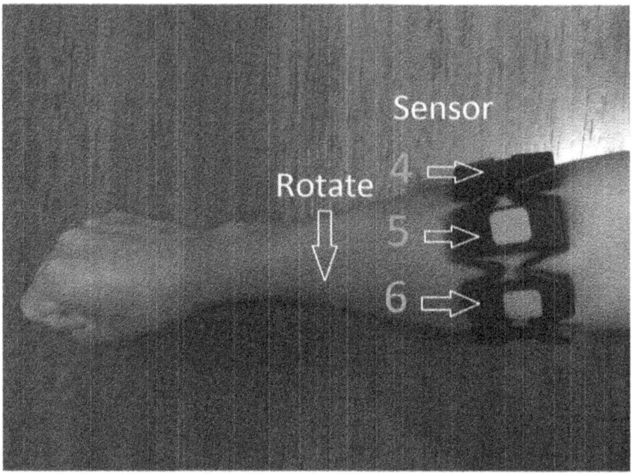

Fig. 2. Myo armband rotated from the base position suggested by manufacturer.

$$\begin{aligned} Dts_{general} &= [(W_1, E_1), \ldots, (W_i, E_i)] \\ Dts_{general} &\in R^{1000x8}, W_i \in [-1, 1]_{1000x1}, E_i \in [1, 8] \end{aligned} \quad (2)$$

2.3 Traditional Method

Traditional gesture recognition systems using the Myo armband need to be trained before they are used [7]. Commonly this methodology works well; however, after people take the armband off they must train the system again if they want to use it with good accuracy. The gestures performed and recorded during a session are shown in Fig. 3.

To process the data, a matrix organized per sensor, user and categories is created (Eq. 3). Ms_i is the transposed matrix with 15 repetitions for each gesture. Ms_i is the total training matrix for user i and has a dimension of 90 rows and 1000 columns. The 90 rows is the result of 15 repetitions multiplied by 6 gestures. Data training matrix for user i is described as follows:

User i:

$$Emg(user_i, category_j) = [Ms_1, Ms_2, Ms_3, Ms_4, Ms_5, Ms_6, Ms_7, Ms_8, Y] \quad (3)$$

$$Ms_i \in R^{15x1000} Y \in \{out, in, close, thumb, relax, tap\}$$

$$\begin{aligned} Dtrain_{user_i} = [&Emg(user_i, out); \\ &Emg(user_i, in); \\ &Emg(user_i, close); \\ &Emg(user_i, thumb); \\ &Emg(user_i, relax); \\ &Emg(user_i, tap)] \end{aligned}$$

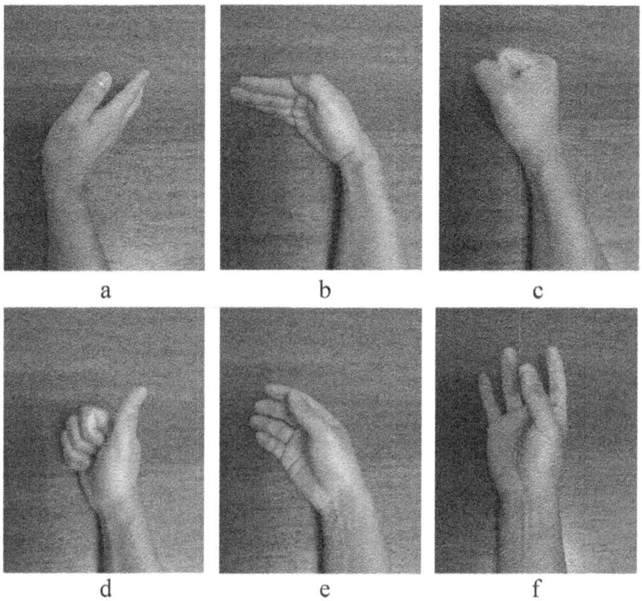

Fig. 3. Gestures to be recognized with both traditional and proposed method. (a) wave out, (b) wave in, (c) close, (d) thumb, (e) relax, (f) tap

Notice that the data has been transposed to be handled and organized according to each gesture and sensor. A table was created by gesture and sensor with the data transposed to work with Matlab. The total matrix for the 40 users is shown below and notice that for each user the data is concatenated.

Total training data for 40 users:

$$Dtrain_{total} = [Ms_1, Ms_2, Ms_3, Ms_4, Ms_5, Ms_6, Ms_7, Ms_8, Y]$$

where:

$$Ms_i \in R^{3600x1000} \text{ and } Y \in \{out, in, close, thumb, relax, tap\}$$

Each data matrix coming from the armband in a time t always come in the same sequence even if the armband is located in a different position from the position suggested by the manufacturer.

The EMG data sequence coming from the armband by default is organized in the following order $Emg_{default}(t) = [s_1(t), s_2(t), s_3(t), s_4(t), s_5(t), s_6(t), s_7(t), s_8(t)]$ where s_1 represents the sensor number 1. When the orientation of the armband is changed, the data order is the same even though the armband was rotated. For $user_1$ the signals were recorded with sensor number *2* matching with the position that was defined as base (see Fig. 1), getting a default matrix order $Emg_{default}$.

Regarding the testing $data_1$ the process to organize the data was followed as the previous one and a total matrix was defined for 40 users too, taking into account that these test data were recorded taking as reference the position suggested by the manufacturer.

Total testing matrix $data_1$ for 40 users:

$$Dtest_1 = [Ms_1, Ms_2, Ms_3, Ms_4, Ms_5, Ms_6, Ms_7, Ms_8, Y]$$

where:

$$Ms_i \in R^{3600x1000} \text{ and } Y \in \{out, in, close, thumb, relax, tap\}$$

Regarding the testing $data_2$, the test $data_2$ matrix represents the set of recordings with different rotations of the armband for each user. It is worth mentioning that the test $data_2$ has different sensors taken as a reference and their distribution activity is not equal. In Fig. 5 the different gestures activity distribution for four users is shown.

Total testing matrix $data_2$ for 40 users:

$$Dtest_2 = [Ms_1, Ms_2, Ms_3, Ms_4, Ms_5, Ms_6, Ms_7, Ms_8, Y]$$

where:

$$Ms_i \in R^{3600x1000} \text{ and } Y \in \{out, in, close, thumb, relax, tap\}$$

For practical reasons the number of columns to work with were reduced from 1000 to 900, because when recording the signals not always the same amount of data was gotten. To avoid inconveniences when concatenating the data only 900 points were taken. For *training data* as well as test $data_1$ and test $data_2$ the same extractors were applied. A SVM $classifier_1$ with original *training data* was trained and tested.

2.4 Proposed Method

The proposed method is based on the maximum amplitude detection (MAD). After that, the data matrix is rearranged according to the sensor with the highest mean amplitude detected. The sensor with highest amplitude is identified using the movement *"wave out"*, this movement allows the maxim values data be concentrated mainly in one sensor S_x which is taken as reference for the new order.

$$Emg = [V_1, V_2, V_3, V_4, V_5, V_6, V_7, V_8], Emg \in R^{200x8} \text{ and } Vi \in [-1, 1]_{200x1},$$

$$Emg_{mean} = mean(Emg) \tag{4}$$

$$S_x = max(Emg_{mean}) \tag{5}$$

$$S_x = max([V_{1mean}, V_{2mean}, V_{3mean}, V_{4mean}, V_{5mean}, V_{6mean}, V_{7mean}, V_{8mean}]) \tag{6}$$

where *max* function represents the maximum value of the vector. After the sensor is identified the new sEMG matrix is organized and described according to the Eq. (7):

$$Emg_{new} = [S_x, S_{mod((x+1),8)}, S_{mod((x+2),8)}, ..., S_{mod((x+7),8)}, S_{mod((x+8),8)}] \quad (7)$$

For $user_{20}$ the MAD sensor is located in the sensor number *6 (s6)*. According to the proposed method the new matrix is organized as follows:

$$EMG_{new} = [s6, s7, s8, s1, s2, s3, s4, s5]$$

For $user_{30}$ the MAD sensor is located in sensor number *5 (s5)*. According to the proposed method the new matrix is organized as follows:

$$EMG_{new} = [s5, s6, s7, s8, s1, s2, s3, s4]$$

Applying the MAD algorithm for the original *training data*, test $data_1$ and test $data_2$ new matrices labeled as *training data**, test $data_1^*$ and test $data_2^*$ were gotten. It should be noticed that the new training matrix is organized according to the maximum amplitude sensor and does not imply that as a result of applying MAD algorithm the same reference sensor must be gotten for all recordings. However, the result of the sensor detection should give similar sensors like the original one obtained when the data were recorded using the position suggested by the manufacturer.

In Table 1 the result of applying the maximum amplitude detection algorithm in the original data for training and testing is shown. This method allows to have greater robustness to rotation as well as greater independence in the placement of the armband, also this allows to have higher performance and avoid the necessity to record the signals every time the systems is going to be used. Table 1 shows the reference electrode calculated for test $data_1^*$ and test $data_2^*$ using MAD sensor activity.

In Fig. 5. the data for four users whose data have been recorded using different orientation of the armband is showed. The distribution of the EMG activity is different and not concentrated in the same region although all the recordings correspond to the same gesture labeled with different colors respectively. The Fig. 5 shows each group of data separately according to the gesture performed for each user. For all users the *wave out* gesture is represented in dark blue. For $user_{17}$, the concentration of the highest sEMG activity is detected over sensors 1, 2 and 3. For $user_{18}$, the greatest concentration of activity during the *wave out* gesture is located in sensors 4, 5 and 6. For $user_{19}$, the greatest EMG activity is concentrated on sensors 4, 5 and 6. Similarly for $user_{20}$, the greatest activity is detected over sensors 5, 6, 7 and 8. It can be appreciated that the concentration of activity for the same gesture is different for each user and this is logical since

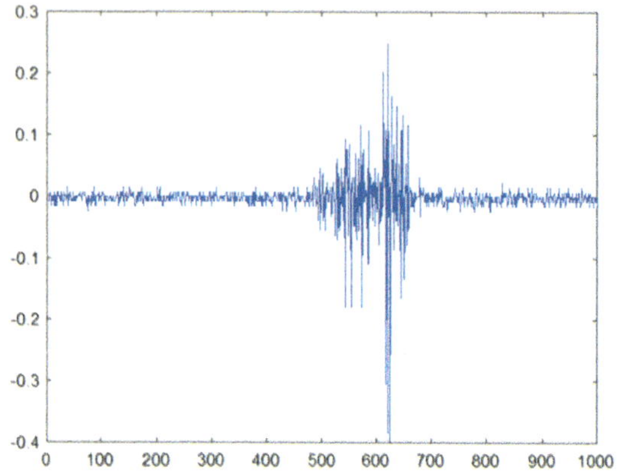

Fig. 4. sEMG for *user3* on sensor 7 s_7 while *wave out* gesture was performed.

each user placed the armband arbitrarily. For other hand gestures, for example for the *wave in* gesture labeled in orange the concentration of activity by sensor is not homogeneous in the same way.

The Fig. 6 shows the same data for the four previous users whose data have been recorded using different positions of the armband. However, to this data the orientation correction using the MAD algorithm was applied.

After applying the MAD algorithm the activity distribution is similar and concentrated in the same region. The recordings correspond to the same gesture labeled with different colors respectively for the four users using the armband placed in different positions.

It can be verified that using the MAD algorithm the data of the users 17, 18, 19 and 20 have been aligned and now these data could be used in any classifier, improving the accuracy because of the new data organization.

After making this correction in the orientation, the data entered into a new $classifier_2$ always have the same order, regardless the position where the user use the armband. It is not necessary to rotate a specific angle to be able to perform the compensation for the rotation. The proposed method always searches for the sensor with the highest activity.

The Fig. 6 shows how the data is automatically aligned, since it takes as reference a sensor that has been calibrated during the beginning of the test session. A summary of the reference electrodes calculated by MAD algorithm applied to the original data is shown in the column 2, 3 and 4 of the Table 1.

Table 1. Reference electrodes calculated by MAD algorithm.

User	$Train*$	$Test_1^*$	$Test_2^*$
$user_1$	2	2	1
$user_2$	3	2	4
$user_3$	7	7	1
$user_4$	2	2	1
$user_5$	2	2	4
$user_6$	2	2	4
$user_7$	2	2	5
$user_8$	8	8	4
$user_9$	2	2	4
$user_{10}$	1	1	1
$user_{11}$	1	1	1
$user_{12}$	1	1	4
$user_{13}$	1	1	1
$user_{14}$	1	1	1
$user_{15}$	2	2	2
$user_{16}$	1	1	3
$user_{17}$	2	2	2
$user_{18}$	2	1	4
$user_{19}$	2	2	5
$user_{20}$	1	1	6
$user_{21}$	2	2	5
$user_{22}$	1	1	7
$user_{23}$	2	2	6
$user_{24}$	1	1	7
$user_{25}$	2	2	1
$user_{26}$	1	1	1
$user_{27}$	1	1	2
$user_{28}$	2	2	5
$user_{29}$	1	1	1
$user_{30}$	1	1	5
$user_{31}$	1	7	7
$user_{32}$	2	2	6
$user_{33}$	3	3	6
$user_{34}$	7	2	2
$user_{35}$	1	1	6
$user_{36}$	2	2	8
$user_{37}$	1	2	2
$user_{38}$	2	2	8
$user_{39}$	2	2	2
$user_{40}$	2	2	2

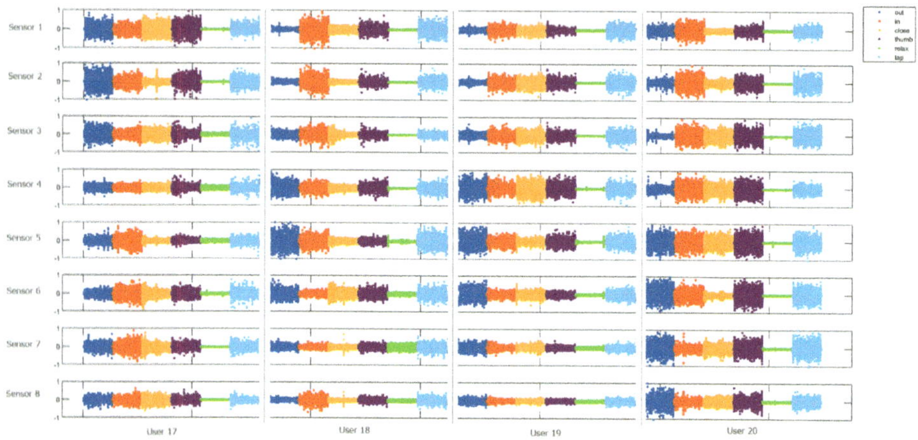

Fig. 5. Normal testing $data_2$ distribution activity for 4 users. All gesture activity is concentrated in different sensor when recording data in different positions

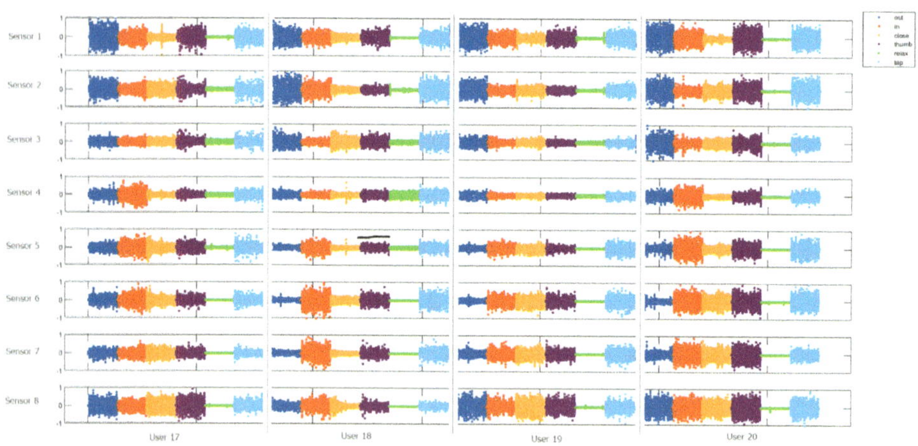

Fig. 6. Testing $data_2^*$ distribution activity with correction to the armband rotation for 4 users. All gesture activity is concentrated in the same sensors (sensors 1, 2, 3) when recording data in different positions

2.5 Features Extractors

sEMG curve envelope, windowing, sEMG energy, mean absolute value and standard deviation are used for both methods as features extractors. Where features extractors are defined as follows:

Mean absolute value:

$$\mid \mu \mid = \frac{1}{N} \sum_{i=1}^{N} \mid V_i \mid$$
$$V_i \in [-1, 1]_{Nx1} \tag{8}$$

where, N denotes the number of points recorded per channel. Being $N = 1000$ points during 5 s.

Standard deviation:

$$S = \sqrt{\frac{1}{N-1}\sum_{i=1}^{N} \mid V_i - \mu \mid^2} \qquad (9)$$
$$V_i \in [-1, 1]_{Nx1}$$

The same features extractors are applied to both methods with the exception that in the proposed method the data is organized differently. The data matrix with the characteristics used to train $classifier_1$ and $classifier_2$ is described below.

$$Emg_{(user_i, feature)} = [Ms_1, Ms_2, Ms_3, Ms_4, Ms_5, Ms_6, Ms_7, Ms_8]$$

where $Ms_i \in R^{15x1000}$ and $feature \in \{std, envelope, welch, absmean, energy\}$

$$Emg_{(user_i, std)} = [Ms_1, Ms_2, Ms_3, Ms_4, Ms_5, Ms_6, Ms_7, Ms_8] \qquad (10)$$

$$Emg_{(user_i, envelope)} = [Ms_1, Ms_2, Ms_3, Ms_4, Ms_5, Ms_6, Ms_7, Ms_8] \qquad (11)$$

$$Emg_{(user_i, welch)} = [Ms_1, Ms_2, Ms_3, Ms_4, Ms_5, Ms_6, Ms_7, Ms_8] \qquad (12)$$

$$Emg_{(user_i, absmean)} = [Ms_1, Ms_2, Ms_3, Ms_4, Ms_5, Ms_6, Ms_7, Ms_8] \qquad (13)$$

$$Emg_{(user_i, energy)} = [Ms_1, Ms_2, Ms_3, Ms_4, Ms_5, Ms_6, Ms_7, Ms_8] \qquad (14)$$

$$Features_{user_i} = [Emg_{(user_i, std)}, Emg_{(user_i, envelope)}, Emg_{(user_i, welch)}, Emg_{(user_i, absmean)}, Emg_{(user_i, energy)}]$$

where $Emg_{(user_i, feature)} \in R^{15x8}$

$$Matrix_{(user_i, category_j)} = [Features_{user_i}, Y] \qquad (15)$$

where $Features_{user_i} \in R^{15x40}$ and $Y \in \{out, in, close, thumb, relax, tap\}$

The total matrix for training is as follows:

$$TrainMatrix_{total} = [Matrix_{(user_1, category_j)}; ...; Matrix_{(user_{40}, category_j)}] \qquad (16)$$

where the size $TrainMatrix_{total} \in R^{3600x40}$

The total matrix for test $data_1$ and test $data_2$ is as follows:

$$TestMatrix_{total} = [Matrix_{(user_1)}; ...; Matrix_{(user_{40})}] \qquad (17)$$

3 Experiments

In this section four experiments have been carried out and two SVM classifiers have been designed in order to check the operation of the proposed system in situations of rotation of the armband. Two SVM classifiers were trained using data for the traditional method and data for the proposed method. $Experiment_1$ and $experiment_2$ are analyzed using the SVM $classifier_1$ that has been trained using the *training data* recorded using the position suggested by the manufacturer. $Experiment_3$ and $experiment_4$ are analyzed using the SVM $classifier_2$. The MAD algorithm has been applied to *training* and *testing data* to reorder accordingly the new reference electrode. It should be noticed that in the Table 1 the *training data** and $test_1^*$ reference electrodes are calculated using the highest potential sensor and do not differ greatly with the position suggested by the manufacturer.

In Table 1 the *training data** as well as the test $data_1^*$ have approximately the same reference sensor after apply the proposed method. These reference sensors indirectly show how the armband was placed by the user. Comparing the two columns it is clear that the data are similar. For column 3, in test $data_2^*$ the algorithm was also applied and the result obtained for the reference sensor is different.

In the Table 1 there are 3 users to have in consideration: users 3, 8, 34. These reference sensors are different compared with the other users, however this is due to the fact that when users made the gesture *wave out* they unwittingly made a strong movement when returning to the relaxation position. This particular situation can be seen in Fig. 4, where the sEMG *wave out* gesture for $user_3$ on sensor number 7 is performed and the reason why MAD algorithm selected sensor number 7 as new reference in Table 1 is showed. The algorithm confirms that the armband was in different positions, but in the same way when confirming the different positions of the armband, the algorithm have changed the order of the sensors.

3.1 Experiment 1

The $experiment_1$ includes training the $classifier_1$ with normal *training data*, then testing the $classifier_1$ with test $data_1$. Both *training* and test $data_1$ are recorded using the recommendations of the armband manufacturer. Users are from 20 to 55 years old. In this experiment, 15 repetitions are performed for each gesture. Users are not asked to calibrate the system, simply they place the armband according to the suggested position.

3.2 Experiment 2

The $experiment_2$ includes training the $classifier_1$ with *training data* and testing the $classifier_1$ with test $data_2$ (armband placed in different positions). In this experiment, the user is previously asked to take the armband off. After this, the user is asked to place the armband in the desired position. In the same way each user is asked to repeat each gesture 15 times. In this experiment the $classifier_1$ is tested with the data recorded in different positions.

3.3 Experiment 3

The $experiment_3$ includes training the $classifier_2$ with *training data** organized according the proposed method and testing the $classifier_2$ with test $data_1^*$ organized according the proposed method too. In this experiment, the MAD algorithm is used to correct the position. The reference sensor for the *training data** is obtained even though these data were recorded using the same position. This data is shown in the column 2 and 3 in the Table 1.

3.4 Experiment 4

The $experiment_4$ includes training the $classifier_2$ with *training data** organized according the proposed method and testing the $classifier_2$ with test $data_2^*$ (armband placed in different positions) organized according to the proposed method. All data for the $experiment_3$ and $experiment_4$ are the same like the $experiment_1$ and $experiment_2$ only with the difference that for $experiment_3$ and $experiment_4$ the rotation correction of the armband has been applied. The correction in the rotation can be seen in Figs. 5 and 6 applied for four users as example.

4 Results, Analysis and Comparisons

The confusion matrix for $experiment_1$ is showed in Fig. 7.

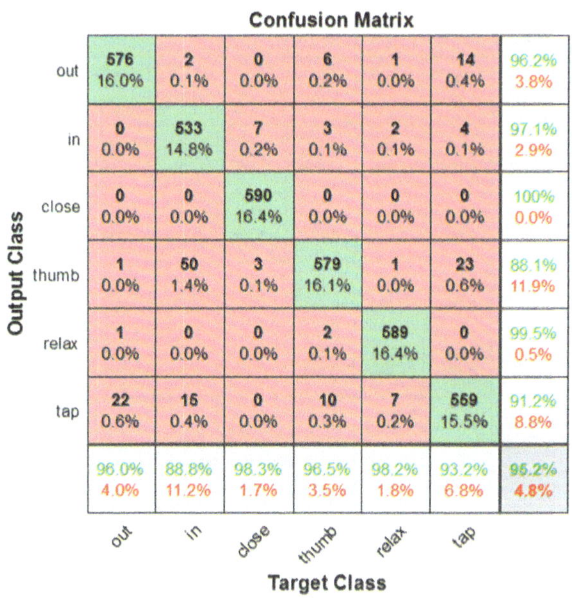

Fig. 7. Confusion matrix with test $data_1$ ($experiment_1$)

The confusion matrix for $experiment_2$ is showed in Fig. 8. Two SVM classifiers were trained and tested using two separate procedures. $Training$ and $testing$ dataset in default order for SVM $Classifier_1$, $training^*$ and $testing^*$ dataset using the proposed method for SVM $Classifier_2$. In Fig. 9 the confusion matrix for $experiment_3$ is showed. Comparing test $data_1$ working with the traditional method versus this novel method, the system accuracy decreases from 95.2% to 93.9% using test $data_1^*$ as input. However, this result is because of the references calculated by MAD algorithm for users 3, 8, 34 are different from the others.

For $experiment_4$ (armband rotated) the confusion matrix using the novel method for test $data_2^*$ is showed in Fig. 10. Comparing test $data_2$ working with the traditional method versus this novel method the system accuracy increases from 59.5% to 92.4% using testing $data_2^*$ as input. With this novel method the recognition system can be used even by new people with great effectiveness and accuracy.

The accuracy system decreases 35.7% when a user uses the armband in different position working with the traditional method. The Fig. 5 shows how is the data distribution for users 17, 18, 19 and 20 when this data is going to be apply for traditional method. Data in dark blue color is related to the movement *wave out*. There are different distributions for all users and the signal power concentration is not located in the same sensor due to users placed the armband in different positions.

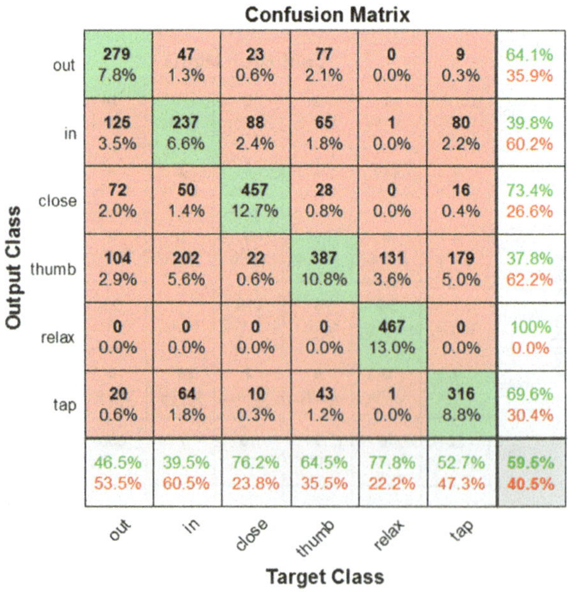

Fig. 8. Confusion matrix with test $data_2$ ($experiment_2$)

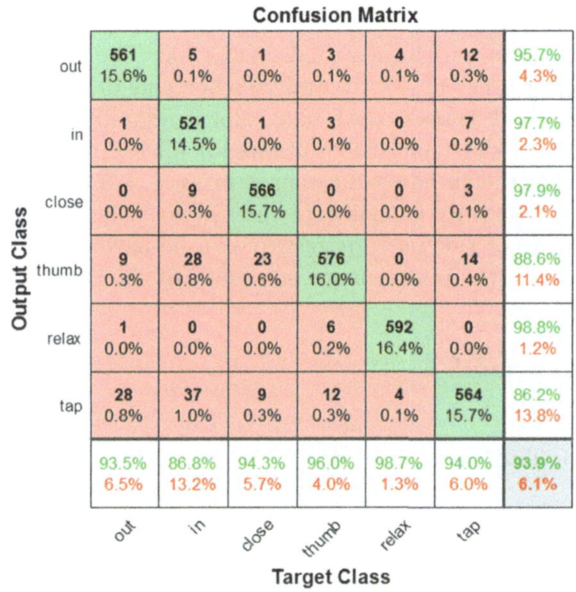

Fig. 9. Confusion matrix with novel method, test $data_1^*$ ($experiment_3$)

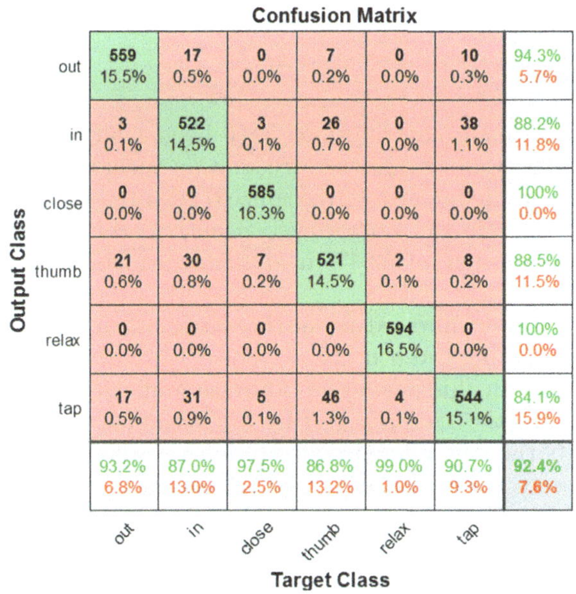

Fig. 10. Confusion matrix with novel method, test $data_2^*$ ($experiment_4$)

The accuracy system decreases only 1.5% when a user uses the armband in different position working with the proposed method. The Fig. 6 shows how is the data distribution for users 17, 18, 19 and 20 when this data is going to be apply using the novel method. Data in dark blue color is related to the movement *wave out*, there are almost the same distributions for all users and the signal power concentration is located in the same sensor even if user place the armband in different positions. Actually to calibrate the system using the MAD algorithm, the *wave out* gesture or the *wave in* gesture can be used.

The Table 2 shows the performance summary of the two systems.

Table 2. Accuracy systems comparison.

Method	$Experiment_1$	$Experiment_2$	$Experiment_3$	$Experiment_4$
Traditional	95.20%	59.5%	–	–
Proposed	–	–	93.9%	92.4%

For consideration and experimentation by anyone interested in the proposed method, the code as well as the dataset of the paper can be found in the following link: https://drive.google.com/drive/folders/1bvWbh-16c4ShFQDP3Q6a8hwBu6UaAW4y.

5 Conclusion

In this paper, three main contributions have been made. The main contributions of this novel method for gesture recognition include, (1) robustness with placement sensors on the forearm to recognize 6 gestures with high accuracy, (2) the low necessity to train the system every time, (3) the recognition algorithm responds with an accuracy of 92.4% in different armband positions using the novel technique.

The system can be calibrated using the *wave out* or *wave in* gesture. The 1.5% decrease can be improved if the system is calibrated at the beginning of the data acquisition. In this paper the calibration at starting the acquisition is not performed. Using the *wave out* gesture to reorganize the data matrix is how the algorithm gets the new reference electrode. Any classifier can be used after the orientation correction. Similarly it is not necessary to use several features over the EMG signals, only 5 features were used in order to have good accuracy.

Future works will include the research for more than 20 hand gestures recognition and the implementation of the system that allows to obtain a response in less than 100 ms with great accuracy. The system response should be the same when the armband is placed on any forearm (right or left). The system will also be tested using an embedded system to make it more portable.

Acknowledgment. The authors gratefully acknowledge the financial support provided by the Escuela Politécnica Nacional and the Corporación Ecuatoriana para el Desarrollo de la Investigación y la Academia (CEDIA) for the development of the research project CEPRA-2019-13-Reconocimiento de Gestos.

References

1. Artemiadis, P.K., Kyriakopoulos, K.J.: EMG-based position and force control of a robot arm: application to teleoperation and orthosis. In: 2007 IEEE/ASME International Conference on Advanced Intelligent Mechatronics, pp. 1–6. IEEE (2007)
2. Tan, D., Saponas, T.S., Morris, D., Turner, J.: Wearable electromyography-based controllers for human-computer interface. U.S. Patent 8,170,656, issued 1 May 2012
3. Muceli, S., Farina, D.: Simultaneous and proportional estimation of hand kinematics from EMG during mirrored movements at multiple degrees-of-freedom. IEEE Trans. Neural Syst. Rehabil. Eng. **20**(3), 371–378 (2011)
4. Edwards, A.L., Dawson, M.R., Hebert, J.S., Sherstan, C., Sutton, R.S., Chan, K.M., Pilarski, P.M.: Application of real-time machine learning to myoelectric prosthesis control: a case series in adaptive switching. Prosthet. Orthot. Int. **40**(5), 573–581 (2016)
5. Tenore, F.V.G., Ramos, A., Fahmy, A., Acharya, S., Etienne-Cummings, R., Thakor, N.V.: Decoding of individuated finger movements using surface electromyography. IEEE Trans. Biomed. Eng. **56**(5), 1427–1434 (2008)
6. Nazemi, A., Maleki, A.: Artificial neural network classifier in comparison with LDA and LS-SVM classifiers to recognize 52 hand postures and movements. In: 2014 4th International Conference on Computer and Knowledge Engineering (ICCKE), pp. 18–22. IEEE (2014)
7. Benalcázar, M.E., Motoche, C., Zea, J.A., Jaramillo, A.G., Anchundia, C.E., Zambrano, P., Segura, M., Palacios, F.B., Pérez, M.: Real-time hand gesture recognition using the myo armband and muscle activity detection. In: 2017 IEEE Second Ecuador Technical Chapters Meeting (ETCM), pp. 1–6. IEEE (2017)
8. Kim, J., Mastnik, S., André, E.: EMG-based hand gesture recognition for real-time biosignal interfacing. In: Proceedings of the 13th International Conference on Intelligent User Interfaces, pp. 30–39. ACM (2008)
9. Zhang, Y., et al.: Wearing-independent hand gesture recognition method based on EMG armband. Pers. Ubiquit. Comput. **22**(3), 511–524 (2018)
10. gForce arm band. http://www.oymotion.com/
11. Kerber, F., Puhl, M., Krüger, A.: User-independent real-time hand gesture recognition based on surface electromyography. In: Proceedings of the 19th International Conference on Human-Computer Interaction with Mobile Devices and Services, p. 36. ACM (2017)
12. http://www.myo.com . Accessed 17 Oct 2018
13. DTing arm band. http://www.dtingsmart.com/
14. eCon arm band. http://econtek.cn/
15. Weissmann, J., Salomon, R.: Gesture recognition for virtual reality applications using data gloves and neural networks. In: Proceedings IJCNN, vol. 3, pp. 2043–2046 (1999)

Industrial Engineering

Optimization of Motorcycle Assembly Processes Based on Lean Manufacturing Tools

Jonnathan Quezada[1], Lorena Siguenza-Guzman[2], and Juan Llivisaca[1(✉)]

[1] Faculty of Chemical Sciences, Universidad de Cuenca, Cuenca, Ecuador
{jonnathand.quezadac, juan.llivisaca}@ucuenca.edu.ec
[2] Department of Computer Sciences, Faculty of Engineering, Universidad de Cuenca, Cuenca, Ecuador
lorena.siguenza@ucuenca.edu.ec

Abstract. Global competition and economic dynamism force the assembly industry to look for productive philosophies of improvement, optimizing processes, and resources. The challenge in the optimization process involves finding tools that will grant an optimal process. This article proposes a process optimization through the application of Lean Manufacturing tools, with the necessary guidelines analyzed and verified by computational simulation in a motorcycle assembler. Firstly, applying a plant distribution, together with the change of productive system, allowed the reduction of distances, elimination of unnecessary movements, and reprocessing. Then, Pull System reduced the stock of both the product in process and the final product in the warehouse, avoiding the generation of future waste and expenses due to overproduction. Finally, the use of the 5S tool entailed maintaining order and cleanliness inside the plant to facilitate the management of internal resources. In the study reported, an average reduction of 40% in assembly time and an average increase in production of 30% of all motorcycle models that were part of the study are achieved.

Keywords: Optimization process · Lean Manufacturing · Assembly industries

1 Introduction

The common denominator of the assembly industries is to work through a rigid capital-intensive production system; these industries waste about 70% of resources on raw materials, labor, and facilities, unnecessary movements, and downtimes that carry excessively high costs [1]. In addition, only 5% of the activities carried out by organizations add real value to the product. These facts, together with constant changes and high competitiveness, have forced organizations to look up production methods and philosophies that generate higher productivity and efficiency, thus managing to respond to market demands [2–4].

The philosophy of Lean Manufacturing (LM), addresses the different wastes related to overproduction, waiting, inventory, transportation, defects, waste of processes, unnecessary movements and underutilization of employees' capacity [5]. For this, LM uses various tools, such as Pull System, 5S, Single Minute Exchange of Die (SMED),

J. Nummenmaa et al. (Eds.): CSEI 2019, AISC 1078, pp. 247–259, 2020.
https://doi.org/10.1007/978-3-030-33614-1_17

Kanban, Kaizen, Heijunka, Jidoka, and Value Stream Mapping (VSM) [6]. LM is used by organizations that want to increase their market share, achieving positive results, and reducing resources since they provide small and frequent improvements. This allows companies that adopt this philosophy to increase their competitiveness, positioning, and profitability in their products and services [7, 8]. Socconini expresses that the real power of LM lies in continually discovering those opportunities for improvement that are hidden because there will always be waste that can be eliminated [9].

In the literature, there are several success stories of the LM application. One of the most documented techniques in the metalworking sector is the plant distribution of areas that allows a better production flow [10]. For example, Castillo exposes the design of plant distribution in order to reduce operational costs, increasing the fulfillment of delivery orders to customers [11]. The study achieved a cost reduction of 15.79% and the increase in the percentage of fulfillment of delivery orders of 51.68%. Salazar and others proposed a plant distribution through a hierarchical analytical process [12]. The authors based their hypothesis that the formation of families and cellular assembly allowed for better material flow, eliminating excessive product movements in process and waiting times generated mostly by the inadequate location of machinery and materials. After their study, an improvement in the use of machinery was achieved by 50%, and a reduction in distances traveled from 237 m to 163.7 m, and thus, a decrease in cycle times per product.

On the other hand, 5S is a tool resulting from Japanese words: classification (seiri), sorting (seiton), cleaning (seiso), discipline (seiketsu) and standardization (shitsuke) [8]. The goal of the 5S tool is to improve and maintain organizational conditions, order and cleanliness in the workplace, safety, working climate, staff motivation, and efficiency and, consequently, quality, productivity, and competitiveness of the organization [13]. Palomino, in its application of LM in the packaging line of a lubricant plant, mentions that with the use of 5S, a substantial reduction in process times was achieved, generating a decrease of 27% in preparation times and 36% in cleaning times [14]. Likewise, Hernandez describes the impact of 5S on small and medium-sized enterprises, SMEs, achieving a positive performance in terms of productivity, quality, organizational climate, and industrial safety, as well as, an improvement of the working climate and a reduction of occupational risks within their workshops [15].

Finally, the Pull System is the option to reduce unnecessary storage and overproduction. Nahmias defines it as the system where the market triggers a pull of materials throughout the production system [16]. Mora and others in their study, focused on the analysis of Pull type production systems by simulation, mention that, when this operation was carried out, an increase of 144 units was achieved, and thus, an increase in the productivity of the organizations [17]. Likewise, Tamayo and Urquiola mentioned that with this system, a decrease of product in the process was achieved and also provided a solution to the problems of production flow [18].

2 Materials and Methods

The main objective of the present investigation was to analyze a case study for the proposal of the process optimization within an assembly company. For that, all motorcycle models assembled in 2018 were considered; thus, the population analyzed consisted of seven motorcycle models, encoded as M1 to M7, respecting the confidentiality agreement established with the company. The analysis was based on the information from the process study, obtained by [19], and the time study, reported by [20].

The company develops its assembly process under the stock production system, i.e., the final product is inventoried for later sale. The production flow consists of four processes: (A) Assembly, (B) Quality Control, (C) Upholstery, and (D) Packing. Assembly is the critical production process of the company, where all the parts are coupled to the chassis of each motorcycle. Quality Control comprises the processes of reviewing the electrical and mechanical operation of motorcycles. Upholstery is the process in which the sponges and seat are placed to each motorcycle that has passed the quality control stage. Finally, the packaging is the last stage, in which the protections to the final product (motorcycle) are placed for later maintenance in the warehouse, until its sale. Within this production system, 18 operators were considered: 12 in cellular assembly (two for each cellular, six cellulars), three quality control officers, one carrier, one in the upholstery area and one in the packaging area. For discrete event simulation, the software "FlexSim 2018 Update 2", under academic license, was used to determine the application success of the LM philosophy. A methodology was configured based on four phases: a study of the current situation, determination of the ME tools to be used, a proposal of optimization, and simulation of the future situation.

2.1 Study of the Current Situation of the Assembly Plant and Determination of the Main Problems (Waste)

Several tools were used to analyze the current state of the company, such as: in situ observations, interviews, spaghetti diagram, time and movement analysis, simulation, flow process chart – worker type and material type as suggested and developed in [19]. The production system corresponds to work in cellular assembly, which is independent of each other. It could be noticed the extensive route of each motorcycle from the raw material area to the assembly area and, from there, to the finished product area. Using the flow process chart – worker type, performed by [19], problems related to distances traveled and over-processing were identified. The main problems identified and their interpretation in terms of LM, were: overproduction, transportation, excess inventory, defects, and movements. In the case of *overproduction,* because the company has a production philosophy of work for the stock, that is, it accumulates inventory in warehouses, often unnecessary of both raw material and products, either in-process or final product. On the other hand, a flow process chart – material type was developed. It represents the nature of the current production process, showing that many unnecessary movements were generated. This because the *transport of material* is carried out on several occasions from one point to another. About *excess inventory*, the current manner of work within the assembly plant produces a warehouse dedicated only to material in process, being the largest warehouse within the organization. In the case of *defects*, countless reworks are generated because there is no quality philosophy in the workplace, which increases production time. Finally, the space of unnecessary tools causes *movements* that do not add value to the product; reflecting a lack of learning in motion economics. As evidenced, internal problems were directly related to two factors: the production philosophy and the nature of the production system. Indeed, the current distribution of the plant generates an excess of transports since the flow of material goes from the assembly cellular to a pre-assembly warehouse; of this to quality control and then returns to the warehouse. The ideal flow would be: cellular assembly – quality control and then to the following processes, eliminating the pre-assembly cellular that generates spatial waste.

To analyze and illustrate the proposal, a simulation of the current situation was carried out with the most representative motorcycles in sales during 2017, and thus, determine the variation of the actual information obtained with the simulated. Figure 1 shows that the most representative models were the first three, which contributed more than 70% in sales within the organization.

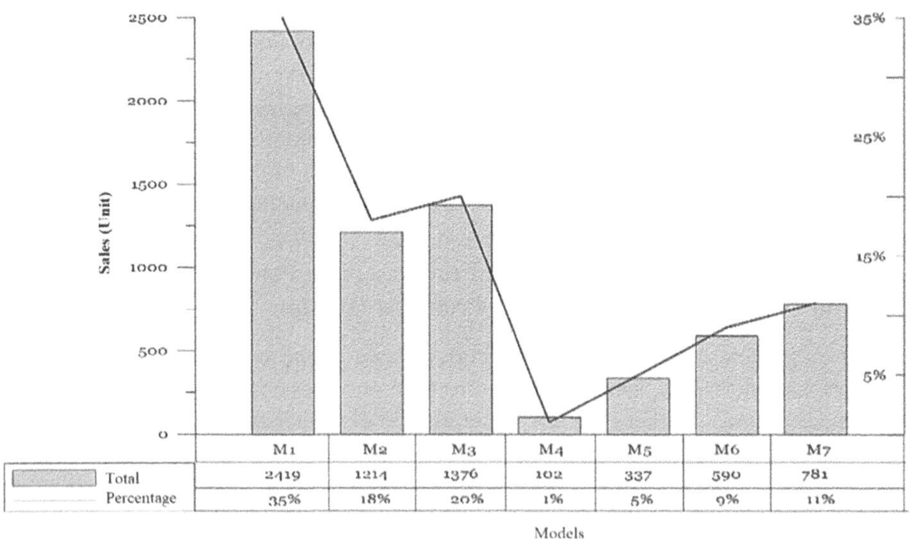

Fig. 1. Analysis of sales distribution of motorcycles.

After the simulation of the current situation (Fig. 2), the data obtained were similar to those observed *in situ* (visits that took place in 2018); this was done to determine the variance presented by the actual simulated information. Of the selected models, a production of 6.7 motorcycles/day per work cell was obtained, i.e., six motorcycles since a WIP (work in process) of 0.7 is generated, and thus, a total of 36 motorcycles (throughput). The average occupancy was 95% in the Assemblers, being the missing 5% of 25 min of idle time in the day. In the case of Quality Control, the average occupancy was 100%; whereas, for Upholstery and Packaging, there is a time of use of

93% and 85% respectively. It should be noted that the simulation did not consider whether there are sales orders on the day. Therefore, the operator does not stop, and production is previously scheduled to have the main models for production. In addition, the percentage of occupation does not only correspond to specific activities, but they also influence rework such as bolt adjustment (lack of movement economy), and unnecessary activities such as transport, resulting in a loss 43 min a day per operator. Figure 2 presents some examples of simulation of the current situation of the assembly line: assembly process, quality control and packing and upholstery stations.

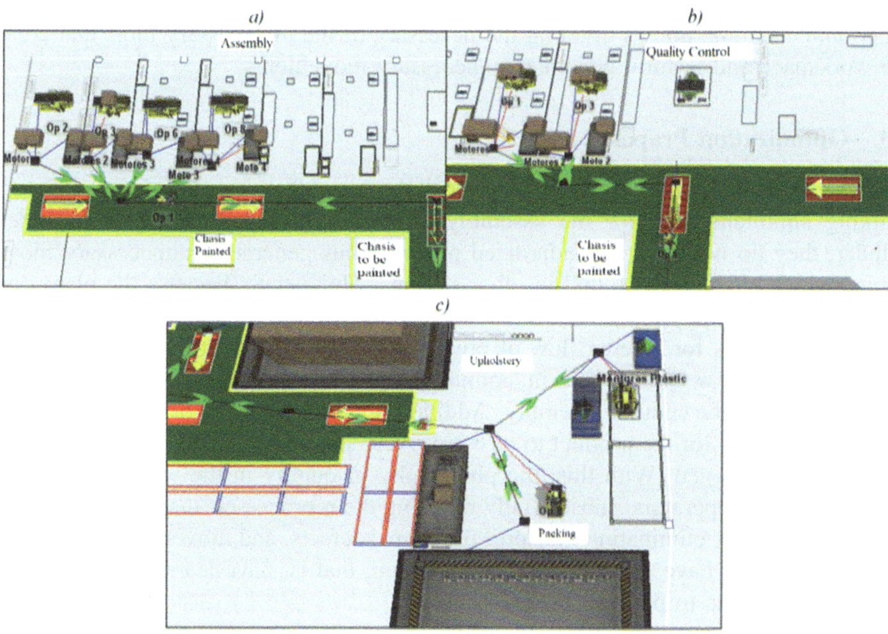

Fig. 2. Simulation of the current production process. (a) Simulation of assembly. (b) Simulation of the quality control. (c) Simulation of upholstery and packing stations.

2.2 Determination of the LM Tools Needed to Solve the Problems Related to the Identified Waste

Once the waste was recognized, the LM tools that best suit these problems were identified in the literature. As a result of this analysis, 5S, Pull System, and SMED were selected [3, 7, 13, 21, 22]. It is necessary to mention that SMED would have no functionality within the present case because the process is manual, and the tool encompasses matrix changes in machines. Therefore, in conjunction with the plant distribution, the first two tools, 5S and Sistema Pull, were used to: (1) allow better material flow, (2) reduce the distances to be traveled by both workers and products, (3) eliminate unnecessary transport, (4) effectively use available spaces, and (5) improve working conditions and safety of operators. These tools were useful in addressing Phase 1 waste issues. The distribution of LM tools to solve the waste issues found was: 5S to solve the problem of Movements, plant distribution for Transport, Pull System for Excess Inventory and Overproduction, and 5S together with Pull System for Defects.

The Pull System, which starts as a reaction to demand, allows to eliminate overproduction by folding to market requirements; it also allows to lower inventories of products in process, thus attacking the problem of excess inventory. Plant distribution allows the redesign of the production system to an assembly line, eliminating unnecessary transports, increasing productivity due to decreased processing times, and accelerating flow. Additionally, through the use of this tool, delays are reduced because working times are balanced. Finally, the 5S tool that focuses on organization, sorting, order, and cleaning, allows dividing the necessary of the unnecessary tools that create the workspace and in turn, generates unnecessary movements.

2.3 Optimization Proposal

The current production system is based on cellular assembly made up of two operators, working simultaneously for the assembly of each motorcycle. Being independent cellular, they do not have a standardized process, thus generating unnecessary movements, which affect faults found in other stations. This occurs because the plant does not have a philosophy of quality in the position or economy of movements, creating rework. Therefore, for a better flow of raw material, a total redistribution was chosen, implementing an assembly line in conjunction with a conveyor belt in the space intended prior to the cellular assembly. Additionally, because the work is sequential on the assembly line, for the product to pass from one station to another, they must fulfill the activities assigned. With this, the philosophy of quality in the workplace will be respected by the operators, substantially reducing the reprocesses, that is, the defective products and, also, eliminating transport problems, defects, and movements. Moreover, the operator must have a slender working method, that is, have learned the necessary activities over time to perform the assembly.

The plant distribution consisted of the following jobs: (A) Part classification, consisting of two operators. At this stage, the unpacking of parts and engine is carried out, as well as, the preparation of bolts and parts for distribution in the four assembly stations. (B) Pre-assembly, consisting of two operators. This position is responsible at the macro level of the assembly of motorcycles, that is, activities such as the connection of the engine, rims, battery and electrical cables to the chassis. (C1) Upholstery, with an operator, the same responsible for placing the sponge, seat and leather upholstery on the motorcycle. (C2) Assembly line, consisting of eight operators, distributed in four stations, responsible for the complete assembly of the motorcycle. (D) Quality control and packaging, with an operator, which has responsibilities that ensure the proper operation of the motorcycle, i.e., the mechanical and electrical system, in addition to the packing and transfer to the final product hold. Furthermore, an assembly line balancing was proposed; thus, the assembly line is more fluid, and there are no stops or agglomeration at the different stations. This swing is done by taking each activity of the motorcycle assembly process identified in Phase 1.

Once the plant distribution was proposed, the 5S tool was implemented, which consists of the stages of project socialization, initial audit, material requirement, staff training, tool execution, and auditing terminate. The *initial audit* serves as a fundamental basis for determining the starting point and the main aspects to be improved within the assembly area; a *list of resources* for the correct implementation of the 5S is subsequently drawn up. *Training of staff* is essential at the time of execution of the plan since the operators will be in charge of the care of their work area; this training will feature the objective of the 5S, a brief description of the tools and the strategy for their implementation. Once the *implementation of 5S* is completed, it is necessary to verify the measure of compliance. For this, a continuous review of each technique used is carried out, through a *quarterly audit* of the state of the plant, using a 5S audit format.

Finally, for the development of the Pull System, it was necessary to have an adequate forecast, based on only manufacturing what will be sold. For this purpose, the Holt-Winters method was used, which is a softening exponential triple forecasting method, that considers both levels, trend, and seasonality of a given time series [23]. Once predicted the sales quantities, it was necessary to calculate the capacity of the plant to determine the number of operators needed to meet demand in peak periods, through aggregate planning. With the use of the production master plan in conjunction with the above, the units and time periods to be produced will be reflected, allowing better control of the assembly line.

2.4 Simulating the Future Situation

For the development of this phase, since it is a simulation of discrete events, the following steps were performed: (1) formulation of the problem and determination of the objectives; (2) system modeling, which consists of creating the system design for simulation; the model must have the characteristics of the system (e.g., entity, attributes, and relationships); (3) implementation of the model on the computer using the simulation software; (4) program verification; (5) model validation; (6) simulation design and pilot tests, where the number of iterations and input variables used are determined; (7) simulation execution; (8) results analysis; and, (9) documentation [24].

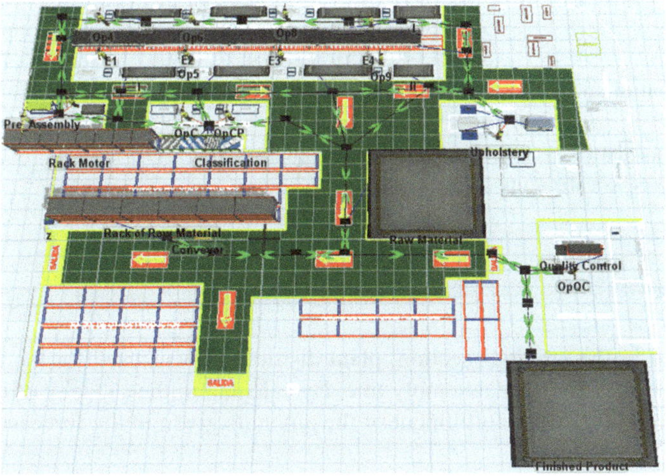

Fig. 3. Simulation of the future state of the motorcycle assembly line.

Figure 3 presents the simulation of the future situation of the assembly line, based on the three most representative motorcycles.

Table 1 reflects the time distributions for the entire assembly line. The times used for the simulation were based on the times observed by [20], with which a distribution fitting and statistical data analysis was performed for simulation. As a result, it was obtained in a normal, exponential, and logarithmic normal statistical distribution, granted by the Stat::Fit software [24]. The simulation considered a normal working day within the assembly plant, which works nine hours a day (540 min) for five days a week. To model the future state of the plant, 30 replicas were made for each motorcycle taken as a case study, obtaining a daily total of 57 motorcycles (throughput) model M1, 45 of the model M2 and 54 of the model M3, which allowed to verify the system stability decreasing variability presented by the times of each model. It should be mentioned that for testing were considered slender working methods, and that the operator already has an accepted experience to perform the work without a problem.

Table 1. Distributions time for the proposed simulation.

Localization		Cycle time probability distribution (Second)		
		M1	M2	M3
Classification		Exponential (405, 5.36)	Exponential (377, 46.1)	Lognormal (576, 2.13, 1.52)
Pre-assembly		Lognormal (386, 3.6, 0.198)	Normal (488, 19)	Normal (504, 6.9)
Assembly line	Station 1	Normal (455, 5.62)	Lognormal (491, 2.72, 0.114)	Lognormal (569, 3.41, 0.383)
	Station 2	Lognormal (432, 2.4, 0.611)	Lognormal (449, 2.82, 0.21)	Lognormal (612, 3.11, 0.464)
	Station 3	Lognormal (449, 2.08, 0,721)	Lognormal (494, 3.16, 0.138)	Normal (558, 19.4)
	Station 4	Normal (441, 7.46)	Lognormal (0.178, 6.16, 0.00464)	Lognormal (427, 4.38, 0.171)
Upholstery		Lognormal (370, 4.14, 0.25)	Lognormal (514, 4.37, 0.408)	Lognormal (313, 5.43, 0.211)
Quality control and packaging		Normal (459, 23.2)	Lognormal (198, 5.86, 0.093)	Lognormal (534, 3.66, 0.688)

Note: Lognormal (γ, μ, σ); Normal (μ, σ); Exponential (γ, μ); μ = Mean; σ = Standard deviation; γ = Localization

3 Results

As mentioned in the previous section, plant distribution was used for the change from the production system to the assembly line. By calculating the capacity of the plant and the hours necessary for the fulfillment of the amounts given in the forecast for the year

2018, it was determined that only 14 operators are needed, and not the 18 currently present in the plant. Operators would be distributed as follows: two operators for the classification area, two for pre-assembly, eight for assembly (two operators per station), one operator for upholstery and one for quality control and packaging. In addition, remaining operators must be trained in the working method.

Comparing the two production systems (cellular work assembly and assembly line) an average decrease of 40% of the standard assembly time of the three models, presented in Table 2, could be determined. This reduction is mainly due to the elimination of unnecessary movements and rework; this since activities that do not add value, within each process, were eliminated and considered in the simulation of slender processes. To determine the times of the four remaining models, the percentage decrease in average time was considered when changing the production system, which is 40%. This reduction was applied to the standard time of the cellular assembly given by [20], thus theoretically determining the standard times in minutes of the missing models, that is, M4 = 90,24, M5 = 87,31, M6 = 73,18, and M7 = 91,07.

Table 2. Decreased time in the assembly line.

Model	Standard time cellular (min)	Standard time line (min)	Percentage of decrease
M1	103,69	59,25	43%
M2	114,87	76,60	33%
M3	119,29	67,25	44%

Figure 4 shows a comparison among the capacity of motorcycles assembled daily in both work cellular assembly and assembly lines and the increase obtained by the change of the production system. By decreasing the standard time, an increase in the daily capacity of motorcycle assembly can be observed. Therefore, as a result of the considerable reduction in standard time, an increase in plant capacity with an average of 30% of assembled units is attributable. The aggregate planning details that with the times obtained for the assembly line and the capacity of the future plant (14 operators), the company can work under the concept zero inventories, since there is no need to obtain several motorcycles in stock to meet the requirements of the coming months.

In the current situation of the selected models, a production of 6.7 motorcycles/day per work cell was obtained, i.e., a WIP of 0.7 since at the stations there are half-assembled motorcycles in the day, which also generate an inventory. With the assembly line, this WIP will not be created since the motorcycles that start on the line will end their assembly process and subsequent storage in the warehouse; therefore, it entails the increase of production and the decrease of time. In addition, by improving the throughput in the production of motorcycles, there was an increase of 11 units per day, in contrast to six units per day (current throughput).

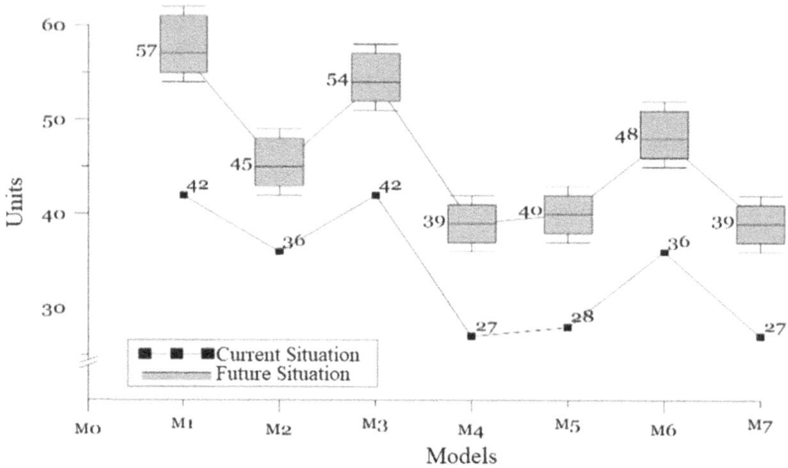

Fig. 4. Comparison of assembled motorcycles between cellular assembly (current situation) and assembly line (future situation).

4 Discussion

Competitiveness is one of the most critical challenges facing organizations today. However, for companies to be robust, they cannot rely solely on low production costs, but on seeking continuous improvement, which can result in the use of information technologies, product differentiation and innovation in the portfolio of products. In addition to factors of reduction of inventories, costs, increase in productivity and quality, this study managed to eliminate unnecessary activities within the operational process, obtaining a significant decrease in waiting times, the reduction of process times and increased assembly capacity of the plant. For example, by standardizing the working method, it was possible to reduce assembly operating times by up to 40%. This result allows the company extra time available to make more products, and thus, increase productivity, as this significant reduction achieved a 30% increase in motorcycle production in the same period and a WIP decrease. In addition, it is shown that the total disposal of products-in-process and stock is possible thanks to the implementation of the Pull System; thus, reducing spaces intended for their storage, as well as the costs generated. This method also allowed for savings in working capital since there are no fixed resources because the stored inventory is solely intended to meet market demand.

Thus, the advantages present in the plant when working through an assembly line are summarized in: control in production times (shorter times), higher productivity by increasing production with lower resources, better workflow and reduction to 11 min/motorcycle of takt time, i.e., the average time between the start of production of one product and the start of production of the next. While, by decreasing the processing time leads to an increase in the available time, which transforms into increased productivity, since, taking as an indicator the production capacity of the same, an increase of 30% is achieved. This, in turn, directly influences the monetary benefits of the organization. On average, a motorcycle profit of $3K is currently earned, while, with the new production system, by getting a cost reduction and by working at full capacity, this benefit would increase up to $4K per motorcycle.

The implementation of the 5S tool, as a practice of LM, represents a significant reduction in the waste within the plant, since the application of this tool allows to have an orderly workplace and use of machinery strictly necessary for the production process. It is important to note that, unlike several authors, who base their studies on the 5S tool as initial part of a LM implementation [22], this study focused its first changes in plant distribution, since a modification of the productive system is proposed by moving from cellular assembly to assembly line, and then, once the end-floor distribution is determined, focus on the principles of 5S.

By applying the Pull System, it was able to reduce stock levels and product inventories in the process (WIP) to the minimum possible. This system is only based on the production of customer demands. Therefore, it is essential to determine the needs of the market through forecasts. For this study, the Holt-Winters method was used, which is a highly trusted method for developing forecasts in short time periods. The results of the simulation show a 90% reduction in the inventory levels of products in process, thus achieving better management of resources, as well as, the release of the space that the warehouse area of work-in-process occupied. These spaces were reorganized by the distribution of the plant for the quality control area, obtaining a better flow of parts. This solves the sequential and interdependence problems among the different jobs mentioned above. This is because when working in an assembly line, it is possible to identify process gaps at the exact point, reducing over-processing.

In addition, and differing from studies found in the literature, the validation of the LM tools of the present case study, except the 5S tool, were given by simulating discrete events with real data raised in the company. For instance, through the simulation of the current situation of the company, it was possible to determine the number of iterations, obtaining 30 replicates and with a period of a warmup for each motorcycle, taking for granted the stability within the modeled system. Also, executing the model taking into account a normal working day within the assembly plant, that is, nine hours a day (540 min) during the five days of the week. These fundamentals give confidence that the results obtained are a real guide as indicators of the improvement of the organization. This also allows demonstrating the theoretical results to a situation close to the real, since, considering the statistical distribution of the times of each workstation (Table 2), the system presented stability in the replicas performed in the simulation, delivering standard times for the future situation without varying samples. However, simulation should be considered as a scenario where the stability of product orders is constant, and operators are fully trained and with the necessary expertise.

5 Conclusions

At the end of this study, it was possible to verify that the optimization proposal, based on LM, presents positive results within the assembly plant case study. This is achieved with the help of simulation, because it can make decisions without modifying the physical plant, taking into account the company's data. The plant distribution served as the basis for optimization since an assembly line was implemented, with a sequence in the product flow. This eliminated movements that do not add value and a reduction of processes and defective products, in addition to reducing the distances traveled, thus achieving a substantial reduction in production times and reducing the WIP.

By implementing the Pull System as a basis for production, the amount of WIP was reduced because only motorcycles will be assembled according to market requirements. In this manner, the return on investment will be achieved in a shorter time, since the company's resources will be allocated only to the assembly of the products that it has planned to sell according to historical data forecasts and market trends. Similarly, working under this system will eliminate the products-in-process and in turn, the space used for this purpose within the production plant, reducing the cost of over-inventory storage. In addition, product quality problems will be reduced because they will be assembled into an ideal minimum lot size (piece by piece), and greater flexibility will be obtained by responding appropriately to market changes.

As limitations within the proposal, it is found that the impact of the 5S tool can only be evaluated or found when implementing it in the company. Similarly, the Pull System as a demand-based system cannot be verified until its application, but using the forecast performed with the data delivered by the company, its success is estimated. The study is a basis for future analyses, where other tools such as Total Productive Maintenance (TPM), SMED and Kanban can be included, and even total development as is the case with the Pull System, it can be converted in its entirety to a JIT (Just in Time) production philosophy for better results. Together, these tools can significantly reduce all waste that afflicts most industries daily.

Acknowledgments. This study is part of the research project "Modelo de Gestión para la Optimización de Procesos y Costos en la Industria de Ensamblaje," supported by the Research Department of the University of Cuenca (DIUC). The authors gratefully acknowledge the contributions and feedback provided by the IMAGINE Project team; similarly, to the management and operational staff of the motorcycle assembly for their willingness to being analyzed as a case study.

References

1. Taj, S., Berro, L.: Application of constrained management and lean manufacturing in developing best practices for productivity improvement in an auto-assembly plant. Int. J. Prod. Perform. Manage. **55**, 332–345 (2006). https://doi.org/10.1108/17410400610653264
2. Melton, T.: The benefits of lean manufacturing: what lean thinking has to offer the process industries. Chem. Eng. Res. Des. **83**, 662–673 (2005). https://doi.org/10.1205/cherd.04351
3. Baluis Flores, C.A.: Optimización de procesos en la fabricación de termas eléctricas utilizando herramientas de lean manufacturing. Pontificia Universidad Católica del Perú, Peru (2013)
4. Pedraza, L.M.: Mejoramiento productivo aplicando herramientas de manufactura esbelta. Universidad EIA, Colombia (2010)
5. Tejeda, A.S.: Mejoras de lean manufacturing en los sistemas productivos. Ciencia y Sociedad **36**, 276 (2011). https://doi.org/10.22206/cys.2011.v36i2.pp276-310
6. Vargas Hernández, J.G., Muratalla Bautista, G., Jiménez Castillo, M.T.: Lean manufacturing ¿una herramienta de mejora de un sistema de producción? Ingeniería Industrial. Actualidad y Nuevas Tendencias **17**, 153–174 (2016)

7. Arrieta, J.G., Muñoz Domínguez, J.D., Salcedo Echeverri, A., Sossa Gutiérrez, S.: Aplicación lean manufacturing en la industria colombiana. Revisión de literatura en tesis y proyectos de grado. In: 9th Latin American and Caribbean Conference, Engineering for a Smart Planet, Innovation, Information Technology and Computational Tools for Sustainable Development, Colombia (2011)
8. Rajadell Carreras, M., Sánchez García, J.L.: Lean manufacturing: la evidencia de una necesidad. Díaz de Santos, Madrid (2010)
9. Socconini, L.: Lean manufacturing paso a paso, 1st edn. Norma, México (2008)
10. Sortino, R.A.: Radicación y distribución de planta (layout) como gestión empresarial. Invenio **4**(6), 125–139 (2001)
11. Castillo Pulgarin, J.: Propuesta de redistribución de planta para la reducción de costos operacionales y aumento en la tasa de cumplimiento de órdenes de entrega en una empresa metalúrgica, Pontificia Universidad Javeriana (2016)
12. Salazar, A.F., Vargas, L.C., Añasco, C.E., Orejuela, J.P.: Propuesta de distribución en planta Bietapa en ambientes de manufactura flexible mediante el proceso analítico jerárquico. Revista EIA **7**, 161–175 (2013). https://doi.org/10.24050/reia.v7i14.427
13. Cardona Betancurth, J.J.: Modelo para la implementación de técnicas Lean manufacturing en empresas. Universidad Nacional de Colombia, Manizales (2013)
14. Palomino Espinoza, M.A.: Aplicación de herramientas de lean manufacturing en las líneas de envasado de una planta envasadora de lubricantes. Pontificia Universidad Católica de Perú, Perú (2012)
15. Hernández, J.C., Vizán, A.: Lean manufacturing: conceptos, técnicas e implantación. Fundación EOI, Madrid (2013)
16. Nahmias, S., Arrioja, R., Nuding, B., Yescas, J.: Análisis de la producción y las operaciones. McGraw-Hill Publishing, México (2007)
17. Mora Barón, A., Tobar López, J., Soto Mejía, J.A.: Comparación y análisis de algunos sistemas de control de la producción tipo «pull», mediante simulación. Scientia Et Technica, vol. XVII (2012)
18. Tamayo García, A., Urquiola García, I.: Concepción de un procedimiento para la planificación y control de la producción haciendo uso de herramientas matemáticas. Revista de Métodos Cuantitativos para la Economía y la Empresa **18**, 130–145 (2014)
19. Benavídez Vera, E.X., Segarra Farfán, E.M., Colina-Morles, E., Siguenza-Guzman, L., Arcentales-Carrion, R.: Levantamiento de procesos como base para la aplicación de sistemas de costeo basado en actividades en empresas de ensamblaje. Revista Economía y Política **30**, 40–71 (2019). https://doi.org/10.25097/rep.n30.2019.03
20. Ramírez Vargas, J.: Predicción de tiempos estándar en líneas de ensamble usando mínimos cuadrados en modelos lineales multivariables. Universidad de Cuenca, Cuenca (2018)
21. Beltrán Rodríguez, C.E., Soto Bernal, A.D.: Aplicación de herramientas lean manufacturing en los procesos de recepción y despacho de la empresa HLF Romero S.A.S. Universidad de La Salle, Colombia (2017)
22. Sarria Yépez, M.P., Fonseca Villamarín, G.A., Bocanegra, C.C.: Modelo metodológico de implementación de lean manufacturing. (83) 51-71. Revista EAN (2017). https://doi.org/10.21158/01208160.n83.2017.1825
23. Mira Segura, L.L., Trejo Martínez, A., López Cruz, D.: Aplicación de Hol-Winters para pronósticos de inventarios. Revista Ciencia UANL. **21**(90) (2018). https://doi.org/10.29105/cienciauanl21.90-2
24. García Dunna, E., García Reyes, H., Cárdenas Barrón, L.E.: Simulación y análisis de sistemas con ProModel, 1st edn. Pearson Educación, México (2006)

A Study on Modeling and Simulation of Automobile Painting Process Based on Flexsim

John Reyes [1(✉)], Darwin Aldas[1], Homer Castelo[2], Rommel Velasteguí[1], Nancy Rodríguez[3], Cristian Suarez[1], and Kevin Alvarez[1]

[1] Universidad Técnica de Ambato, Ambato, Tungurahua, Ecuador
{johnpreyes,darwinsaldas, rs.velastegui,asuarez0935, kalvarez0229}@uta.edu.ec

[2] Universidad Tecnológica Indoamérica, Tungurahua, Ecuador
hcastelo@uti.edu.ec

[3] Instituto Superior Tecnológico Guayaquil, Tungurahua, Ecuador
nprodriguez@institutos.gob.ec

Abstract. This research seeks through the simulation software FlexSim 2017 to model, analyze, visualize and optimize the automobile painting process in a mechanical workshop whose areas are: preparation, sanding, painting, baking, washing and storage of automobiles. The methodology used is field research, applied, qualitative and explanatory, obtaining statistical data that is analyzed through the use of software libraries such as Fluid, Time Tables, Process Flow, Expert-Fit and Experimenter. The results of Experimenter identify the bottleneck and on the basis of the simulation, productivity is improved by 27.45%, which represents the processing of another car in the sanding area. These results can be replicated to other industry workshops, achieving substantial improvements in productivity.

Keywords: Process simulation · FlexSim · Bottleneck · Automobile painting · Productivity

1 Introduction

Nowadays, due to the high competitiveness in the industry and in business, it is necessary to have efficient processes to generate high quality products delivered on time and at the lowest possible cost. To achieve this goal, different alternatives are used such as experimenting in the real system, which generates very high costs due to not experimenting with a model through simulation [1].

Computer simulation allows us to understand the reality and complexity of a production system through the construction of artificial objects and dynamic experimentation before interacting with the real system. For this it is necessary to build models that represent reality and that can be interpreted by a computer [2].

J. Nummenmaa et al. (Eds.): CSEI 2019, AISC 1078, pp. 260–273, 2020.
https://doi.org/10.1007/978-3-030-33614-1_18

The simulation results allow to compare the performance of a system within several scenarios for decision making. The use of a model represents an appropriate didactic support in subjects related to production management [3] since it allows to analyze cases of companies with an efficient management of their processes with data and scientifically proven facts, it is possible to identify improvement opportunities in their processes [4]. The simulation links effects in the process analysis and allows to rationally manage the development of the manufacturing system in different time frames [5].

The simulation technique is used as a tool for making decisions regarding the optimization of resources and its favorable impact on the current level of productivity [6]. The decision making related to the production management of a manufacturing system requires the development of a simulation model that allows the user to verify the current production capacity and the future proposal [7]. The simulation models are built taking into account the conditions and parameters of each production system to compare with those obtained at the end of the simulation of each model [8]. To show a concrete scenario of simulation from a random perspective in practical models using, for example, the FlexSim software requires the understanding of the elementary concepts that make up this analogy, which will help to avoid failures [9, 10].

The objective of this work is to simulate the automotive painting process to allow the company to model, analyze, visualize and optimize the behavior of its processes. Fictitious representation of this reality makes it possible to improve productivity by implementing improvements in processes, resources, products and services in a dynamic model. With the use of Flexsim 2017, the simulation model has the ability to consider complex interrelated tasks and project them through the realization of many alternative combinations in seconds, facilitating the decision making of the model that better fits the objectives set.

2 Theoretical Framework

Flexsim is an object-oriented software used to develop, model, simulate, visualize and monitor a real system or process. It allows the modeling of objects in 2D and 3D with the focus on the development of industrial systems and warehouses, also uses objects from other design software packages that includes AutoCAD, ProE, Solid Works, Catia, 3D Studio, AC3D, Rivit, Google Sketch-Up, etc. [11, 12]. FlexSim is used by leading companies in the industry to simulate production processes before taking them to real execution. It has a variety of object libraries to simulate industrial methods, each object has its own graphical user interface that is used to add data and logic. Compared with other simulation software, it is the only one that creates an environment that is as close to reality [13], this program allows us to interpret, understand and analyze the reality and complexity of a manufacturing process by providing several alternatives for decision making in order to reduce waste in the process [14]. FlexSim offers the option to simulate different scenarios and variables with different changes and all simultaneously. This is the reason why it becomes the main choice of companies to plan and create efficient organizational manufacturing strategies [15]. The FlexSim software is established as a good alternative to simulate due to its large section of libraries and objects that allow to deal with much more complex situations without having to write software code [12].

The use of tools in Flexsim as Process Flow presents a faster and more accurate construction process of the model, for this reason it allows the development of all the details, parameters and estimation of the main indicators that characterize the production process [16]. Taking this application approach the simulation technique using FlexSim, allows to develop precise and efficient simulation models for manufacturing processes, queue systems, inventory systems, investment projects, flight simulations and aerodynamic tests, driving simulations and car crash, natural disasters, defense strategies, layout design, chemical reactions, etc. [17].

3 Methodology

The development of the research and the experimental design is carried out based on work study and statistical analysis of variables.

3.1 Experimental Design

First, a statistical study is carried out to validate the effective execution of the arrivals to the simulation model for this case study. Second, using the AutoCAD software, the layout of the company is identified and elaborated. Third, in Flexsim's 3D design, the statistical values are entered in the objects that simulate the processes of the production cycle. Then several elements that were exported to FlexSim are designed using program file extensions such as stl or sk. Then, with a graphic structure of the simulation already implemented, the different entities and locations that simulate the different processes of the cycle time are incorporated. With the defined entities and locations, events and system states are programmed. Finally, the resources, attributes and variables that represent the cycle time are defined and the simulation is executed with the different experiments focused on optimizing the bottleneck of the system.

3.2 Work Study

Based on the process diagram in Fig. 1, the process variables are determined: waiting time and production capacity.

Based on Eq. 1, time study is carried out in the different processes that comprise the manufacturing of the product and the sanding process is identified as the bottleneck of the system, these data serve as the basis for modeling the simulation. To do this, the total value of the cycle time must be established, considering the necessary supplements as basic fatigue that represents 4%, 5% for personal needs, 2% of slack for standing, 2% for abnormal position uncomfortable (flexed), 2% of use of force (lifts 15 lb), 5% related to atmospheric conditions and 2% for strong intermittent noise, values established using the general electric table [18].

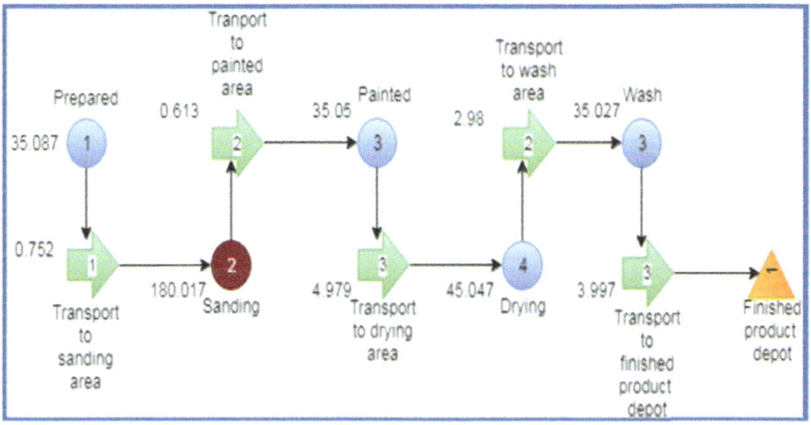

Fig. 1. Path diagram of the automobile painting process

$$Standard\,Time = Normal\,Time * (1 + Allowances) \tag{1}$$

Once the calculation of the time is done, the total time of the work cycle whose value is of 7.01 h is determined. The throughput is calculated dividing one for the standard time, which in this case is equivalent to 1.14 daily units or 6.27 units per week. Using Eq. 2 from Little's law, work in process (WIP) will be obtained through simulation.

$$Avg.Lead\,Time = \frac{Work\,in\,process}{Throughput} \tag{2}$$

3.3 Statistical Analysis

The study is based on an initial sample with pseudo random numbers which are the basis of the design to be simulated. Then the mean and the standard deviation are calculated for a larger vector that represented approximately 1000 numbers using the Excel calculation software; thus, the greater the sample, the statistical study has better precision. The Kol-mogorov-Smirnov test (also known as the K-S test) is performed on these values in order to determine the uniformity and independence. The results were positive and identify the best distribution for the process under study.

When applying the Expertfit tool of FlexSim, the generated sample is compared with the different types of existing distributions, initially showing 3 models that could have an acceptable relationship level. The tool will place in first position the one that best represents the data model entered, in addition to other parameters, which are the basis of information for the programming of locations or attributes as appropriate for the statistical distribution of arrivals. Some libraries like ExpertFit, Experimenter among others need the activation of the software license.

Based on a comparison between the different experiments, we identify that the Weibull E, Beta and Johnson distributions adapt to the data, and through the Kolmogorov test we obtain that the optimal distribution type is Beta. At the end the software shows the parameters that will be used in Flexsim for the arrivals, see Fig. 2.

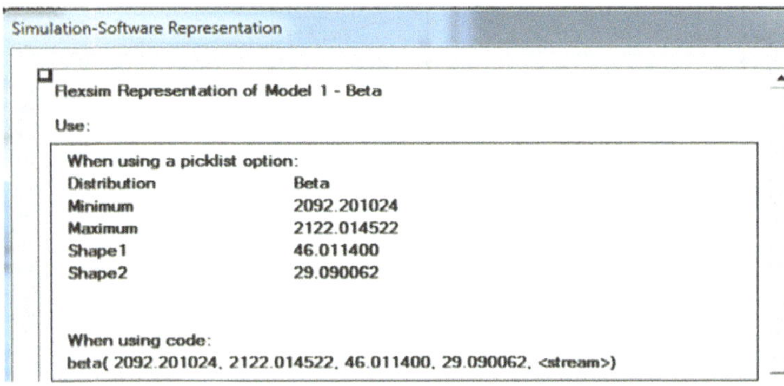

Fig. 2. Statistical calculation provided by Expertfit 2017

4 Results

4.1 Modeling on FlexSim

Using the AUTOCAD software from a 2D plane, all 3D objects are designed, see Figure 3. The design of the car painting workshop in each work area is done by placing the necessary elements for the simulation imported from AUTOCAD, then the icons extracted from the FlexSim library are changed by others, so that the simulated process is as close as possible to a real environment, see Fig. 4.

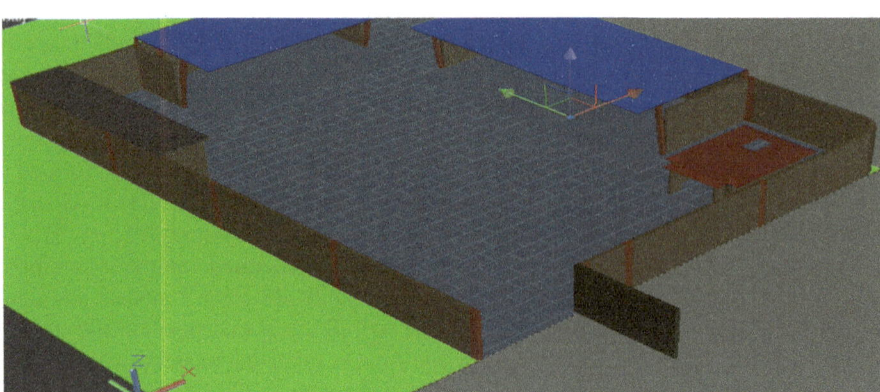

Fig. 3. Design of facilities, from AUTOCAD 2013.

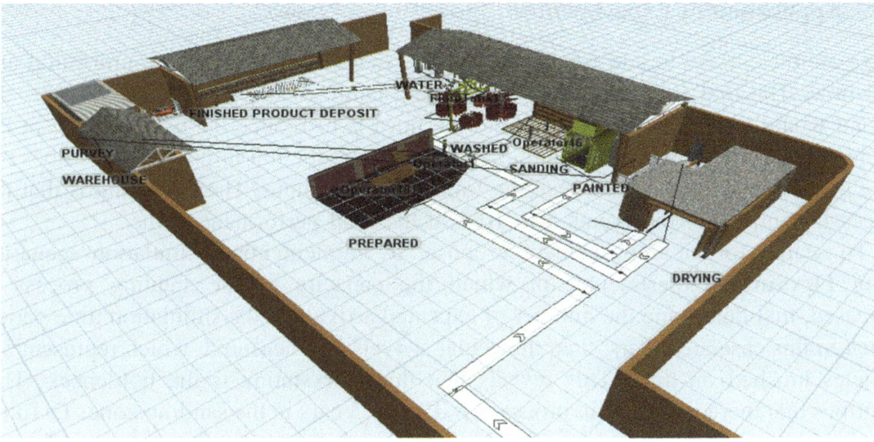

Fig. 4. Design of facilities, from FlexSim 2017.

For the Flexsim implementation, the standard time of each process is needed, the same ones obtained through the time study. With the identification of the bottleneck that restricts the flow of the system a possible solution will be considered by means of the Experimenter tool that shows a series of possible scenarios of which in some of them a leveling of the work load will be shown. This improvement can be interpreted as the increase in machinery or labor force, as shown in Fig. 5.

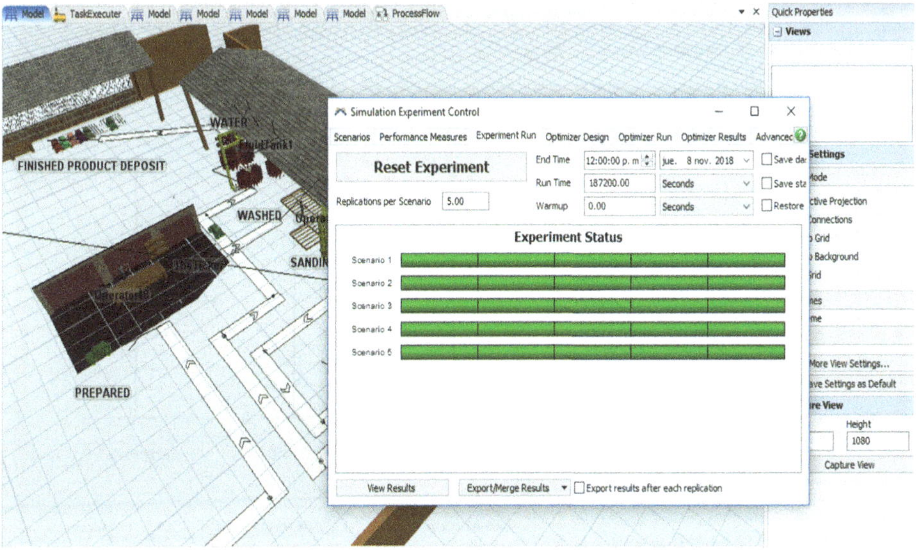

Fig. 5. Implementation of improvement scenarios, from FlexSim 2017.

4.2 Application of Experimenter

There are three ways to add capacity to a process that will help reduce utilization and lead time. The first is to reduce the stochastic nature of the product arrival rate, the second would be to add a machine, and the third is to reduce the waiting time in the machine. There is a persistent notion that a smaller reduction in process time does not have a significant influence on waiting time and throughput. The reasoning is that the processing time only represents 5 to 10% of the total performance time.

To search for an improvement proposal by implementing simulation scenarios using FlexSim's Experimenter in which the capacity of the sanding process is increased, first determine the number of expected arrivals in the simulation time period which in this case are 7 cars, then the actual arrivals are measured, which represent 5.5 vehicles finished on time. This reveals that this workstation is the bottleneck. This identifies that there is a work in process (WIP) of 1.5 cars in the sanding zone. To fulfill the objective, an assumption is used: 2 entities can be attended at the same time for one week (8 working hours from Monday to Friday and 4 extra hours on Saturday) and finally the experiment is carried out. Figure 6 shows, in the second simulation scenario, that when adding capacity for machines in the painting process, the maximum production capacity is stabilized, passing from 5 to 6 automobiles. This indicates that the painting process has ceased to be the bottleneck. This result will be the new base, and then more experiments will be done until finally reaching the expected arrivals at the beginning.

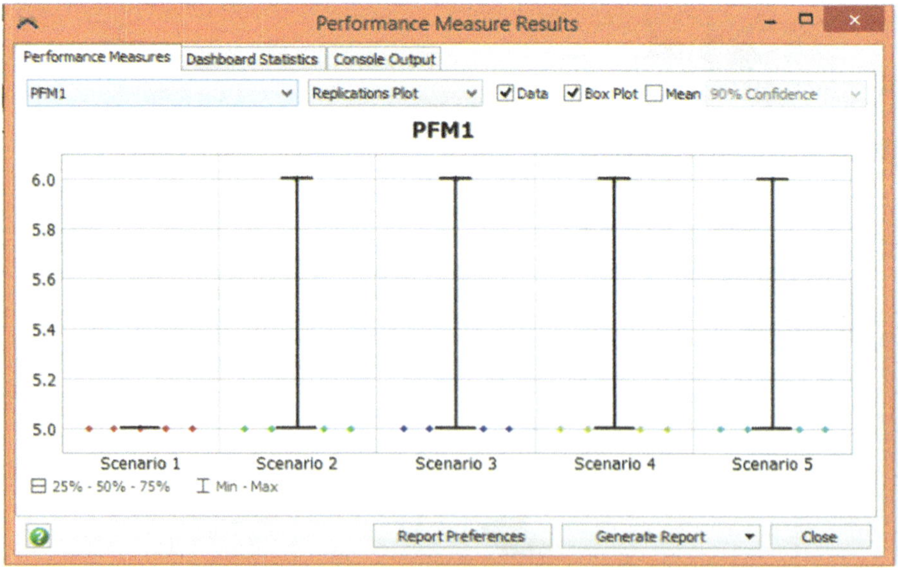

Fig. 6. Experiment scenarios from FlexSim 2017.

4.3 Application of Time Tables

To show a simulation that is much closer to reality, we set up a work schedule whose working time starts at 8:00 a.m. to 5:00 p.m. with a break time from 12:00 a.m. to 1:00 p.m. it is observed in Fig. 7.

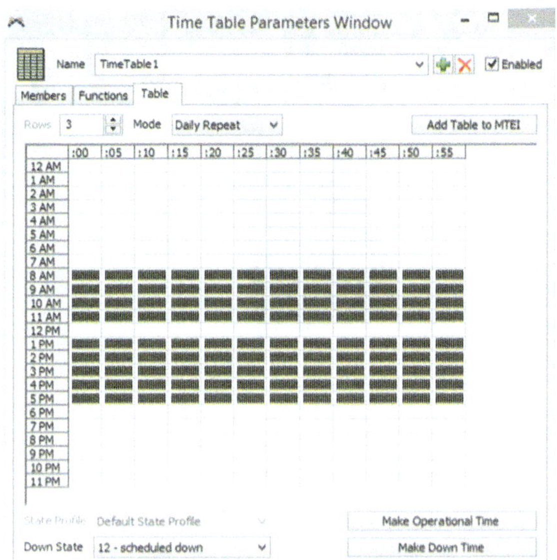

Fig. 7. Work schedules, from FlexSim 2017

To create the schedules in Flexsim, in the tool box section choose the time table option and proceed to select all those elements that will be influenced by the desired schedule. Finally, the table option indicates the work schedule to which the workers are governed.

As a result, it is obtained that the machines are marked from a yellow box as shown in Fig. 8, indicating a work break that is at a resting time or outside working hours.

4.4 Application of Process Flow- TaskExecuter

To create a new Process Flow, the Task Executer option is selected to show the paint transport flow of one of the operators, such as the Source. The Wait for event option keeps the token until a certain event is activated, this activity verifies if that event occurs in the simulation model. When that event occurs, the tracking that is given to the selected process flow (token) is released.

In order to clarify the coding, some operations will be briefly explained below. In the first Wait for Event the properties are modified as indicated in Fig. 9. In the Event option, the dropper is clicked and in the 3D model the operator is selected and when a sub-menu is displayed, the option OnRe-sourceAvailable is selected, and in row 3 of

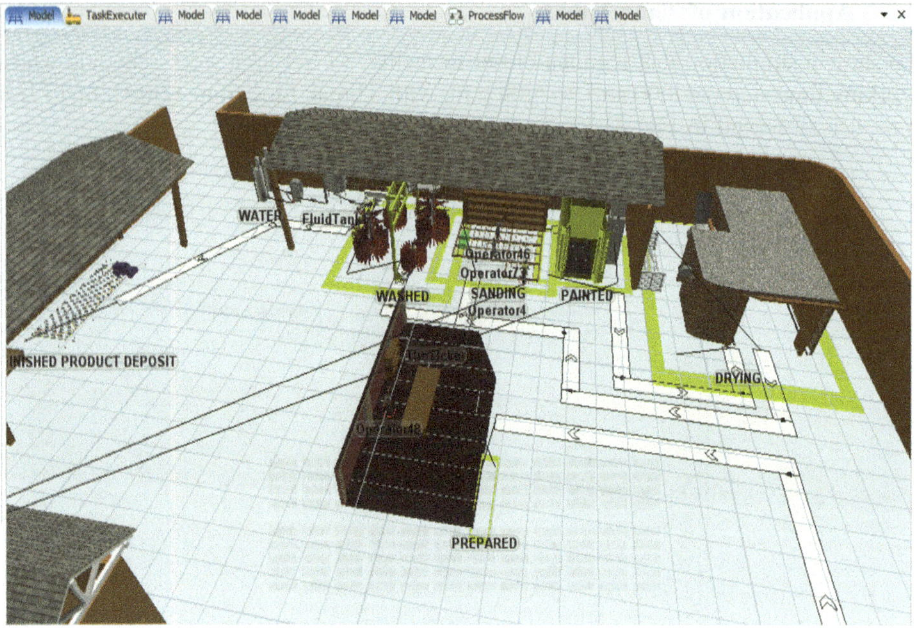

Fig. 8. Showing resting times, from FlexSim 2017

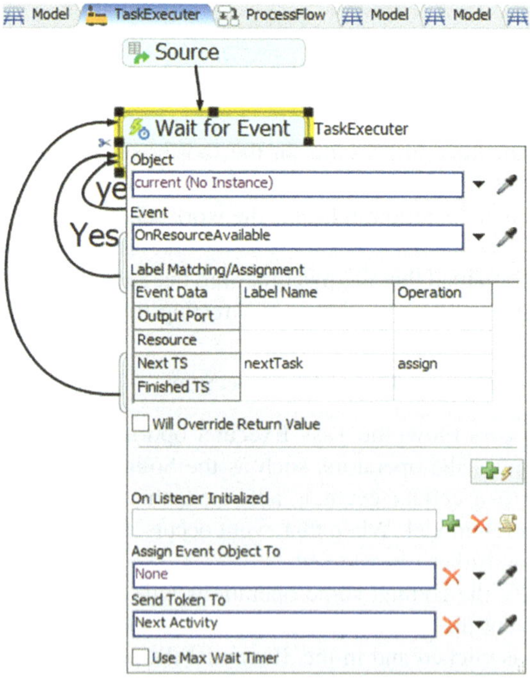

Fig. 9. Property window for Wait for event, from FlexSim 2017

the table in Next TS the Label Name is used as nextTask and in operation the assign option is selected. In the option Decide, the Send Token To box is modified, as shown in Fig. 10.

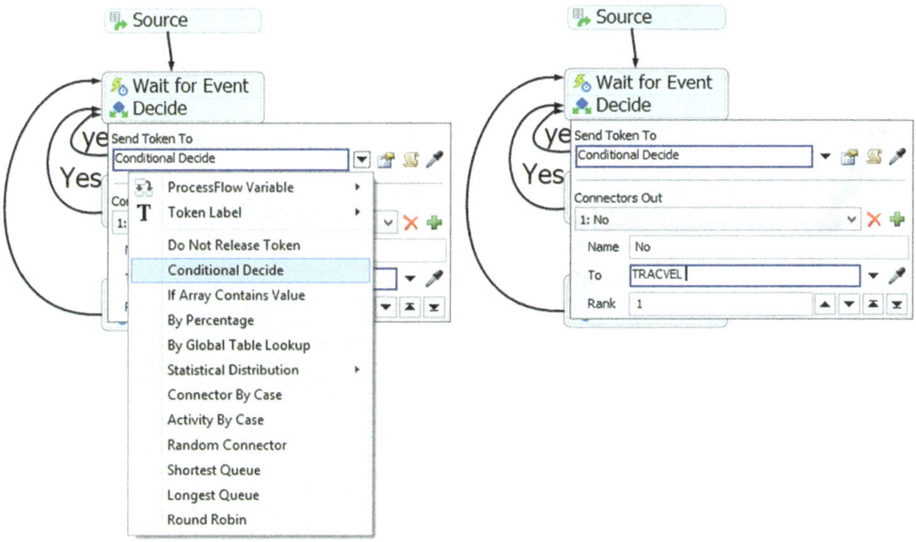

Fig. 10. Configuration of elements that shape the process Flow from FlexSim 2017

To track the flow of the Tokens of the Travel that is made to the worker, the dropper is taken and the queue is selected, which in this case is the paint shop and the Wait until complete option is active. In the second Wait for Event the waiting characteristics are changed, selecting the object that the operator is in charge of transporting, but in the sub menu OnStartTask is selected, it is given a time of 10 s to rest in case of no material.

In Fig. 11. we have the left view of the 3D model and the right view of the TaskExecuter. When there is a car in the baking process, the operator moves carrying the paint and rests when the assigned transport activity ends and a new order is awaited. The right view shows the Process Flow of the model´s Token.

4.5 Application of Fluid Library

In many real systems the materials are not necessarily discrete pieces but fluid, however "A Fluid Model" presents the fundamental concepts to build a hybrid model (discrete material and fluids) in FlexSim. In many real systems the materials are not necessarily discrete pieces but fluid, however "A Fluid Model" presents the fundamental concepts to build a hybrid model (discrete material and fluids) in FlexSim. In this case study, two

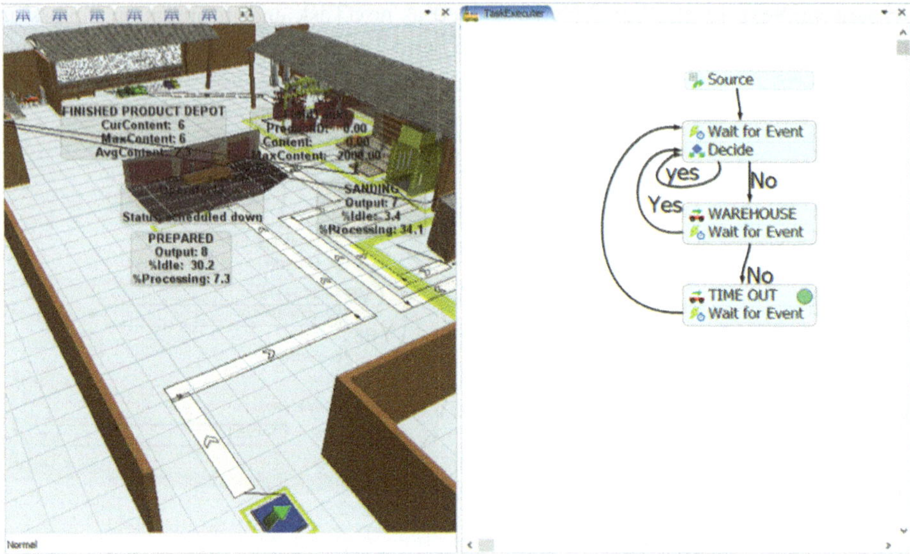

Fig. 11. Run of the task execution, from FlexSim 2017

fluid generators are used, which are for the containment of water and soap, as shown in Fig. 12. The generators have a capacity of 500 L and 300 L as it is necessary to wash a car an approximate 500 L [1].

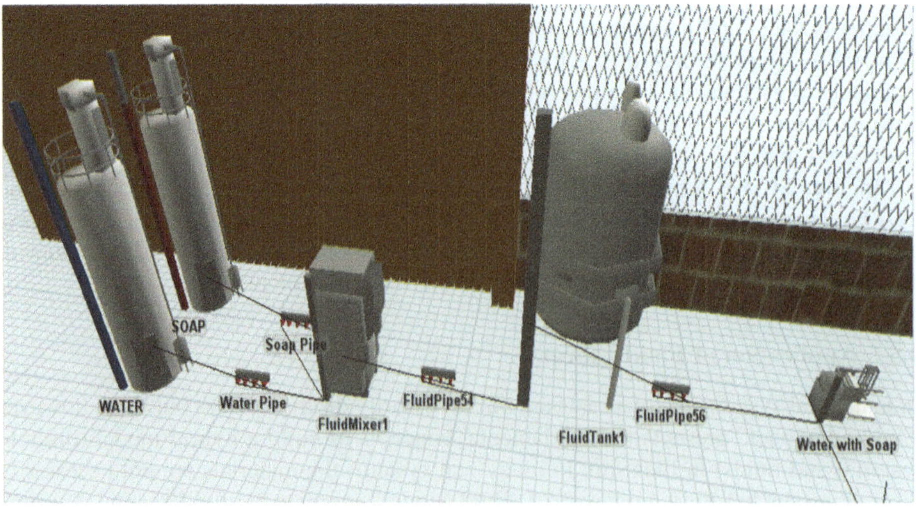

Fig. 12. Design of water and liquid soap storage tanks, from FlexSim 2017

Pipes are used that carry the fluids to a mixer that defines the amount of water and soap for the mixture. The coding is shown in Fig. 13. Then the liquids that are passed to a storage tank with a capacity of 2,000 L are mixed where the mixture will be dispensed as the car arrives at the auto-wash zone.

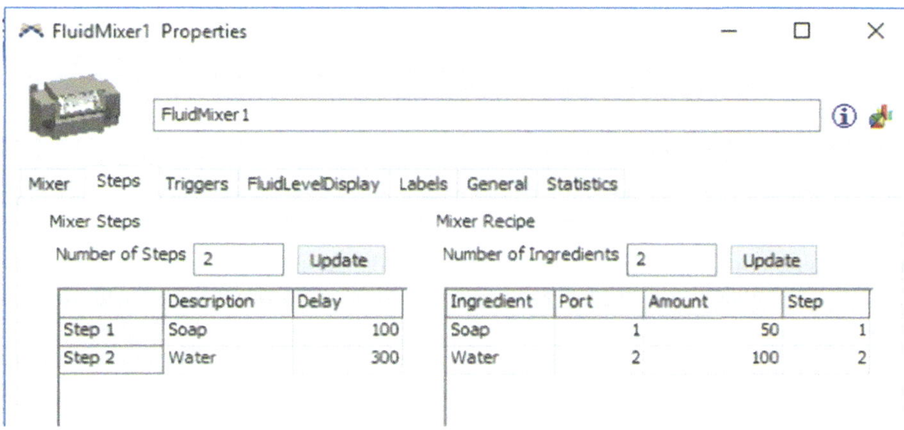

Fig. 13. Fluid Mixer properties - mixes-part 2 Configuration, from FlexSim 2017.

Then a Fluid to Item is used whose output connects to a Combiner which takes the car plus the water that it generates as an output product. The final design for a car wash is shown below in Fig. 14.

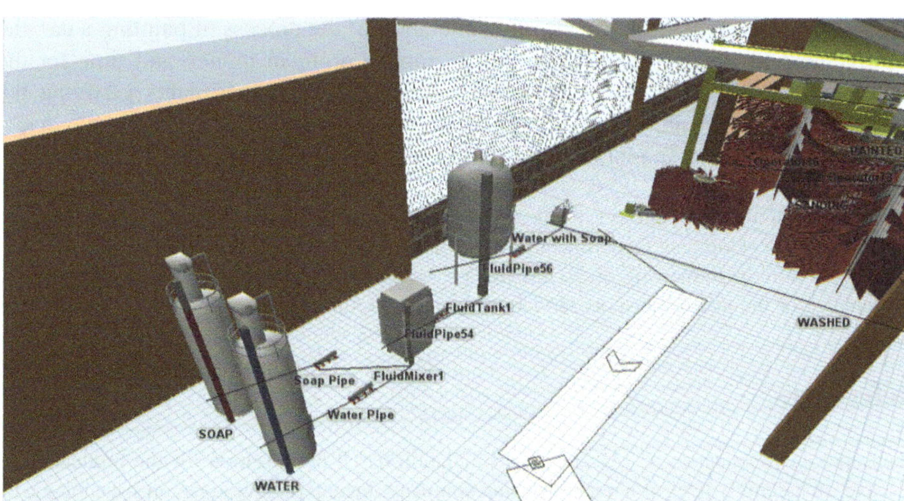

Fig. 14. Final design of the automatic washing area through the mix of water and liquid soap, from FlexSim 2017

4.6 Improvement Proposal

Based on the bottleneck found in this system, the simulation is performed by placing an additional operator in the process. Then the standard time is calculated again using Eq. 1, obtaining a decrease of the cycle time to 5,175 h. Then, Eq. 2 is applied, obtaining that as a result the company shows a production capacity of 1.54 units per day or 7.72 units per week, which implies an increase of 27.45% in relation to the starting initials.

5 Conclusions

Through Flexsim it is possible to design different types of simulation models that allow to make correct decisions in the production systems according to the requirements of the company. An immediate benefit is that it avoids having economic losses since the simulation is done in a virtual way and allows to compare results with the real environment.

In the study carried out, it is determined that the bottleneck is in the sanding process, by proposing a second simulation scenario in which the area is increased to two stations, generating an increase in capacity of approximately one and a half car. In addition, in the second scenario, the flow of production capacity is leveled, since for a third and fourth scenario production is at the same level, that is, 6 cars in the week due to the established time of work in each of the areas.

The process of bottleneck (sanding) was experienced by increasing a worker. The results show an increase in production capacity from 6.27 to 7.72 units per week, which implies an increase of 27.45%. By inserting more scenarios for the sanding area with more processes, productivity is maintained in all scenarios, since arrivals are around 1 car per day.

Based on the study of modeling and simulation of the process of painting a car, the improvement of this tool is evident, which implies profit in money and time for the company. In this way different industries could carry out experiments verifying the bottleneck and giving an immediate solution to it without losing money and time in real experimentation when implementing a proposal without certainty of the benefit.

Acknowledgments. The authors thank the Technical University of Ambato (UTA) and the Research and Development Department (DIDE)for the support provided during the execution of this work within the framework of the research project called "Socio-environmental impact of the externalities of the urban transport service in Ambato. Optimization model". Code DIDE10.

References

1. López, A., González, A., Alcaraz, S.: Simulation-based optimization for the production of axes in assembly lines of a manufacturing company. Ingenieria investigación y tecnología **20** (1), 1–9 (2019)
2. Simón, I., Santana, F., Granillo, R., Piedra, V.: La simulación con FlexSim, una fuente en las operaciones de un sistema híbrido. Científica **17**(1), 1–12 (2013)

3. Forero, Y., Giraldo, J.: Simulación de un Proceso de Fabricación de Bicicletas: Aplicación Didáctica en la Enseñanza de la Ingeniería Industrial. Formación Universitaria **9**(3), 39–50 (2016)
4. Reyes, J., Aldas, D., Alvarez, K., García, M., Ruíz, M.: The factory physics for the scheduling: application to footwear industry. In: Proceedings of the 7th International Conference on Simulation and Modeling Methodologies, Technologies and Applications, pp. 248–254. SCITEPRESS - Science and Technology Publications, Madrid (2017)
5. Jurczyk-Bunkowska, M.: Using discrete event simulation for planning improvement in small batch size manufacturing system. Stud. Syst. Decis. Control **198**, 19–43 (2020)
6. Yang, Q., Zhang, D., Zhou, H., Zhang, C.: Process simulation, analysis and optimization of a coal to ethylene glycol process. Energy **155**, 521–534 (2018)
7. Naseem, A., Shah, S., Khan, S., Malik, A.: Decision support system for optimum decision making process in threat evaluation and weapon assignment: current status, challenges and future directions. Annu. Rev. Control **43**, 169–187 (2017)
8. Aldas, D., Reyes, J., Morales, L., Alvarez, K., Portalanza, N., Aman, R.: Manufacturing strategies for an optimal pull-type production control system. case study in a textile industry. In: 2018 Congreso Internacional de Innovación y Tendencias en Ingeniería (CONIITI), pp. 1–6. Bogotá (2018)
9. Zhu, X., Zhang, R., Chu, F., He, Z., Li, J.: A flexsim-based optimization for the operation process of cold-chain logistics distribution centre. J. Appl. Res. Technol. **12**(2), 270–278 (2014)
10. Pongjetanapong, K., O'Sullivan, M., Walker, C., Furian, N.: Implementing complex task allocation in a cytology lab via HCCM using Flexsim HC. Simul. Model. Pract. Theory **86**, 139–154 (2018)
11. Xu, G., Feng, J., Chen, F., Wang, H., Wang, Z.: Simulation-based optimization of control policy on multi-echelon inventory system for fresh agricultural products. Int. J. Agric. Biomed. Eng. **12**(2), 184–194 (2019)
12. Liu, T., Duan, Y., Liu, Y.: Simulation and optimization of the AS/RS based on Flexsim. In: Lecture Notes in Electrical Engineering, vol. 375, pp. 855–863 (2016)
13. Gołda, G., Kampa, A., Krenczyk, D.: The methodology of modeling and simulation of human resources and industrial robots in FlexSim. FlexSim in Academe: Teaching and Research, pp. 87–100 (2019)
14. Hoffa, P., Pawlewski, P.: Simulation of supply chain with disturbances using flexsim - case study. Commun. Comput. Inf. Sci. **524**, 90–101 (2015)
15. Zhang, H., Li, C., Li, Y., Song, H., Li, X.: Study on seed-metering device belt mixed flow assembly line of FlexSim. In: Qi, E., Shen, J., Dou, R. (eds.) Proceedings of the 22nd International Conference on Industrial Engineering and Engineering Management 2015, pp. 633–642. Atlantis Press, Paris (2016)
16. Kęsek, M., Adamczyk, A., Klaś, M.: Computer simulation of the operation of a longwall complex using the "Process Flow" concept of FlexSim software. Adv. Intell. Syst. Comput. **835**, 97–106 (2019)
17. Cantú, J., Guardado, M. del C., Balderas, J.: Simulación de procesos, una perspectiva en pro del desempeño operacional. Revista Iberoamericana de Producción Académica y Gestión educativa **3**(5), 4–6 (2016)
18. García Criollo, R.: Estudio del Trabajo: Ingeniería de Métodos y Medición del Trabajo, 2nd edn. Mc Graw Hill, Mexico (2005)

Ergonomic Postural Evaluation System Through Non-invasive Sensors

Christian Mariño(✉) and Javier Vargas

Facultad de Ingeniería en Sistemas Electrónica e Industrial, Universidad Técnica de Ambato, Ambato, Ecuador
{christianjmarino, js.vargas}@uta.edu.ec

Abstract. The research consists in the study of cutting workplaces, in order to subsequently develop an ergonomic evaluation of the workers in this area and to obtain results of the ergonomic risks associated with the work. It could be determined that the appropriate method for the ergonomic evaluation was RULA. Once these workplaces had been evaluated in the traditional way, the postural risk evaluation system was developed and programmed. This system works with the Kinect V2 sensor, specifically with the depth, skeleton tracking, and color sensors; the system determines angles and measurement points between each joint of the body, each of them referenced to the value established in the RULA evaluation matrices at certain times.

Keywords: Ergonomic evaluation · Sensor · RULA

1 Introduction

Nowadays, the inclusion of ergonomic principles in the design of production processes is an important activity that is taking on relevant importance in companies; the main reason being to reduce the Musculoskeletal Disorders (MSDS) that are generated in jobs [1]. It is necessary to evaluate and measure ergonomic risk factors in order to obtain a comfortable workplace. There are ergonomic risk assessment methods classified [2], as observational or indirect and direct measurement methods that present disadvantages such as [3]: invasive, low precision, complex data analysis, being applied only by expert assessors, high cost, among others.

Several ergonomists find them unsuitable for measuring risk in the actual workplace. There are also semi-automatic assessment methods using 3D sensors and high-tech software [4], where more and more research is being developed for these methods. The use of sensors such as Kinect [5], can help to solve these problems because they collect data more reliably (see Fig. 1), avoid measurement errors, the sampling frequency can be increased due to the automation of data collection, making the risk estimation more accurate and non-invasive [6].

It is possible to track body parts through the manipulation, control and monitoring of a computer through the Kinect sensor applied to ergonomic risk assessment [7]. The use of the Kinect 2 sensor presents the results with measurement accuracy and efficiency in real time, being a real progress for computer vision tasks, allowing qualitative and quantitative evaluation models [8].

J. Nummenmaa et al. (Eds.): CSEI 2019, AISC 1078, pp. 274–286, 2020.
https://doi.org/10.1007/978-3-030-33614-1_19

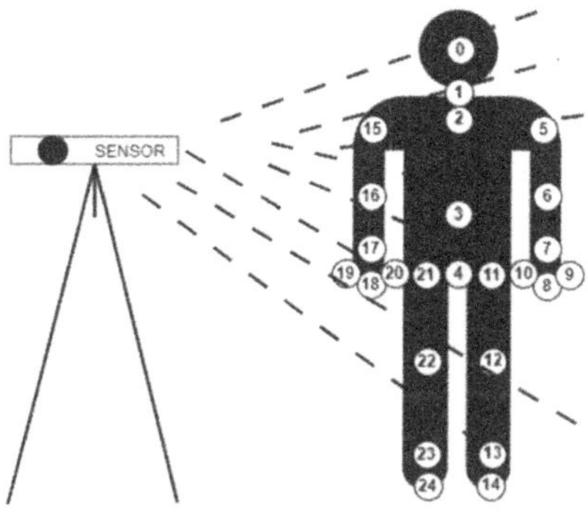

0 Head	5 ShoulderLeft	15 ShoulderRight
1 Neck	6 ElbowLeft	16 ElbowRight
2 SpineShoulder	7 WristLeft	17 WristRight
3 SpineMid	8 HandLeft	18 HandRight
4 SpineBase	9 HandTipLeft	19 HandTipRight
	10 ThumbLeft	20 ThumbRight
	11 HipLeft	21 HipRight
	12 KneeLeft	22 KneeRight
	13 AnkleLeft	23 AnkleRight
	14 FootLeft	24 FootRight

Fig. 1. Person in the evaluation environment with sensor location

The capture of people's movements and even manage a computer program with total independence of cables or special clothes mounted with multiple sensors at different points of the body [9], make all these movements are captured by the sensor in real time, at the lowest cost, allowing subsequent analysis of the data obtained and therefore determine the level of ergonomic risk to which workers are exposed in the production of footwear [10].

In the business environment, the aim is to determine some musculoskeletal disorders [11], for example, the evaluation of the workplace in driving a car, where it depends primarily on the repetitive movements of the driver and response times in each activity [12]. On the other hand, there is the evaluation of jobs in general, since people use the computer daily and the sitting position is not adequate, these symptoms are created called "Office Workers Syndrome" [13]. Which is analyzed if a position of the worker is inadequate this thanks to the non-invasive sensor of evaluation. Because the data are taken from people in their respective activities and different trades or professions such as: medicine, driving, office, industry, etc. These data are used as a means of analysis to determine potential risks at work [14], in addition to being able to decide whether an activity should be changed or eliminated from the worker. Going further to data mining it has to be that the data taken from a worker can serve as a database of training movements. Either for a rehabilitation system or send that data to robots and thus be able to improve the activity in an appropriate mechanism of work [15].

Taking into account that the industry seeks the best interaction of the worker with the job, methodologies such as RULA, REBA, OWAS, etc. are a case study [16]. Which seek to evaluate each job or activity in certain times of repetition. Now the evaluation depends on the equipment used for data collection, which are classified in two groups [17]: invasive and non-invasive devices. The disadvantage of invasive devices is the discomfort of the worker at the time of being evaluated and the data may not be accurate. On the other hand, if a worker in an assembly industry is evaluated with a non-invasive sensor, the person will perform their activity normally, thus presenting the real problems in the activity or improvements in it.

2 Methods

The purpose of the research is to develop a tool that facilitates and optimizes technical management in the evaluation of ergonomic risks of occupational origin [18], in this sense was taken into account, for the development of this work, the footwear companies specifically in the area of cutting each of their activities.

2.1 Development Study

The first phase of this project is the development of initial planning to determine the procedure and ergonomic assessment method to be used by applying an indirect methodology. With this, the decision was taken to carry out an analysis of the cutting workplace in the production of footwear, in which it was evident that most of the repetitive efforts and movements during the working day were in the upper limbs. For this reason, the Rapid Upper Limb Assessment (RULA) method [19], and the Rapid Entire Body Assessment (REBA) method [20], are the methods initially selected to be used as standardized ergonomic assessment techniques.

The traditional evaluation of work postures is based fundamentally on observation, which is why the following is a list of the various cards applied for the study of work and postural evaluation of cutting activity in footwear companies.

2.2 Convenience Sampling

The evaluation within the system is planned to work with 25 companies is met with the evaluation sample of 25 cutting jobs in footwear companies, where the total population evaluated participant in the research project was 70 people. The type of sampling carried out was intentional or convenient, because for the study only the exclusive personnel in manual cutting workstations were required.

2.3 System Systematization

The first execution parameter is the determination of the base Operating System, within the specifications of Kinect V2, Windows 10 is established in its Pro version. Requirements for Kinect V2: 64-bit (x64) processor, 4 GB memory, USB 3.0 controller dedicated to the Kinect sensor for Windows v2, Graphics adapter with DX11 capacity for Microsoft Kinect v2 sensor [21].

The requirements are the basics for the initial operation of Kinect. For the initial tests of the sensor commercial and free access programs were used such as: Brekel Pro Body v2 [22]; Kinect capture software with a maximum of 6 people within a programming environment with Unity. Kinect Studio and Visual Gesture Builder [23]; Kinect own applications, which analyze and record pre-established movements of the body postures.

These programs and most Kinect applications have the ability to capture the person in a frontal way. This poses a research problem of being able to obtain data from a person in a lateral way (sagittal plane) [24]. The bibliographic research is based on the types of sensors that Kinect has such as Depth, Skeleton and Color (see Fig. 2). The data that can be had with the Skeleton sensor are 25 joints of the body, while the depth sensor allows to save data in captures of 3D frames [25]. The combination of these two sensors makes it possible to obtain data based on the programming of a complete rotation of one of these joints [26].

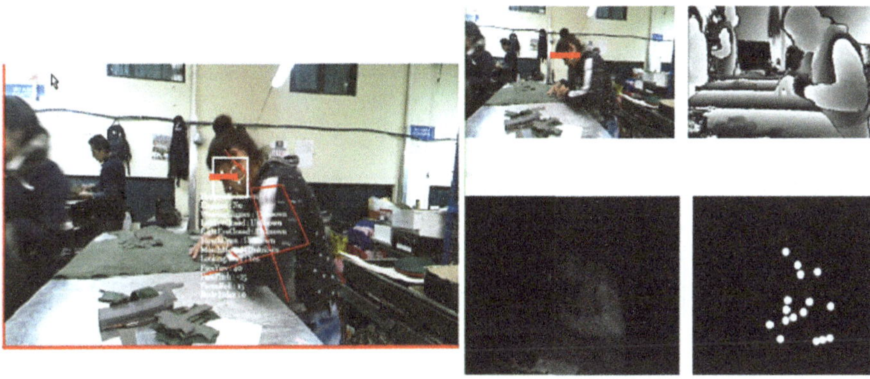

Fig. 2. System implementation in workplaces with different environments.

2.4 System Systematization

Initial Kinect tests, the initialization of the sensor has a procedure of physical recognition of the Kinect, until the ignition of the same within the SDK (see Fig. 3).

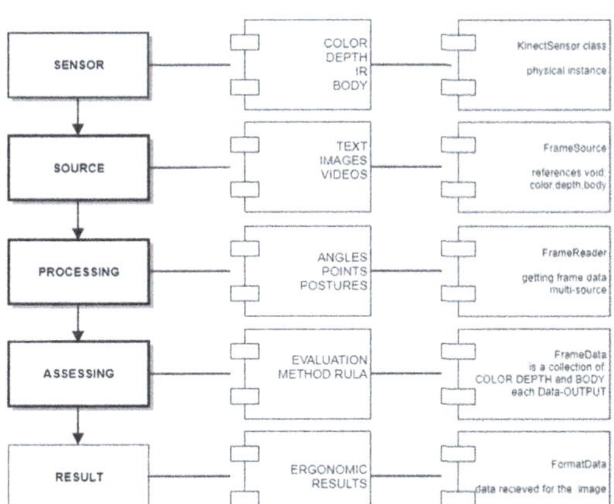

Fig. 3. Sensor processing stack.

Implementation of the system modules with a non-invasive sensor for the evaluation of the workplace [27], to determine the Kinect processing and to improve the use of the GPU a study is made with Kinect V2 and NVidia CUDA [28]. Therefore, for the detection of people and the follow-up of RGB-D data, obtaining as a result that the rates of obtained tables are better with respect to the use of CUDA in the processing. This group of data is processed by different classes (see Fig. 4), which allow from the capture to the evaluation and generation of the reports of each of the participants.

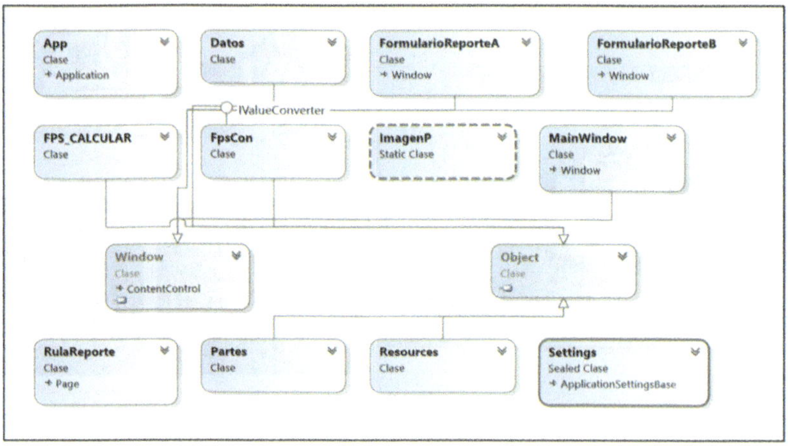

Fig. 4. Sensor processing stack.

The data obtained from each articulation is stored for each frame captured, at certain times. The averages are 52 columns and 1000 records (in rows) that are recorded in Excel, this for an evaluation of 5 min. Each one of these data allows to determine the position and the value to be obtained according to the RULA valuation matrices. In addition, each group of rows is congruent with the frame that belongs to it, the process of evaluation of RULA values [29].

The operation of the depth camera allows to determine the person in a workspace obstructed by objects and different types of lighting. The skeleton sensor completes the depth sensor, which determines the analysis points specified by RULA [30]. This means that the skeleton sensor is mapped over the depth sensor. For the first versions of the system with Kinect, it was used libraries and parameterization between Vitruvius [31] and investigation of angles to obtain points in the axes x, y, z, which allows the determination of the angles of evaluation of certain articulations of the frontal plane.

The interface model is based on a process of switching on the sensors (color, depth, skeleton), the process of rapid analysis is contemplated, accuracy depends on the evaluation time. The signals (green box) that a person is within the capture focus of Kinect depends on a value of >1, correctly traced and <1 outside the evaluation range of Kinect (see Fig. 5).

In the development process each Kinect procedure is based on the division of the data into CSV with each corresponding frame. Having too much information in plain text leads to implement a database in PostgreSQL for better response time in the query of each frame with each of the output data capture sensor.

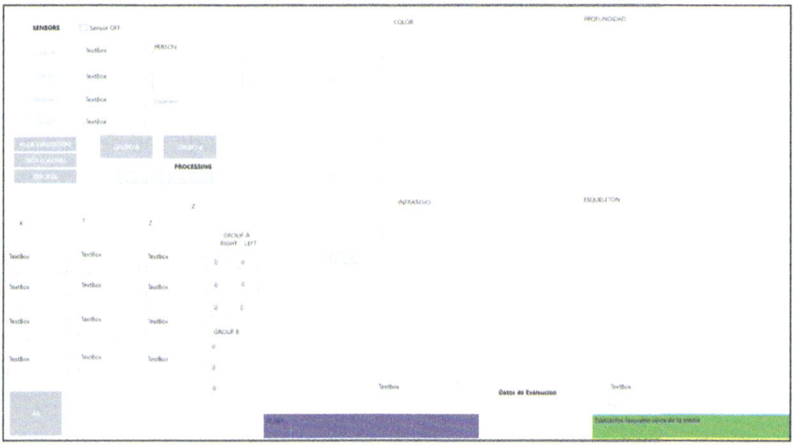

Fig. 5. Prototype implementation schema.

3 System Performance Tests

Initial tests of the system (see Fig. 6), are based on the detection of postures in determined times, each evaluation counts on a sequential one that analyzes the points of each participant until the activity is determined as repetitive or of inadequate posture.

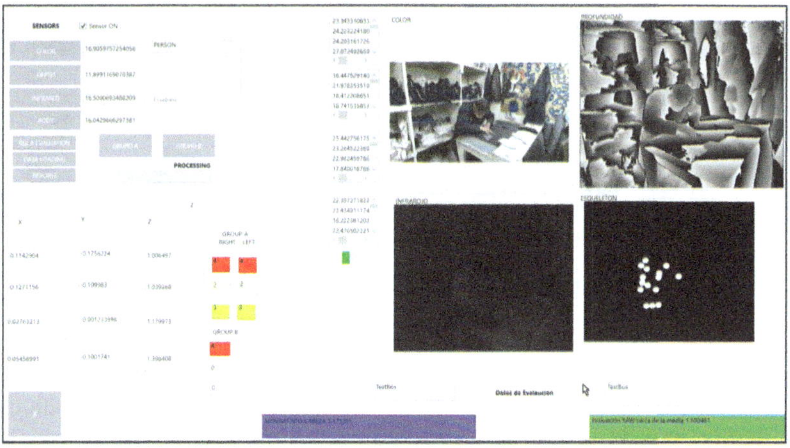

Fig. 6. Implementation of a system in workstations of a cutting company.

The reports are generated based on the highest risk frames, for which the RULA comparison tables are loaded, establishing a score for the final valuation. There are two forms of reports for both group A and group B established by RULA [29] (see Fig. 7), the final generation of the report depends on the evaluation time of the participant.

For group A it is determined according to the RULA methodology that the extremities of the body evaluated are the arm, forearm and wrist. On the other hand, for group B it is determined according to the RULA methodology that the extremities of the body evaluated are the neck, trunk and legs (static value). It must be taken into account that the values obtained are a comparison between body and depth sensor data. These data allow to establish an analogy of the capture for each frame which is created for each second of the repetitive activities.

The system contemplates obtaining three frames that come from the n postures evaluated in certain times. These three frames are managed internally by the data, i.e. each capture entails the depth data that the body adjusts to the mean of evaluation of the inadequate posture. Each inadequate posture or evaluated by the method entails the study of the angles that exceeded for this posture in addition to the time of repetition in the activity.

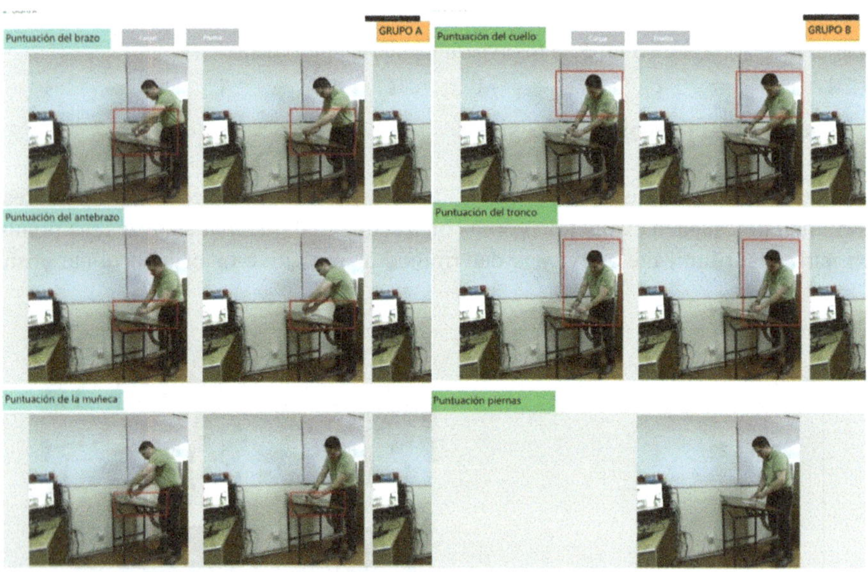

Fig. 7. Implementation of a system in workstations of a cutting company.

As a previous point of investigation, it was established that the data obtained in the assessment of the wrist may contain errors, due to the distance of evaluation of the sensor with the person. In order to parameterize these values and obtain data closer to reality, it was decided to use an additional evaluation sensor which through stable programming a deeper approach to the hands of the person evaluated. In projection the turns of the wrists are determined by a rotation matrix which determines data in depth of the person.

Once the evaluation is completed, we proceed to the creation of reports which, with generated by the system, taking into account that, if there are additional data in the methodology tells us that, if there is any kind of load, if the limb is on one side of the body, if there are forced flexions, if abduction data, and so on. These data allow us to determine the risk level of the posture or activity that the person is performing.

4 Results

The method of development of dynamic systems, allows a restructuring based on the requirements of evaluation of the postures of each person and the type of activity performed (see Fig. 8). These activities are based on a previous study of the RULA methodology, the valuation times are determined by the precision required by the data. As it is a non-invasive system, it allows the worker to perform his task normally. In addition to the evaluation of the system, an expert in evaluation by means of an angle measurement instrument corroborates the data obtained with the sensor in the system.

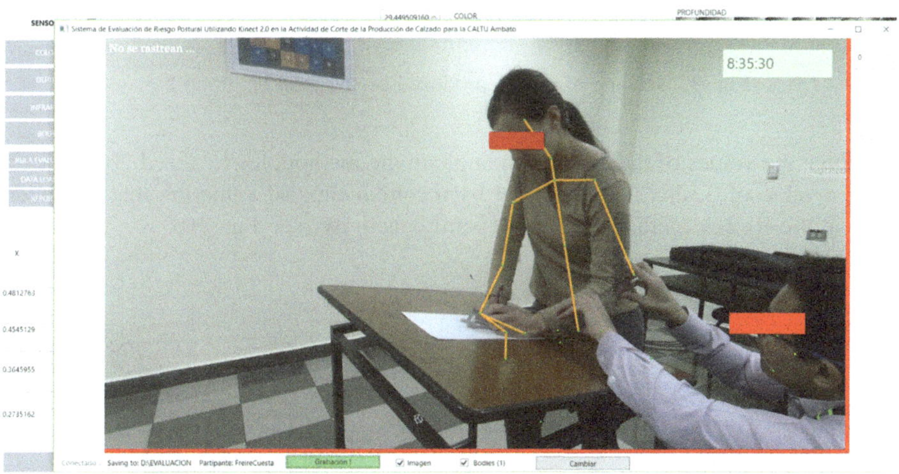

Fig. 8. Development of dynamic systems (real-time evaluation method).

The results are divided into two phases: The first phase, seeks to obtain the joints of the person in a CSV configuration file, the saves an approximate 8000 tuples (rows) of information. Each stored data is processed and tied with the frames corresponding to the evaluation activity. The second phase, analyzes the data based on the tuples of the CSV as an internal process, each frame of the 1000 that can be extracted from each evaluation in a time of 5 min, is saved for the final report and the valuation belonging to the positions determined by RULA [29].

The results were validated in the laboratory tests comparing the direct measurements versus the measurements that the developed system took in the same position and instant of time.

Taking the result of group A, the rating of the arm is 4 for the right side and 3 for the left side. These values are obtained by the frame that establishes that the angle of the arm was 41°, or; this means that the values of the evaluation angle in the methodology exceeds 20° of extension giving a value of 2, after that is added 2 additional points because there is abduction and has the shoulder elevated (see Fig. 9).

Taking the result of group B, the value of the neck is 3. This value is determined by the degree of flexion that the worker evaluated has which for this case is 57°; this

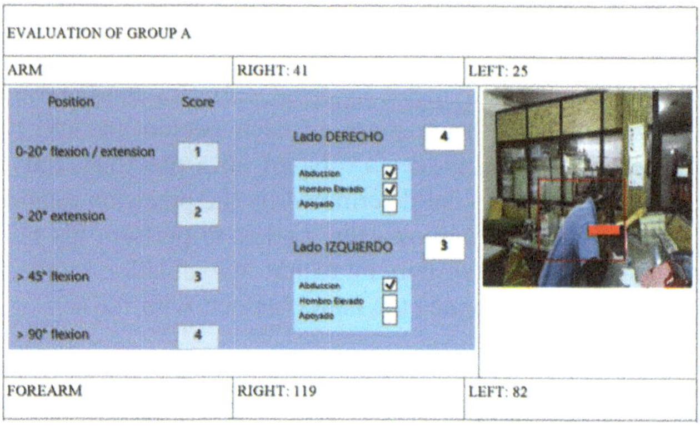

Fig. 9. Scheme of results produced by the system, group A.

means that the values of the evaluation angle in the methodology exceed 20° of flexion giving a value of 3, this is determined by a certain time of valuation that is contemplated in the RULA methodology as repetitive activity (see Fig. 10).

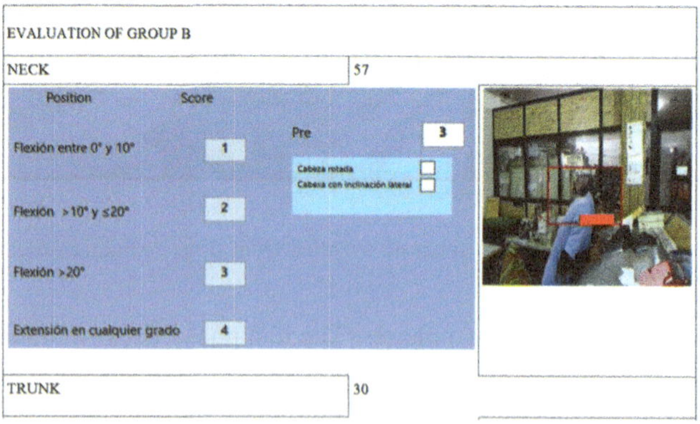

Fig. 10. Scheme of results produced by the system, group B.

For more detailed information, the evaluation parameters of the system are presented. The data obtained from the neck within the template is determined by (see Fig. 11); the activity performed by the person, the evaluation side with respect to the sensor, the percentage time of the frames, the time that the activity lasted with that working angle, the total number of frames within a range of 1 to 5000, time and date of evaluation, the angle, the most representative frame for the report and finally is determined based on the percentage of type and the most inappropriate posture with the frame selected for a previous result of the risk level.

Therefore, the final score of the RULA assessment method within the system tells us that for the right side we have a risk level 4 and for the left side we have that the risk level is 3 (see Fig. 12). These determinations and values are born from the RULA score

NECK

ACTIVITY:	CUT _M	ACTIVITY:	CUT _M	ACTIVITY:	CUT _M
SIDE:	RIGHT	SIDE:	RIGHT	SIDE:	RIGHT
TIME FR(%):	69	TIME FR(%):	44	TIME FR(%):	79
MIN/SEG:	7 / 08	MIN/SEG:	6 / 39	MIN/SEG:	8 / 01
PHOTOGRAMS EVALUATED (5000):	3777	PHOTOGRAMS EVALUATED (5000):	3587	PHOTOGRAMS EVALUATED (5000):	4877
TIME:	2018-04-05 10:52:26	TIME:	2018-04-05 10:57:14	TIME:	2018-04-05 10:54:10
ANGLE:	55	ANGLE:	36	ANGLE:	57
NUM_ PHOTOGRAM:	1992	NUM_ PHOTOGRAM:	6444	NUM_ PHOTOGRAM:	3549
NUM_ PHOTOGRAM SELECTED:					3549

Fig. 11. Summary of results after valuation.

tables. Each value was from the systematization of the data corresponding to each person evaluated. Once the results are presented, they are validated with respect to the visual perspective of the frame found with the highest level of risk, which additionally allows access to the other frames obtained in that evaluation time.

SCORES FOR GROUPS A AND B			
GROUP A			
ARM R	4	ARM L	3
ANTEBRAZE R	3	ANTEBRAZE L	1
WRIST R	2	WRIST L	1
TURN	1	TURN	1
SCORE A ON THE RIGHT SIDE	4	SCORE A ON THE LEFT SIDE	3
ACTIVITY _M	1	ACTIVITY _M	1
LOAD / FORCE	0	LOAD / FORCE	0
SCORE C RIGHT SIDE	5	SCORE C LEFT SIDE	4
GROUP B			
NECK	3		
TRUNK	4		
LEGS	1		
SCORE B	5		
ACTIVITY _M	1		
LOAD / FORCE	0		
SCORE D	6		
FINAL SCORE RULA			
RIGHT	7	LEFT	6
	LEVEL: 4		LEVEL: 3
Se requieren cambios urgentes en la tarea		Se requiere el rediseño de la tarea	

Fig. 12. Format of the report of the result of the evaluation and ergonomic risk assessment

5 Conclusions

The semi-automatic system of ergonomic evaluation with the developed sensor proves to be a non-invasive method that allows the worker to carry out his activities in a natural way making the data more real and reliable, allowing also the discovery of the ergonomic risk associated to the postures in a task with precision, thus reducing the errors of estimation of measures taken by the evaluators with a direct measurement; in addition that reduces considerably the time in the development of an ergonomic evaluation.

The ergonomic evaluation method recommended for the development of the software in the cutting activity is the RULA method, because the activity has a high level of repeatability and requires the movement of the upper extremities, in addition its application in real working environments is more recommended.

In manual cutting, of the companies studied with this type of cut, 33% have level of critical risk in the workplace, so it is recommended according to the methods of postural evaluation that changes are made in the task urgently. And 67% of the companies have a high level of risk in the work area, which indicates that a redesign of the task or job is required. Determining based on these results that manual cutting activity is riskier than die cutting activity. Therefore, the ergonomic evaluation system with the non-invasive sensor found performance levels of manual jobs at 13% for jobs that may require task changes, 65% for jobs that require task redesign and 22% for jobs that require urgent changes in the task of 23 evaluated manual cut jobs.

Acknowledgments. The authors would like to thanks to the Technical University of Ambato (UTA) for financing the project "System of Evaluation of Postural Risk Using Kinect 2.0 in the activity of Cutting of the Production of Footwear for the CALTU Ambato", for the support to develop this work.

References

1. Nasl Saraji, J., Ghaffari, M., Shahtaheri, S.: Survey of correlation between two evaluation method of work related musculoskeletal disorders risk factors REBA & RULA. Iran Occup. Health **3**(2), 5–6 (2006)
2. Kong, Y.K., Lee, S., Lee, K.S., Kim, D.M.: Comparisons of ergonomic evaluation tools (ALLA, RULA, REBA and OWAS) for farm work. Int. J. Occup. Saf. Ergon. **24**(2), 218–223 (2017). https://doi.org/10.1080/10803548.2017.1306960
3. Öztürk, N., Esin, M.N.: Investigation of musculoskeletal symptoms and ergonomic risk factors among female sewing machine operators in Turkey. Int. J. Ind. Ergon. **41**(6), 585–591 (2011). https://doi.org/10.1016/j.ergon.2011.07.001
4. Li, X., Han, S., Gül, M., Al-Hussein, M., El-Rich, M.: 3D visualization-based ergonomic risk assessment and work modification framework and its validation for a lifting task. J. Constr. Eng. Manage. **144**(1), 1–13 (2018). https://doi.org/10.1061/(ASCE)CO.1943-7862.0001412
5. Tee, K.S., Low, E., Saim, H., Zakaria, W.N.W., Khialdin, S.B.M., Isa, H., Awad, M.I., Soon, C.F.: A study on the ergonomic assessment in the workplace. In: AIP Conference Proceedings, pp. 1–12 (2017). https://doi.org/10.1063/1.5002052

6. Mariño, C., Santana, R., Vargas, J., Morales, L., Cisneros, L.: Reliability and validity of postural evaluations with kinect v2 sensor ergonomic evaluation system. In: Advances in Intelligent Systems and Computing, pp. 86–99. Springer, Cham (2018). https://doi.org/10.1007/978-3-030-02828-2_7
7. Plantard, P., Shum, H.P.H., Le Pierres, A.S., Multon, F.: Validation of an ergonomic assessment method using kinect data in real workplace conditions. Appl. Ergon. **65**, 562–569 (2017). https://doi.org/10.1016/j.apergo.2016.10.015
8. Tsakiraki, E.: Real-time head motion tracking for brain positron emission tomography using microsoft kinect v2. In: KTH Royal Institute of Technology and Health, Stockholm (2016)
9. Asteriadis, S., Chatzitofis, A., Zarpalas, D., Alexiadis, D.S., Daras, P.: Estimating human motion from multiple kinect sensors. In: International Conference on Computer Vision/Computer Graphics Collaboration Techniques and Applications, pp. 1–6 (2013). https://doi.org/10.1145/2466715.2466727
10. Geoffrey, D.: Ergonomic methods for assessing exposure to risk factors for work-related musculoskeletal disorders. Occup. Med. **55**(3), 190–199 (2005). https://doi.org/10.1093/occmed/kqi082
11. Radwin, R.G., Lee, S., Li, K., Lieblich, M., Park, B.K.D.: Discussion panel on computer vision and occupational ergonomics. In: Proceedings of the Human Factors and Ergonomics Society, pp. 957–959. Human Factors an Ergonomics Society Inc. (2016). https://doi.org/10.1177/1541931213601220
12. Zhao, M., Beurier, G., Wang, H., Wang, X.: In vehicle diver postural monitoring using a depth camera kinect. In: SAE Technical Papers, pp. 1–9. SAE International, United States (2018). https://doi.org/10.4271/2018-01-0505
13. Paliyawan, P., Nukoolkit, C., Mongkolnam, P.: Office workers syndrome monitoring using kinect. In: Proceedings of the 20th Asia-Pacific Conference on Communication, APCC 2014, pp. 58–63. Institute of Electrical and Electronics Engineers Inc, Thailand (2015). https://doi.org/10.1109/APCC.2014.7091605
14. Buisseret, F., Dierick, F., Hamzaoui, O., Jojczyk, L.: Ergonomic risk assessment of developing musculoskeletal disorders in workers with the microsoft kinect: TRACK TMS. IRBM **39**(6), 1–3 (2018). https://doi.org/10.1016/j.irbm.2018.10.003
15. Bai, J., Song, A., Xu, B., Nie, J., Li, H.: A novel human-robot cooperative method for upper extremity rehabilitation. Int. J. Soc. Robot. **9**(2), 265–275 (2017). https://doi.org/10.1007/s12369-016-0393-4
16. Kee, D., Karwowski, W.: A comparison of three observational techniques for assessing postural loads in industry. Int. J. Occup. Saf. Ergon. **13**(1), 3–14 (2007). https://doi.org/10.1080/10803548.2007.11076704
17. Seo, J., Alwasel, A., Lee, S., Abdel-Rahman, E.M., Haas, C.: A comparative study of in-field motion capture approaches for body kinematics measurement in construction. Robotica **37** (5), 928–946 (2017). https://doi.org/10.1017/S0263574717000571
18. Intranuovo, G., De Maria, L., Facchini, F., Giustiniano, A., Caputi, A., Birtolo, F., et al.: Risk assessment of upper limbs repetitive movements in a fish industry. BMC Res. Notes **12**, 2–7 (2019). https://doi.org/10.1186/s13104-019-4392-z
19. Corlett, E.N., McAtamney, L.: RULA: a survey method for the investigation of world-related upper limb disorders. Appl. Ergon. **24**, 91–99 (1993)
20. McAtamney, L.: Rapid upper limb assessment (RULA). In: Handbook of Human Factors and Ergonomics Methods. CRC Press, Boca Raton (2004). https://doi.org/10.1201/9780203489925
21. Chen, C., Yang, B., Song, S., Tian, M., Li, J., Dai, W., et al.: Calibrate multiple consumer RGB-D cameras for low-cost and efficient 3D indoor mapping. Remote Sens. **10**(2), 1–28 (2018). https://doi.org/10.3390/rs10020328

22. Leong, C.W., Chen, L., Feng, G., Lee, C.M., Mulholland, M.: Utilizing depth sensors for analyzing multimodal presentations. In: International Conference on Multimodal Interaction, pp. 547–556 (2016). https://doi.org/10.1145/2818346.2830605
23. Mgbemena, C.E., Oyekan, J., Tiwari, A., Xu, Y., Fletcher, S., Hutabarat, W., et al.: Gesture detection towards real-time ergonomic analysis for intelligent automation assistance. In: Advances in Intelligent Systems and Computing, pp. 217–228 (2016). https://doi.org/10.1007/978-3-319-41697-7_20
24. Manghisi, V.M., Uva, A.E., Fiorentino, M., Bevilacqua, V., Trotta, G.F., Monno, G.: Real time RULA assessment using kinect v2 sensor. Appl. Ergon. **65**, 481–491 (2017). https://doi.org/10.1016/j.apergo.2017.02.015
25. Dutta, T.: Evaluation of the kinect sensor for 3-D kinematic measurement in the workplace. Appl. Ergon. **43**, 649–659 (2011)
26. Rayme, N.S., Kamat, S.R., Shamsuddin, S., Wan Mahmood, W.H., Azizan, N.: Ergonomics study of working postures in manual hand layup process. In: Lecture Notes in Mechanical Engineering, pp. 15–26. Springer, Singapore (2018). https://doi.org/10.1007/978-981-10-8788-2_2
27. Mariño, C., Vargas, J., Aldas, C., Morales, L., Toasa, R.: Non-invasive monitoring environment: toward solutions for assessing postures at work. In: Iberian Conference on Information Systems and Technologies CISTI, pp. 1–4. IEEE Computer Society, Spain (2018). https://doi.org/10.23919/CISTI.2018.8399204
28. Vargas, J., Mariño, C., Aldas, C., Morales, L., Toasa, R.: Kinect sensor performance for Windows V2 through graphical processing. In: International Conference on Machine Learning and Computing: ACM International Conference Proceeding Series, pp. 263–268. China (2018)
29. MacLeod, D.: Rapid upper limb assessment (RULA). In: The Rules of Work, 2nd edn. CRC Press, Florida (2013)
30. Yan, X., Li, H., Wang, C., Seo, J.O., Zhang, H., Wang, H.: Development of ergonomic posture recognition technique based on 2D ordinary camera for construction hazard prevention through view-invariant features in 2D skeleton motion. Adv. Eng. Inform. **34**, 152–163 (2017). https://doi.org/10.1016/j.aei.2017.11.001
31. Vangos, P.: Angle Calculations with Kinect|Vitruvius. https://vitruviuskinect.com/angle-calculations. Accessed 15 July 2019

Energy Supply of a Hybrid System of Biomass and Wind Turbines of the Pichacay Landfill Towards an Intelligent Network for the City of Cuenca-Ecuador

Daniel Icaza[1(✉)] and David Borge-Diez[2]

[1] University of Leon, León, Spain
dicaza00@estudiantes.unileon.es
[2] Department of Electrical, Systems and Automation Engineering, University of Leon, León, Spain
david.borge@unileon.es

Abstract. The modeling, simulation and analysis of the energy conversion equations describing the behavior of a hybrid system of wind turbine and biomass system for power generation are presented in this paper. A numerical model based on the aforementioned equations, encoded in MATLAB and the results were compared with the experimental data. The model predicted quite interesting results compared to the experimental data provided by the EMAC EP under various conditions. This study seeks to contribute to the new demands of future energy for the City of Cuenca, its possibilities for expansion according to the Territorial Planning Plan and with a vision that this electric energy be part of a reliable and growing smart grid.

Keywords: Renewable energy · Wind turbine · Biomass · Modeling · Simulation · Smart grid

1 Introduction

Renewable and non-conventional energy generation methods such as wind, solar, hydropower, biomass, geothermal, thermal storage and waste heat recovery power supply solutions for remote areas such as the localities surrounding the Pichacay Landfill in Cuenca - Ecuador, which are not directly accessible by the Electricity network. The hybrid renewable energy system is an integrated system of two or more renewable energy systems, and can complement each other, provides a higher quality and reliable power source independent of the utility's grid network [1–11]. Plants are wind/biomass become an increasingly attractive option, as the price of fossil fuels and land increase and the cost of thermal technology [5]. Although biomass power plants can operate continuously, they can be costly, supply chain security and mass transport [9].

In Cuenca-Ecuador, due to its agricultural nature, residual biomass is a renewable source of energy with a high potential for use. Bioenergy or biomass energy, is a type of renewable energy from the use of organic matter formed in some biological process. Another study was proposed by the references [12] for the implementation of hybrid

J. Nummenmaa et al. (Eds.): CSEI 2019, AISC 1078, pp. 287–307, 2020.
https://doi.org/10.1007/978-3-030-33614-1_20

systems in the rural area disconnected from the electrical network. In our case we will analyze and compare the theoretical part of the system proposed with the experience in the field, precisely in the sanitary landfill of Pichacay, a place where solid wastes are treated mainly of the City of Cuenca in Ecuador.

The system consists of a motor element, alternator, heat recovery system and auxiliary elements. It would involve the construction of an installation to consume the gas and wood waste for the production of electric energy and the use of the power of the flue gases to generate the steam with which to heat the water for the dryers of wood. A long term can save on consumption and the cost of energy [20].

Cogeneration is defined as the sequential production of electrical and/or mechanical energy and the energy of energy is used in industrial processes from the same primary energy source, and is today an alternative as a method of energy conservation for the industry according to the policies of regional economic globalization and the international policy aimed at achieving sustainable development.

A highly efficient and low emitting concept that has been considered for the future is the hybrid gas turbine high temperature [12–18]. The use Elliott, P. and Booth, R. as well as Organic Rankine Cycle (ORC) plants depends on the temperature and state of flue internal gas flow [19].

Use trash waste to generate energy another alternative for remote houses or a different way to generate energy, where the electric grid does not reach. But it should be noted that in Ecuador there are very few places where the biomass resource is really constant and there is a demanding treatment, this is the case of Pichacay in Ecuador.

The Sanitary Landfill is one of the components of the Pichacay human and environmental development complex, which is located in the parish of Santa Ana, 21 km from the city of Cuenca, on the edge of the Parish of Quingeo.

Its operation began on September 3, 2001, in compliance with strict standards for this type of sanitary equipment. On December 14, 2002, the ministry of environment granted the environmental license.

Due to its technical, environmental, occupational health and safety characteristics, it maintains an integrated management system based on international standards ISO 9001, ISO 14001 BSI OHSAS 18001.

It is a large opportunity the generator of energy in rural areas in isolation in the electricity network of the company supplying energy [18, 23], in these areas it is convenient to install hybrid generation systems especially in Pichacay is considered to be a procedure more than Biomass generation system can be summoned with solar and wind systems, in this case in particular has been considered the wind system of the geographical area it is possible to implement small wind turbines as mentioned in [26, 27]. The viability and importance of solar energy use in global electrification also have been presented in that review and analyzed [21]. Another study was proposed by Damen [22] for implementation in rural areas disconnected from the grid in Sao Paulo. The code HOMER was used to optimize that hybrid system. In addition, other studies were presented on PV-wind-battery hybrid and PV-wind-diesel-battery hybrid intended for rural electrification in Europa [23–25]. For the purpose of validating this simulation model, the energy conversion equations were coded with MATLAB.

This paper is concerned with the main heat and mass transfer mechanisms taking place in a hybrid system of biomass based gas turbine and wind turbine for power generation and heating of the urban centers adjacent to the Pichacay landfill, such as the Patrimonial Center of Quingeo, Urban Center of Santa Ana and urban center of the Valle. A numerical simulation using one dimensional model is presented hereby to describe the process as well as thermal behaviour of associated system.

2 Location of Research

The sanitary landfill is located in the Parish of Santa Ana very close to the territory of the Quingeo Parish. The zone of delimitation between these two sectors is the River Quingeo. See Figs. 1 and 2.

Fig. 1. Location of the Pichacay Sanitary Landfill.

Fig. 2. Pichacay Landfill geographic environment.

In addition to generating electricity from the biomass, also takes advantage of the wind resource of the place since from the high peaks of the river Quingeo and its surrounding areas cause large wind currents and form a wind tunnel downstream overcoming for several hours of the day the 4.5 m/s, for this reason is the production of wind energy in the sector, for the moment on a low scale that may in future be better exploited, nevertheless we consider the model of a hybrid system for the present study.

Next, in the Fig. 3 we can identify the power station where it receives the energy contributions, both the biomass, wind and future can increase other contributions of another type of renewable energy, however it should be clear that adjustments and calibrations must be made to receive new contributions. In our case study we contemplate it without connection to the electrical network of the local trading company.

Fig. 3. Power conversion center.

3 Mathematical Modeling

In the following sections, the energy conversion equations of the biomass energy and wind energy into an electrical energy are presented for locality Pichacay.

3.1 Biomass Simulation

The main function of the bio-gas heater is to satisfy the buildings heating requirements and operate as a back up system in case of maintenance or malfunction of the main geothermal based heating subsystem. Thus, an energy balance on the bio-gas burner and heater, gives [10, 11]. The oil loop is comprised of a tank heat exchanger; ORC waste heat boiler, piping and pump as well as control valves, as seen by EMAC EP technicians in the Fig. 4.

The radiation is the major heat transfer by-product because of the high temperature of the gas. However, other heat transfer mechanisms are present in the combustion chamber such as convective, evaporation and combustion and must be taken in

Fig. 4. Regulation of control valves by EMAC EP technicians.

consideration in order to solve the following energy conversion equations of biomass process [28]:

$$Energy_{in} = Energy_{out} + Losses \tag{1}$$

Where Eq. (1) can be written in the following form:

$$\frac{dT_{wh}}{dt} = -\frac{mL_w}{\rho_w V_{heater}}(T_{wh} - T_L) + \frac{Q_{add}}{\rho_w V_{heater} C_{pw}} \tag{2}$$

$$Q_{add} = 4.18 * CV_{bio} * m_{bio} * \eta_{heater} \tag{3}$$

Where T_{wh} = outlet temperature of water from the biogas heater
V_{heater} = volume of the heater
CV_{bio} = calorific value of bio-gas
m_{bio} = mass flow rate of bio-gas fuel
η_{heater} = heater efficiency

The heat load and capacity of hot water radiators could be modeled as follows [12]. The heat load for a building changes linearly with the difference between indoor and outdoor temperatures [28], like in the Eq. 4.

$$Q = \frac{Q_o * (T_{in} - T_{out})}{(T_{in} - T_{out,0})} \tag{4}$$

Where Q The heat load (kW).
T_{in} Indoor temperatures (°C).
T_{out} Outdoor temperatures (°C).
Q_o The design heat load (KW).
$T_{out,0}$ The design outdoor temperature (°C)

$$Q = Q_o(\frac{\Delta T_{in}}{\Delta T_{in,0}})^n \tag{5}$$

$$\Delta T_{in} = \frac{T_s - T_{re}}{ln[\frac{T_s - T_{in}}{T_{re} - T_{in}}]} \tag{6}$$

Where Ts The supply temperatures of the water (°C).
T_{re} The return-temperatures of the water (°C).
$\Delta T_{lm,0}$. The value of LMTD that is calculated under design conditions.

$$WGT = \eta\, m_{flue\,gas}(h_1 - h_2) \tag{7}$$

Where,

η: is the gas turbine efficiency that takes account various losses during the combustion process.
h_1, h_2: enthalpies of flue gas at inlet and outlet of gas turbine, respectively.

The hot flue gas emitted from the gas turbine is coupled with a thermal oil loop and Organic Rankine Cycle (ORC) to generate refrigerant vapor at waste heat boiler as shown in Fig. 5 [16, 17].

The following thermodynamic equations can be written to evaluate the performance of the ORC:

$$W_{ORC} = m_{ref}(h_1 - h_2) \tag{8}$$

$$Q_{WHB} = m_{ref}(h_1 - h_4) \tag{9}$$

$$\eta_{ORC} = \frac{W_{ORC}}{Q_{WHB}} \tag{10}$$

h_1, h_2, h_4: enthalpies of refrigerant mixture at inlet, out of vapor turbine, inlet and outlet to waste heat boiler of ORC, respectively.

3.2 Wind Turbine Simulation

The power of a particular wind turbine is given by [26, 29]:

$$P_W = 0.5 * C_{p1} * \rho_{air} * A * v^3 * \eta_a \quad (11)$$

Where: P_W = Wind power sweep produced by the blades per unit area. C_{p1} = Betz power coefficient. ρ_{air} = Air density, A is the Area swept by the blades of the wind turbine and v is the wind velocity.

Taking into account the internal performance of the wind turbine, the following can be written:

$$\eta_a = \eta_f.\eta_g.\eta_{box} \quad (12)$$

Where: η_f, η_g are mechanical friction and generator efficiencies respectively and the efficiency speed multiplication box is η_{box} [29].

The power output of the wind turbine in Eq. (11) can be expressed in single-phase power AC as:

$$P_{1f} = \sqrt{3}.\eta_{c1}.V_{line}.I_{line}.Cos\varphi \quad (13)$$

With single phase AC power is P_{1f}, line current I_{line}, represents power factor $Cos\varphi$, and the electric conversion efficiency is referred to as η_{c1}.

3.3 Controller

Generally, the controller power output is given by [29, 30]:

$$P_{Cont-dc} = V_{bat} * I_{rect} \quad (14)$$

Where: V_{bat} is multiplication of the nominal voltage DC in the battery for any particular system and I_{rect} represent the output current of the rectifier in DC.

3.4 Battery Charging and Discharging Model

The battery stores excess power going through the load charge controller (Fig. 6). The battery keeps voltage within the specified voltage and thus, protects over discharge rates, and prevent overload [29, 30].

During the charging period, the voltage-current relationship can be described as follows [28]:

$$V = V_r + \frac{I\left(\frac{0.189}{(1.142-soc)+R_i}\right)}{AH} + (soc - 0.9)\ln\left(300\frac{I}{AH} + 1.0\right) \quad (15)$$

And:

$$V_r = 2.094[1.0 - 0.001(T - 22\,^{\circ}C)] \quad (16)$$

The current and voltage during discharge can be described in terms of the state of charge (SOC) of the cell [29–31]. The battery state of charge is the instantaneous ratio of the actual amount of charge stored in the battery and the total charge capacity of the battery at a certain battery current (ranging from 0.3 to 1.0) which is represented in Eq. 17 [28].

$$V = V_r + \frac{I}{AH}\left(\frac{0.189}{soc} + R_i\right) \tag{17}$$

And R_i is given by:

$$R_i(\Omega) = 0.15[1.0 - 0.02(T - 22\,^{\circ}C)] \tag{18}$$

Where,

V_r, I: the terminal voltage and current respectively
$R_i(\Omega)$: Internal resistance of the cell and T is the ambient temperature.
AH: Ampere-hour rating of the battery during discharging process

Finally, the power produced by the PV array can be calculated by the following equation,

$$P = VI_{rect} \tag{19}$$

Where I_{rect} represent the output current of the rectifier in DC (14).

3.5 Inverter

The characteristics of the inverter are given by the ratio of the input power to the inverter P_{in} and inverter output power P_{in} [29–31]. The inverter will incur conversion losses and to account for the inverter efficiency losses, η_{inv} is used:

$$P_{in}.\eta_{inv} = P_{out} \tag{20}$$

The AC power of the inverter output P(t) is calculated using the inverter efficiency η_{inv}, output voltage between phases, neutral V_{fn}, for single-phase current I_f and $cos\varphi$ as follows:

$$P(t) = \sqrt{3}\eta_{inv}V_{fn}I_f\, cos\varphi \tag{21}$$

Finally, the hybrid system energy conversion efficiency for harnessing energy from wind turbine and Biomass/CHP-ORC is given by:

$$\eta_{sistem} = \eta_{gas} * \eta_{wind} \tag{22}$$

$$\eta_{sistem} = \frac{(Qin + Qadd)H1}{P(gas)} * \frac{P_{1f}}{P_{WT}} \tag{23}$$

Where Q_{in} and Q_{add} are the wind energy and gas turbine combustion heat, respectively. *H1* is the altitude difference.

4 Numerical Procedure

Figures 5 and 6 show the equipment that integrates the network and the corresponding connections for the process of transforming energy from methane gas and wind to electric energy. We describe the Eqs. (1) to (23) which are solved according to the flow diagram presented in Fig. 7, where the biogas input parameters, the wind conditions as well as the ORC are defined. They are integrated into the system but are independently calculated through an iterative analysis until the system gives us an error that is as small and acceptable as possible.

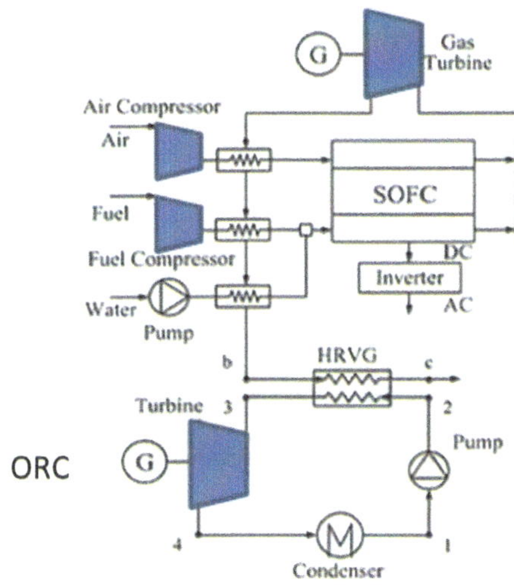

Fig. 5. Proposed hybrid system Biomass and CHP- ORC subsystem

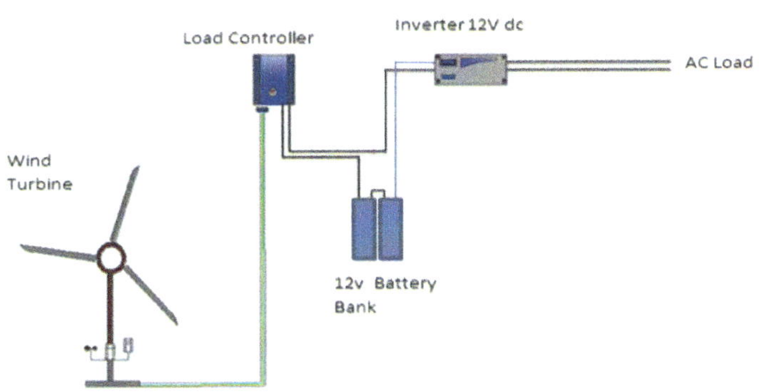

Fig. 6. Proposed hybrid system wind subsystem.

The numerical procedure begins with the use of the wind and biogas flow conditions for a transformation process to convert this energy into electrical energy. The feed and homogenization of the input substrates varies according to the characteristics of the same (pumping, augers, etc.). The digesters are cylindrical deposits (concrete or steel) equipped with agitation and heating equipment that ensure optimal conditions of the process of biomethanization. The generated biogas accumulates in a gasometer (which can be installed directly on top of the digesters or as a separate unit). Once the hydrogen sulphide (H2S) is removed by a desulfurization and condensate system, the biogas is conducted to a cogeneration unit where it is transformed into electricity and heat. The electricity generated can be sold to grid or be self-consumed, the heat covers the plant's own demand and the surplus can be used for heating or external industrial systems. It is also important to note that in parallel there is an additional source of electric power generation that is from the wind, we consider this source as a very interesting alternative in the place given the acceptable conditions of wind flows from the heights of the Green Sector of Quingeo and bright places. It is sought that the data introduced in the generation model give us graphical results and based on iterations a pattern of behavior of the system is established with the use of MATLAB supporting us with the equations of energy transformation. Finally, the total efficiency of the hybrid system is calculated in each input condition.

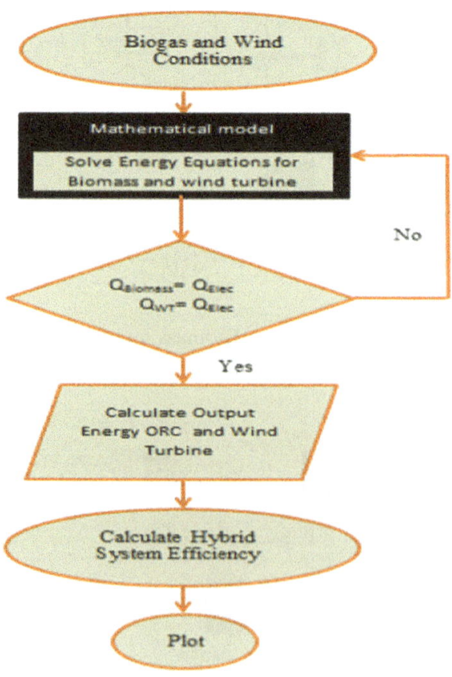

Fig. 7. Flow diagram of Hybrid system calculation.

5 Results and Discussion

The purpose of this manuscript is to enter in the same system the Eqs. (1) to (23) corresponding to the hybrid system and taking into account that the generation of energy is not necessarily going to be simultaneous, and for validation purposes, both the equations and their Data were codified with a numerical integration analysis with finite differences. These data were adjusted according to the simulations carried out in order to establish tolerance limits between theory and experimental curves. The use of software such as Matlab allows the establishment of maximum permissible error limits and we have taken it as a tool for design and optimization of these generation systems very useful for this type of research. In the following sections, we present analyzes and discussions of predicted numerical results, as well as the validations of the proposed simulation model.

In the Fig. 8 presents the typical profiles of environmental insulation at the site during several months of the year 2016 and 2017 at different times of the day. It is important to note that the maximum temperature is at noon, however it is important to incorporate this data so that its results are as accurate as possible and no further adjustments are required to the model used.

5.1 Biomass Simulation

The biomass will continuously feed the reactor, so that when introducing Air at high temperature will produce a gaseous fuel rich in H2 and CO. A Fluidized bed gasifier typically operates at temperatures of 750–900 °C [11], Having chosen a working atmospheric pressure. Figures 9, 10, 11, 12 and 13 show the respective simulations at different temperature values. the Process inside the reactor takes into account the contribution of the bubble and emulsion phases. The velocity and volume fraction parameters are Fundamental to determine the proportion of moles of each phase that exists per floor of the reactor.

For the present study three loads of municipal solid waste incinerators were considered; 100, 150 and 200 t/d with lower heating values (LHV) of 1000, 1700 and 2300 kcal/kg for the simulation. It was segmented to low quality solid waste with 59% moisture, ash 8% and fuel 33% inputs that serve the simulation [28].

In practice, increasing the temperature in a gasification process Increases O2 consumption, resulting in a mayor depletion of char (which It leads to an increase in conversion). An increase in temperature in The typical operating ranges (750–900 °C) favor the enrichment of the Synthesis in CO and H2, due to greater increases in the velocities of Reaction favoring the creation of these compounds, compared to the rest Belonging to the synthesis gas. In the range 750–800 °C, the endothermic nature Reactions producing H2 tend to Increase of this, and a decrease of CH4 as the temperature increases. For the Temperatures 850–900 °C are expected to dominate the reactions of Boudouard and Reformed to steam (R9 and R12, respectively), increasing the content in CO. The increase in temperature favors the destruction of tars (tars) Generated in the volatilization, allowing to increase the yield of the conversion.

In natural form the wet biomass is degraded by bacteria and microorganisms. In conditions of presence of oxygen this process delivers as end-products carbon dioxide

(CO_2), water, sulfate, nitrite, nitrate and sales of ammonium if this process is carried out under anaerobic conditions (absence of Oxygen) is obtained as a flammable gas product which is called biogas and a residue wet of difficult degradation denominated sludge. Depending on the characteristics of the original biomass, this gas may have a composition of between 50 and 75% of methane (CH_4), so it has a good energy potential, reaching around 5000 kcal/m3. The calorific value of biogas depends directly on concentration methane by way of reference the calorific value of 1 m3 of methane is 9.97 kWh. The methane consumption is calculated from 60% to approximately 6 kWh/m3 of biogas.

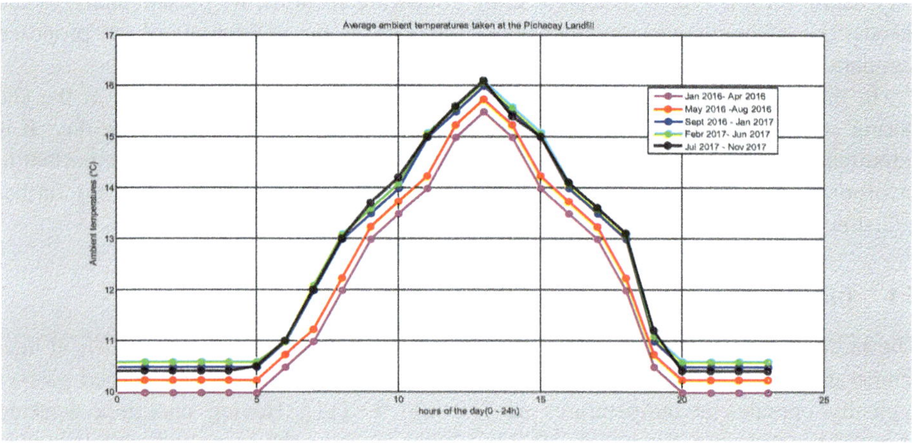

Fig. 8. Ambient temperatures (°C) profile January 2016–November 2017

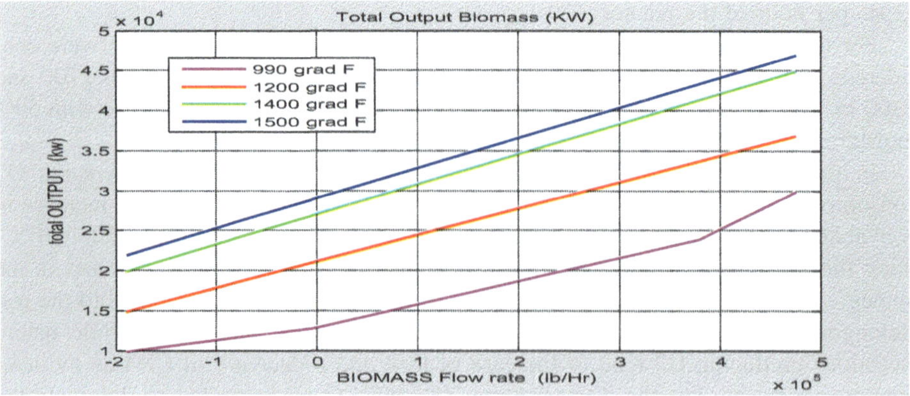

Fig. 9. Biomass output power at different biomass loading

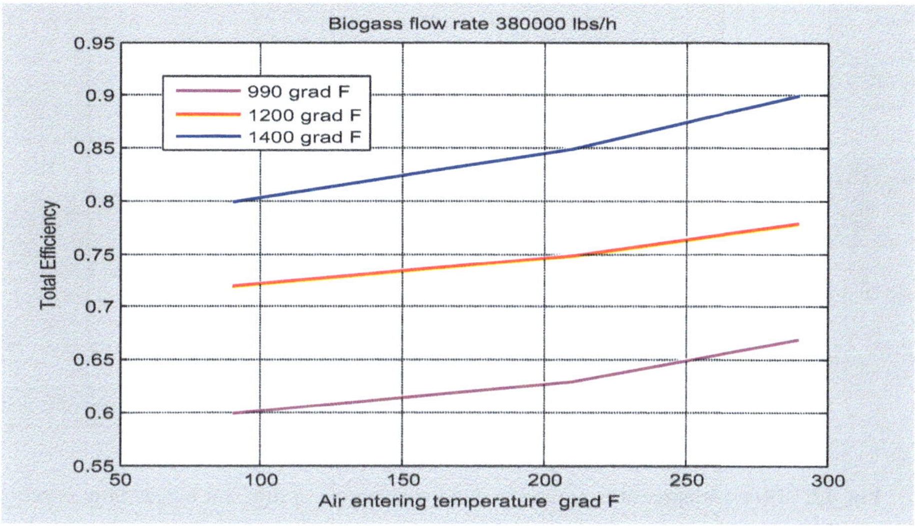

Fig. 10. Biomass output Efficiency at different air entering temperature

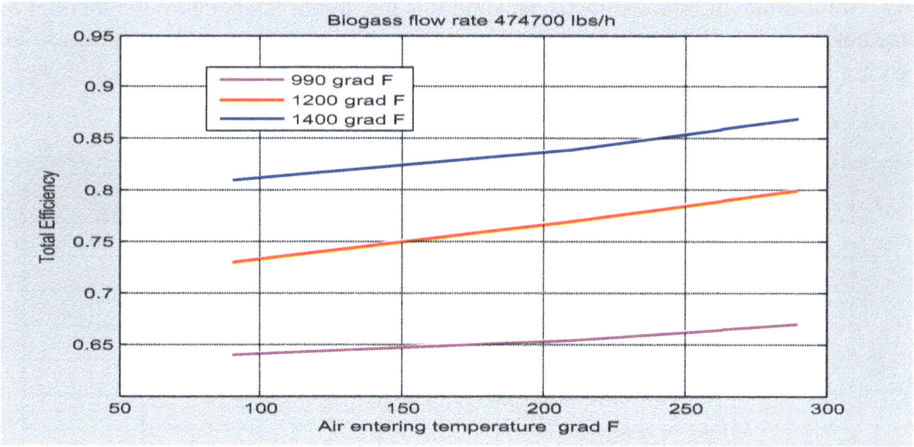

Fig. 11. Biomass efficiency at different biomass different air entering temperature

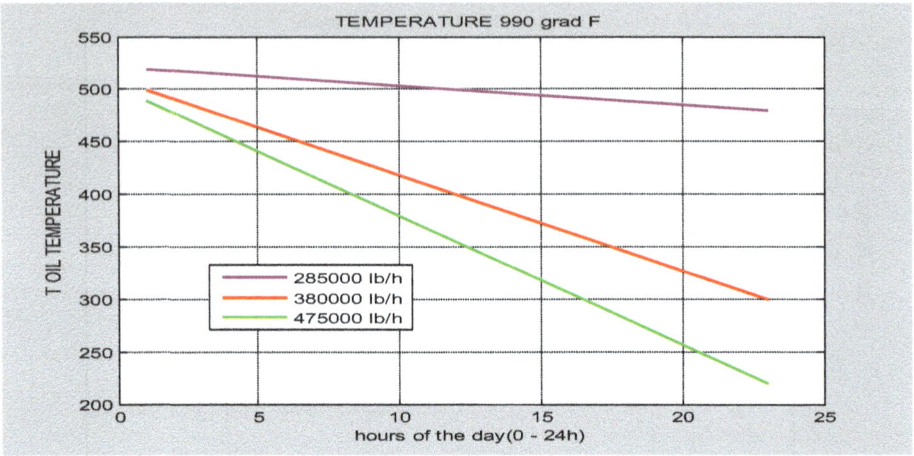

Fig. 12. Time-variation of heat transfer fluid temperature at different biogas flow rates

The dynamic action in a thermal oil medium was used to transport the heat from the biomass combustion gas to the ORC which can be predicted by Eq. (2) and shown in Fig. 10 for different biomass loads. Beyond this maximum temperature the thermal oil was not considered in this simulation since it could disintegrate and compromise the process of heat transfer in the residual heat boiler as shown in Fig. 5.

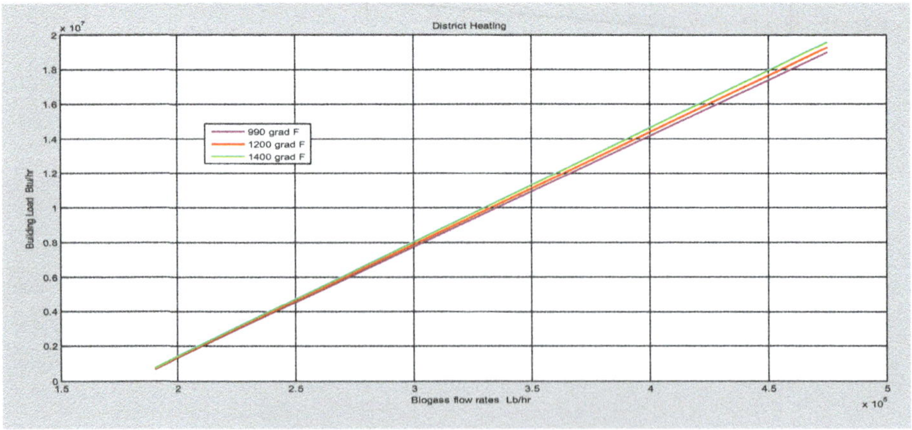

Fig. 13. District heating load Btu/hr.

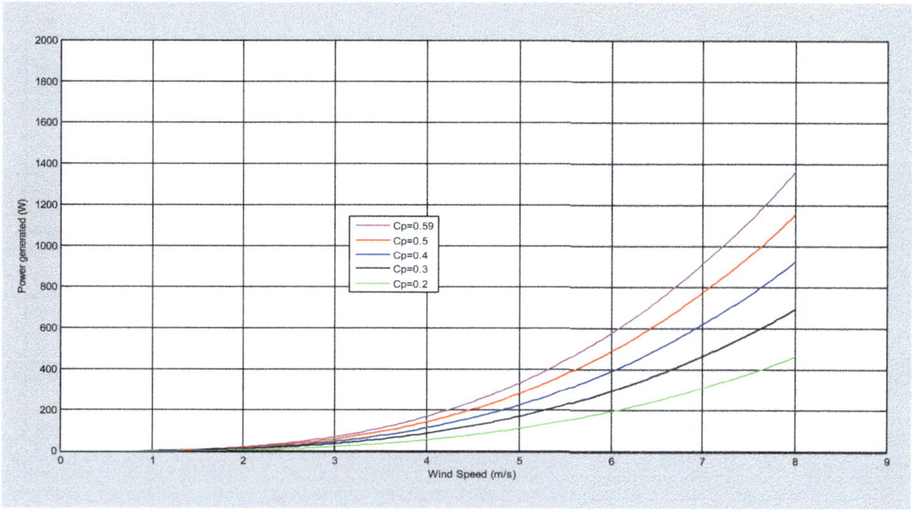

Fig. 14. Power-speed curve for different values of Betz Coefficient.

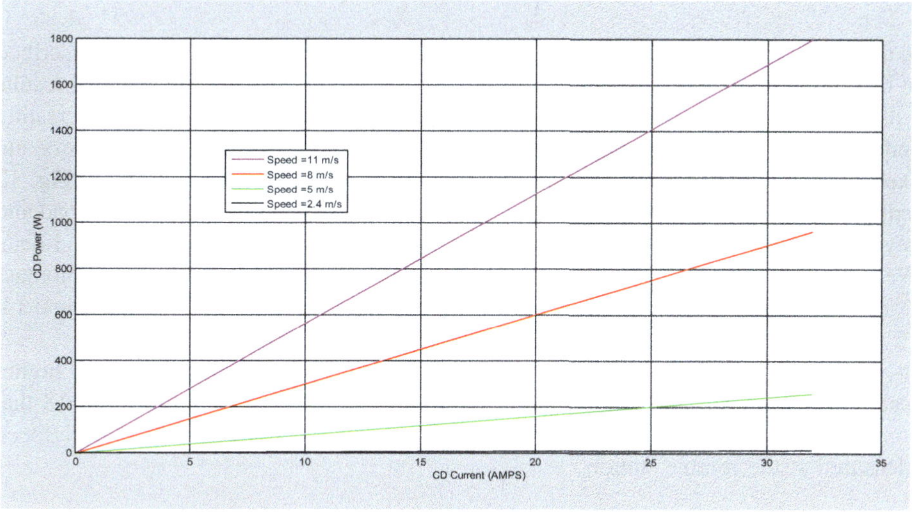

Fig. 15. DC Power- DC Current for wind Speed (m/s).

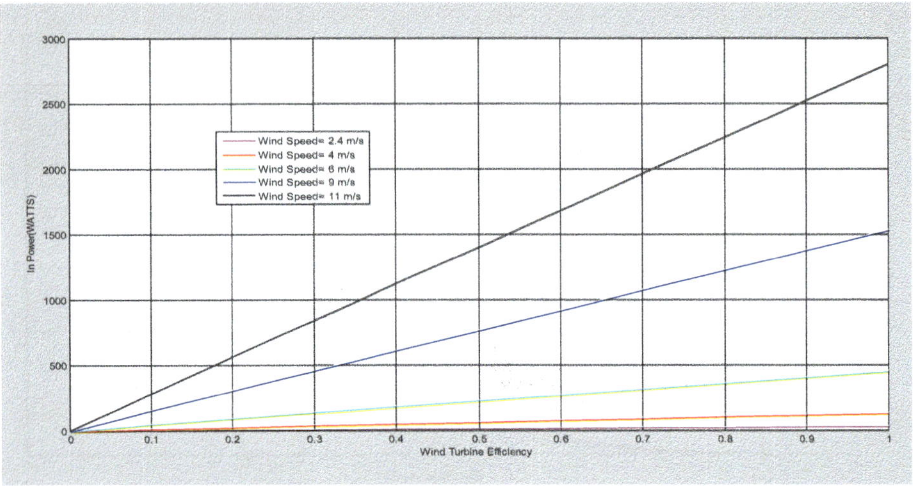

Fig. 16. Energy conversion efficiency at various wind speeds

6 Validation of Simulation Model

In this paper we also sought to validate our prediction of the numerical model described in Eqs. (1 to 23), we have constructed Figs. 17, 18 and 19 to compare expected results with data presented in the literature for biomass and wind turbine. It is very interesting and evident in Fig. 17 that the prediction of the model compares rather with theory and experimental data [28, 29] to various biomass loads. However, the analysis of Fig. 17 indicates that our model predicted very well the biomass data up to the 160 t/d load and beyond that point there were some discrepancies between model prediction and data. We believe that these discrepancies are due to the fact that data on the high biomass load were not fully taken into account heat transfer losses at too large loads (Figs. 14, 15 and 16).

Although the production of electric energy due to the wind source is not of higher incidence, we managed to identify that the theoretical curve indicated in Fig. 18 that comes from Eq. 11 has a pattern of behavior quite similar to the data of power-speed Measured at the remote station.

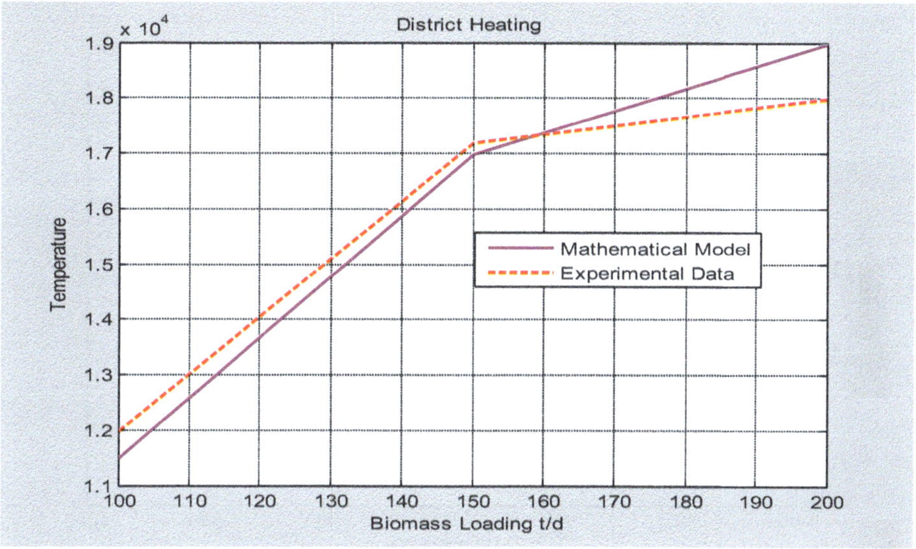

Fig. 17. Comparison with biomass loadings (t/d) at temperature.

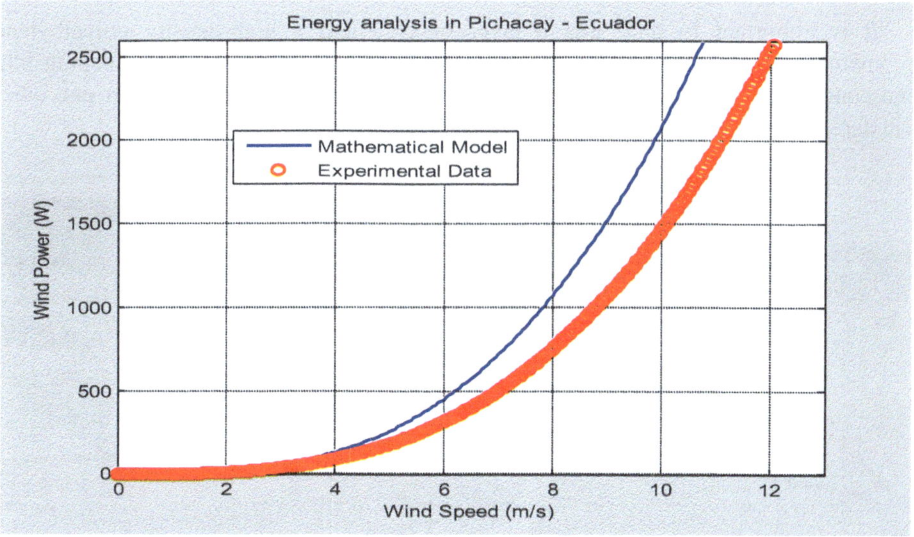

Fig. 18. Comparison of wind speed – wind power.

In Fig. 19 there are good results in the analysis of the model used with the data taken in the field in direct relation to the energy production referring to the variation of temperature, which is why our model is very accurate and very reliable.

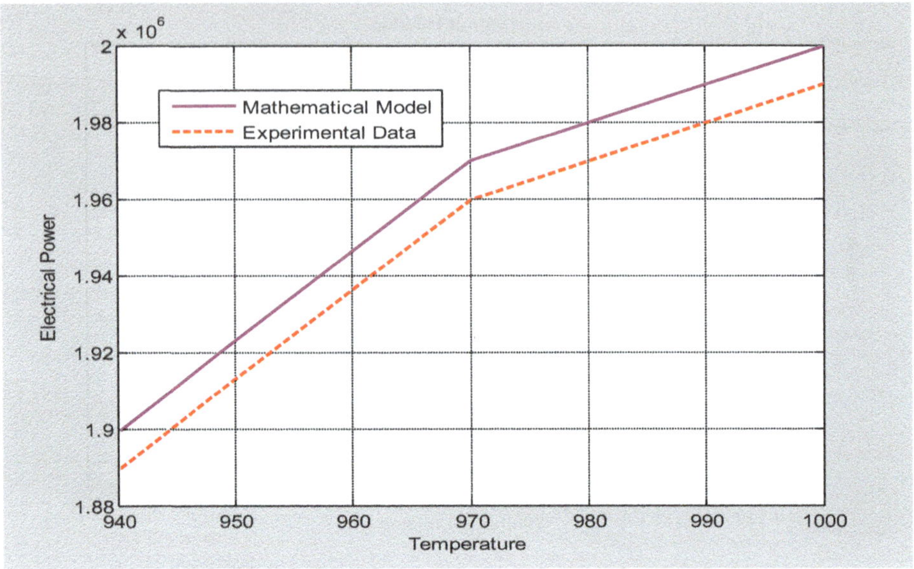

Fig. 19. Comparison of temperature – electrical power.

It is important to note that a layer of clay and a high-density polyethylene geomembrane as shown in Fig. 20 are placed to prevent leachates from seeping into and contaminating ground water or surface waters from the Quingeo River and other tributaries of the sector.

Fig. 20. Conformation and waterproofing of the bottom floor.

7 Conclusions

The energy conversion equations describing the total power generated by a hybrid system of wind turbine, biomass integrated ORC-combined heat and power cycle have been solved and presented. The biomass data also illustrates that the higher the biomass loading the higher the efficiency.

Furthermore, the wind turbine study results showed that the higher the wind speed the accelerated increase in the current. Consequently, the higher the wind speed the higher the electrical power.

This project that starts from the academy, is being developed from the generation of energy from biomass in the Pichacay landfill and that it is possible to include wind generation sources to form hybrid systems.

It is also possible to incorporate other sources of renewable energy generation that can be included progressively, such as solar and be of contribution to supply the growing demand for energy to the City of Cuenca and its surroundings, such as electric vehicles, domestic electric stoves, new road routes, with vision towards an intelligent network.

The analysis carried out in this article in different load conditions traces scenarios of electric power production, however at no time is definitive, it is based on the increasing income of organic waste that comes from the City of Cuenca, to greater amount of greater waste amount of energy produced and the possibility of forming hybrid systems.

It has not been possible to compare with other models used since they have not met the operating characteristics such as Pichacay in the Canton Cuenca with the intention of forming hybrid systems and that have the tendency to form an intelligent network with respect to the load.

It is expected that for a new investigation data will be available in a longer period of time and take advantage of other studies in the maturity phase with which comparisons can be made according to the literature proposed.

Finally, the proposed model predicted results and compared fairly with data under various biomass loading conditions.

Acknowledgement. The research work presented in this paper was made possible through the support of the EMAC EP and officials of the municipal GAD of Cuenca who provided me with the facilities for research.

References

1. Huerta, F., Gruber, J., Prodanovic, M., Matatagui, P.: Powerhardware- in-the-loop test beds: evaluation tools for grid integration of distributed energy resources. EEE Ind. Appl. Mag. **22** (2), 18–26 (2016)
2. Papaspiliotopoulos, V.A., Korres, G.N., Kleftakis, V.A., Hatziargyriou, N.D.: Hardware-In-the-Loop design and optimal setting of adaptive protection schemes for distribution systems with distributed generation. IEEE Trans. Power Delivery **32**(1), 393–400 (2015). https://doi.org/10.1109/TPWRD.2015.2509784

3. Kotsampopoulos, P.C., Lehfuss, F., Lauss, G.F., Bletterie, B., Hatziargyriou, N.D.: The limitations of digital simulation and the advantages of PHIL testing in studying distributed generation provision of ancillary services. IEEE Trans. Industr. Electron. **62**(9), 5502–5515 (2015)
4. Wang, X., Li, Y.W., Blaabjerg, F., Loh, P.C.: Virtual-impedance-based control for voltage-source and current-source converters. IEEE Trans. Power Electron. **30**(12), 7019–7037 (2014)
5. Calis, H.P., Haan, J.P., Peppink, G., Boerigter, H., Van der Drift, B., Venderbosch, R.J., Faaij, A.P.C., Van den Broek, R.: Technical and economic feasibility of large scale synthesis gas production in the Netherlands from imported biomass feedstock – a Strategic Decision Analysis study. Report prepared by: Shell Global Solutions International B.V., Energy Research Center of the Netherlands, Biomass Technology Group B.V., Department of Science, Technology and Society – Utrecht University, Ecofys B.V., sponsored by the Agency for Research in Sustainable Energy (SDE project number P2001-008), Holand (2003)
6. Hoogwijk, M., Faaij, A., Eickhout, B., de Vries, B., Turkenburg, W.: Global potential of biomass for energy from energy crops under four GHG emission scenarios Part A: the geographical potential. The geographical potential. Biomass Bioenerg. **25**(2), 119–133 (2005)
7. Braber, K.: Anaerobic digestion of municipal solid waste: a modern waste disposal option on the verge of breakthrough. Biomass Bioenerg. **9**(1–5), 365–376 (1995)
8. Kaltschmitt, M., Rosch, C., Dinkelbach, L.: Biomass gasification in Europe. Institute of Energy Economics and the Rational Use of Energy (IER), University of Stuttgart. Report prepared for the European Commission, DG XII, EUR 18224 Germany (1995)
9. Hillring, B.: Rural development and bioenergy – experiences from 20 years of development in Sweden. Biomass Bioenerg. **23**(6), 443–451 (2002)
10. Matrakidis, C., Orphanoudakis, T.G., Stavdas, A., Fernandez-Palacios Gimenez, J.P., Manzalini, A.: HYDRA: a scalable ultra long reach/high capacity access network architecture featuring lower cost and power. J. Lightwave Technol. **33**(2), 339–348 (2014)
11. Naber, J. E., Goudriaan, F., Louter, A.S., Biofuel, H., Engineers, S.: Further development and commercialisation of the Shell Hydrothermal Ugrading process for biomass liquefaction. In: Proceedings of the 3rd Biomass Conference of the Americas, Montreal. pp. 1651–1659 (1997)
12. Morris, M., Waldheim, L., Faaij, A. Stahl, K.: Status of large-scale biomass gasification and prospects. In: Kurkela, B., Knoef, H. (eds.) Handbook Biomass Gasification (2005)
13. Rodriguez, M., Faaij, A., Walter, A.: Techno-economic analysis of co-fired Biomass Integrated Gasification/Combined Cycle systems with inclusion of economies of scale. Energy Int. J. **28**(12), 1229–1258 (2003)
14. Solantausta, Y., Bridgwater, T., Beckman, D.: Electricity production by advanced biomass power systems. VTT Technical Research Centre of Finland (1996)
15. Tijmensen, M.J.A., Faaij, A.P.C., Hamelinck, C.N., van Hardeveld, M.R.M.: Exploration of the possibilities for production of Fischer Tropsch liquids via biomass gasification. Biomass Bioenerg. **23**(2), 129–152 (2002)
16. Gustavsson, L., Borjesson, P., Johansson, B., Svenningsson, P.: Reducing CO_2 emissions by substituting biomass for fossil fuels. Energy **20**, 1097–1113 (1995)
17. Howell, J.R., Bannerot, R.B., Vliet, G.C.: Solar-Thermal Energy Systems: Analysis and Design. Mcgraw-Hill College, New York (1982)
18. Azar, C., Larson, E.: Bioenergy and land-use competition in the Northeast of Brazil. A case study in the Northeast of Brazil. Energy Sustain. Dev. **4**(3), 64–72 (2000)

19. Berndes, G., Hoogwijk, M., van den Broek, R.: The potential contribution of biomass in the future global energy supply – a review of 17 studies. Biomass Bioenergy **25**(1), 1–28 (2003)
20. Elliott, P., Booth, R.: Brazilian biomass power demonstration project. Shell Centre (1993)
21. Dornburg, V., Faaij, A.: Efficiency and economy of wood-fired biomass energy systems in relation to scale regarding heat and power generation using combustion and gasification technologies. Biomass Bioenerg. **21**(2), 91–108 (2001)
22. Damen, K.: Future prospects for biofuel production in Brazil – a chain analysis comparison of ethanol from sugarcane and methanol from Eucalyptus in Sao Paulo State. Department of Science. Technology & Society, Utrecht University, NW&S-E-2001-31 68 (2001)
23. White, D.C., Ringelberg, D.B.: Utility of the signature lipid biomarker analysis in determining the in situ viable biomass, community structure, and nutritional/physiologic status of deep subsurface microbiota. In: Microbiology of the Terrestrial Deep Subsurface. CRC Press (2018)
24. Buhler, R.: Fixed bed gasication for electricity generation T application in Europe. In: Kaltschmitt, M., Bridgwater, T. (eds.) Biomass Gasication and Pyrolysis. State of the Art and Future Prospects. PyNE, Newbury (1997)
25. Mukunda, H.S., Dasappa, S., Paul, P.J., Rajan, N.K.S., Shrinivasa, U., Sridhar, G., Sridhar, H.V.: Fixed bed gasication for electricity generation. In: Kaltschmitt, M., Bridgwater, T. (eds.) Biomass Gasication and Pyrolysis. State of the Art and future prospects. PyNE, Newbury (1997)
26. Alvarez, D.O.I.: Modelado, simulación y construcción de una turbina de viento DIAWIND-A2 como una nueva alternativa de generación eléctrica en áreas rurales de Ecuador. Killkana Técnica **1**(3), 9–16 (2017)
27. El Mercurio. http://www.elmercurio.com.ec/342627-electricidad-domestica-con-la-fuerza-del-viento/. Accessed 17 July 2019
28. Fahmy, F.H., Nafeh, A.A., Ahamed, N.M., Farghally, H.M.: A simulation model for predicting the performance of PV powered space heating system in Egypt. In: 2010 International Conference on Chemistry and Chemical Engineering, vol. 1(8), pp. 173–177. IEEE, Japan (2010)
29. Icaza, D., Córdova, F., Toledo, J., Carlos, C., Lojano, A.: Modeling and simulation of a hybrid system solar panel and wind turbine in the locality of Molleturo in Ecuador In: ICRERA 2017, San Diego CA, USA, pp. 620–625 (2017). https://doi.org/10.1109/ICRERA.2017.8191134
30. Icaza, D., Sami, S.: Modeling, simulation and stability analysis using MATLAB of a hybrid system solar panel and wind turbine in the locality of puntahacienda-quingeo In Ecuador. Int. J. Manag. Sustain. **7**(1), 1–24 (2018). https://doi.org/10.18488/journal.11.2018.71.1.24
31. Icaza, D.: Modeling, simulation and construction of the D-ICAZA-A1 wind turbine destined for the rural areas of Ecuador. In: Innovative Smart Grid Technologies Conference - Latin America (ISGT Latin America). IEEE PES, Quito (2017)

Fractal Control Design with Anti-windup Effect for Optimal Operation of a Power Flyback Source

Jesús Rodríguez-Flores(✉), Víctor Isaac Herrera, Andrés Morocho-Caiza, and Christian Merino

Facultad de Informática y Electrónica, Escuela Superior Politécnica de Chimborazo, Riobamba, Ecuador
{jesus.rodriguez, isaac.herrera, andres.morocho, c_merino}@espoch.edu.ec

Abstract. A flyback source was designed together with a control study which allows its safe voltage range operation. The performance of a fractal PID type control was evaluated, starting from full conditions which allow the controller initialization considering lineal behaviors in a nominal point of operation corresponding to a case study of 2000 W flyback source. To achieve the proposed targets, a non-linear model was developed which reproduces the commuted behavior with the elements necessary to close the control loop. The use of the closed-loop controller took place by implementing an "in series" filtering strategy operated by the controller that avoids overcompensation and control hits, introduced by the fractal derivative prediction. To get a better adjustment on each of the controller effects, an individual fractal index was incorporated for each one of them. The source behavior was evaluated with and without the controller, and the performance of the entire and fractal controller was compared with the same values for the Kp, Ti and Td parameters, demonstrating that the fractal structure is able to improve the operating conditions. The evaluation was carried out with a selected set of static load, and with dynamic tests of both gradual and sudden load variation in time. The capacity of the fractal PID control to correct the deviation during the different tests was demonstrated.

Keywords: PID fractal control · Anti-windup · Flyback source · Filters · Feedback

1 Introduction

Progress in the fractal calculation has been significant, as an application, since the 1960s in the XX century [1]. How we see the fractal calculation leads to an integral concept with a certain degree of derivation or vice versa, and even if its form calculated in the past makes its implementation more complicated, approximate forms allow and facilitates it [2].

The fractal calculation approximation usually appears discreetly and rationally for its implementation [3], the reason for which an approximate order must be chosen. In this article, we work with a sixth-grade approximation of the fractal integral using

J. Nummenmaa et al. (Eds.): CSEI 2019, AISC 1078, pp. 308–328, 2020.
https://doi.org/10.1007/978-3-030-33614-1_21

Tustin. This approach leads to a high computational cost; however, with the advancement of technology, computing time is no longer a problem, due to the use of increasingly efficient micro processed systems.

The introduction of the fractal calculation has been investigated by different authors [3–8] in both fields of mathematics and control engineering. This interest in the topic leads to the use of a fractal control in the design of a Flyback switched source, of variable and non-linear nature [9]. Similar work in terms of power electronics is observed in the design of the dc-dc converter of switched capacitors presented in [10]. The investigations [11, 12], approach the switched sources modeling issue, based on the averaged state space. However, this technique suppresses non-linear and high-frequency behaviors that are necessary to take into account when designing the controller in a source with real implementation. For this reason, the presented model reproduces the entire dynamics, replicating the nonlinearities with all their spectral component of the Flyback source.

Ultimately, this article aims to provide strategies for the initialization of the PID controller with the incorporation of an in-series filter both of a fractal nature, as well as the incorporation of a multiple-pole feedback filter with response time equal to the inverse of the commutation frequency. Besides, the methodology in the fractal implementation with the incorporation of a delay for the breaking of the algebraic loop inherent to the Tustin method is proposed. Along with the flyback source design, closed mathematical expressions are provided aiming to give a starting point for the tuning of the PID controller with the incorporation of a serial filter.

2 Methodology

The procedure to develop the fractal controller applied to the Flyback source entails, first of all, the development of the Flyback source, which state of the art is in permanent review to implement different variants in the modeling process. In particular, a non-linear model will be implemented, of a descriptive nature, which is supported in an ideal transformer modeled as controlled sources, and the presence of the exponential model of the diode to guarantee the blocking of current, thus preventing the return of current to the energy source.

2.1 Calculation of Construction Parameters of the Flyback Source Under the Consideration of Constant Current

The development of the Flyback source has been designed under the following considerations: operation in permanent regime, without considering losses in the switching (therefore in the contempt of the transients involved in the transition of states of the semiconductor), average value of the voltage variation and current variation (equal to zero, for both the inductor and the capacitor), average value of constant voltage in the capacitor and average value of constant current in the inductor. These considerations allow establishing the following parameters and Eqs. 1–12, with their corresponding results.

Parameters from Eqs. 1–5 allow calculating the duty cycle of switching by Eq. 6.

$$f = 50\ [\text{kHz}], \tag{1}$$

$$N_1 = 30, \tag{2}$$

$$N_2 = 300, \tag{3}$$

$$V_s = 12\ [\text{V}], \tag{4}$$

$$V_o = 120\ [\text{V}], \tag{5}$$

$$D = \frac{\frac{V_o}{V_s}\frac{N_1}{N_2}}{1 + \frac{V_o}{V_s}\frac{N_1}{N_2}} = \frac{1}{2}[-] \tag{6}$$

Parameters from Eqs. 7–10 allow calculating the magnetizing inductance and capacitor of the source by Eqs. 11 and 12, respectively.

$$P_o = 2000\ [\text{W}], \tag{7}$$

$$R_o = \frac{V_o^2}{P_o} = 7.2[\Omega], \tag{8}$$

$$\frac{\Delta I_{Lm}}{I_{Lm}} = \frac{20}{100}[-], \tag{9}$$

$$\frac{\Delta V_o}{V_o} = \frac{2}{100}[-], \tag{10}$$

$$L_m = \left(\frac{N_1}{N_2}(1 - D)\right)^2 \frac{R_o}{\frac{\Delta I_{Lm}}{I_{Lm}} f} = 1.8\ [\mu\text{H}], \tag{11}$$

$$C = \frac{D}{\frac{\Delta V_o}{V_o} R_o f} = \frac{625}{9}[\mu F], \tag{12}$$

where f is the switching frequency of the PWM, N_1 and N_2 are the windings of the primary and secondary, respectively, V_s and V_o are the input and output voltages of the source and ΔV_o is the variation of the output voltage. L_m is the magnetizing inductance.D is the duty cycle of switching. P_o and R_o are the output power and load resistance. I_{Lm} is the magnetizing current and ΔI_{Lm} is its variation. C is the capacitor of the source.

In the design considerations of the Flyback source, the calculation of a magnetizing inductance is established, with the criterion of current curl. This current curl criterion allows guaranteeing a continued current flow in the transformer, at the loading ideal operating point, which performance guarantees less stress for the power components, in particular concerning the blocking times and reverse recovery of the source diode.

2.2 Simplified Mathematical Model of the Open-Loop Operation of the Flyback Source, Criteria for Sampling Time Selection

The basic development considerations of a Flyback model source are based on two states that constitute the essence that breaks with non-linearity. However, for analytical purposes, it is necessary to omit the non-linearity of elements such as the diode, the element of commutation and others associated with electromagnetic effects in the transformer, which do not add relevant importance to the analysis.

For model development, a passive circuit is considered. However, the behavior of the transformer is reproduced, in a simplified way, taking into account the controlled sources, for each of the operating states. Finally, the operating states are associated with the opening or closing of the power element used in the switching. Figure 1 shows the Flyback source circuit, while Figs. 2 and 3 show the states, under the simplified analysis, to obtain a mathematical model.

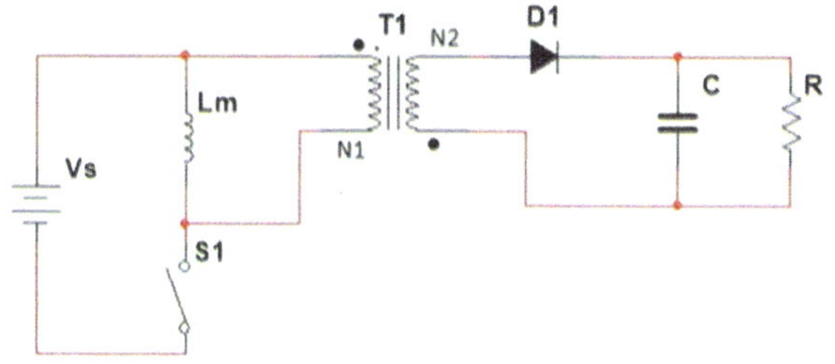

Fig. 1. Flyback source.

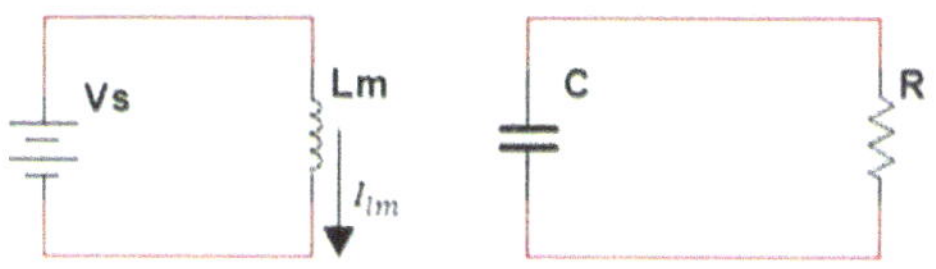

Fig. 2. Flyback source closed switch.

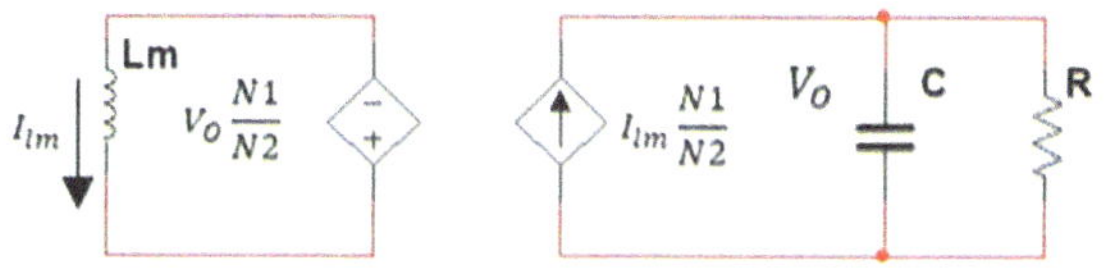

Fig. 3. Flyback source, open switch.

The operation models with bases in the switch state, allow conceiving the following law for the current-voltage. Both equations, both 13 and 14, show the behavior of voltage and current in the periods of 0 to DT and DT to T, respectively.

Equations 13 and 14 present the variables of second-degree systems (time constant of the second-degree system, []), (natural frequency, [rad/s]), (damping frequency, [rad/s]), and (damping constant, [−]), corresponding to the values calculated in Eqs. 15 and 16, with the variables already defined in previous equations.

$$V_O(t) = \begin{cases} V_{O_{off}} \cdot e^{\frac{-t}{RC}}[V] \\ \forall\, 0 \leq t < DT(with\, switch\, closed) \\ \frac{N_1}{N_2} \cdot L_m \cdot I_{Lm_{on}} \cdot \frac{\varpi_n^2}{\varpi_d} \cdot e^{-\sigma(t-DT)} \cdot sin(\varpi_d(t - DT)) \\ + V_{O_{on}}[V] \\ \forall\, DT \leq t < T(with\, switch\, opened) \end{cases} \tag{13}$$

$$I_{Lm} = \begin{cases} I_{Lm_{off}} + \frac{V_S}{L_m} t[A] \\ \forall\, 0 \leq t < DT(with\, switch\, closed) \\ I_{Lm_{on}} \cdot \frac{\varpi_n}{\varpi_d} \cdot e^{-\sigma(t-DT)} \cdot \sin\left(\varpi_d(t - DT) + \tan^{-1}\frac{\varpi_d}{-\sigma} + \pi\right) - \\ \frac{V_{O_{on}}}{L_m} \cdot \frac{N_1}{N_2} \cdot \frac{1}{\varpi_d} \cdot e^{-\sigma(t-DT)} \cdot \sin\left(\varpi_d(t - DT)\right)[A] \\ \forall\, DT \leq t < T(with\, switch\, opened) \end{cases} \tag{14}$$

$$\varpi_n = \frac{N1}{N2} \cdot \frac{1}{\sqrt{L_m C}} \left[\frac{\text{rad}}{\text{s}}\right] \tag{15}$$

$$\xi = \frac{1}{2} \cdot \frac{N2}{N1} \cdot \frac{\sqrt{L_m C}}{RC} [-] \tag{16}$$

where $V_{O_{on}}$ is the initial voltage condition when the source switch is opened. $V_{O_{off}}$ is the voltage initial condition when the source switch is closed. $I_{Lm_{on}}$ is the current initial condition when the source switch is opened. $I_{Lm_{off}}$ is the current initial condition when the source switch is closed.

An iterative process, in which, the condition of the Flyback source state represents the condition of the next state, holding its bases in Eqs. 13 and 14, we obtain the graphs of the current-voltage behavior at the capacitor and inductor levels respectively, based on this information, establishes the criterion for the determination of the sampling frequency of the study.

As shown in Fig. 4, the design condition allows establishing the periodicity of the voltage and current signal of the source under the nominal design conditions. The periodicity of the signals is determined to start from Eq. 17. Where T is the period of the damping frequency of the equivalent system, T_S is the sampling time and f is the switching frequency.

$$T = \frac{2\pi}{\varpi_d} \,[\text{s}] \tag{17}$$

$$400T_S \leq \frac{1}{f} \ll \frac{2\pi}{\varpi_d} \tag{18}$$

To ensure a good sampling time, 400 samples are taken in a quarter of the period of the switching signal. Therefore, the sampling time is firstly determined by using Eq. 18.

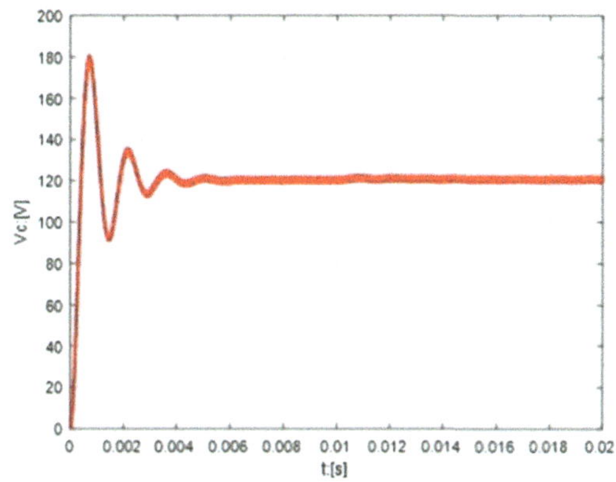

a) Voltage response

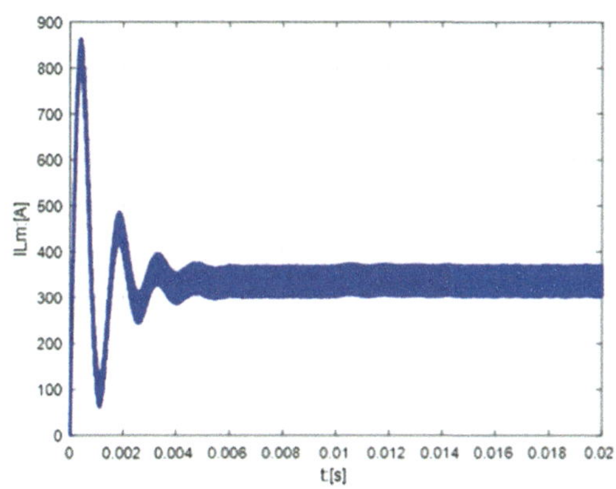

b) Current response

Fig. 4. Attain voltage-current behavior for the simplified model.

2.3 The Simplified Model of the Open-Loop Operation of the Flyback Source, Considering the Non-linear Model of the Diode

The block diagrams of Figs. 5 and 6 reproduce, in a simplified manner, the behavior of the Flyback source when the switch is closed and open, respectively, and when the design load is considered.

The source modeling allows the evaluation of its behavior, before the variation of the duty cycle, and the load variation. This study allows determining the impact on the source behavior, before the variation of these parameters.

Figure 7 shows how the voltage of the source behaves when it varies from a tenth to a thousand times its value. As shown in Eq. 16, the increment of the ohmic value of the load decreases the value of the damping constant. With this behavior appears an increase in the overshoot of the source before it is established. Finally, it is observed that the study requires a logarithmic spacing of the ohmic value in order to show its effects on the tension.

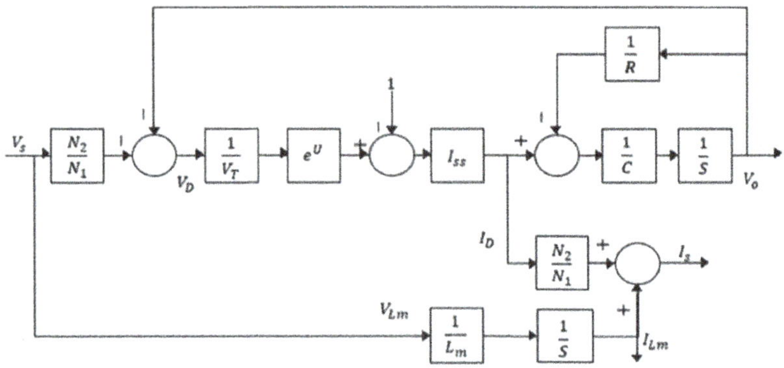

Fig. 5. Model of the Flyback source with the switch closed.

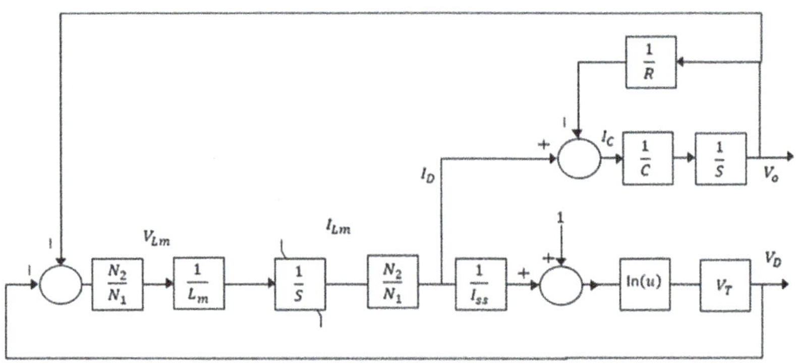

Fig. 6. Model of the Flyback source with the switch open.

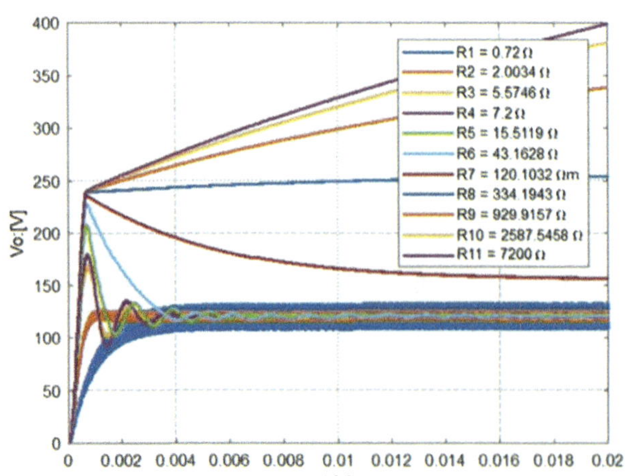

Fig. 7. Behavior of the Flyback source without regulation and with load variation.

When the value of the duty cycle of the switching increases linearly from 20% to 80%, there is a relationship with the overshoot. This allows concluding that the increase in the duty cycle of the switching produces an equivalent increase of the damping of the system and, therefore, decreases the voltage overshoot of the source, as shown in Fig. 8.

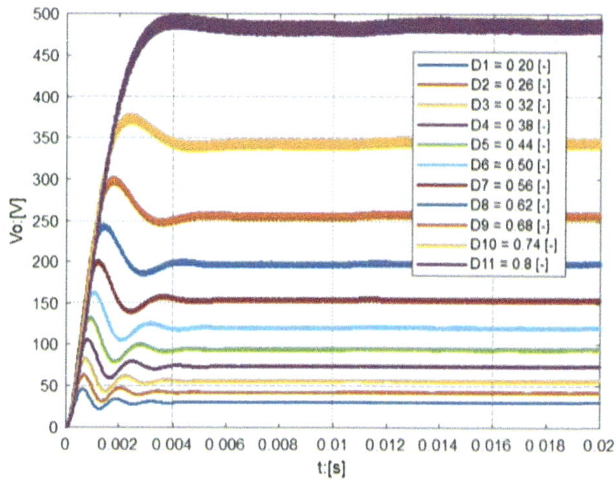

Fig. 8. Output Voltage of the Flyback source for different values of *D*.

2.4 Second Order Reduced Model for the Flyback Source Under the Design Conditions

As a starting point, the performance of the Flyback source is evaluated under the design conditions, the average establishment value is determined, and the gain at frequency 0. The damping coefficient and the value of the simplified model are determined with the transient and spectral study, respectively.

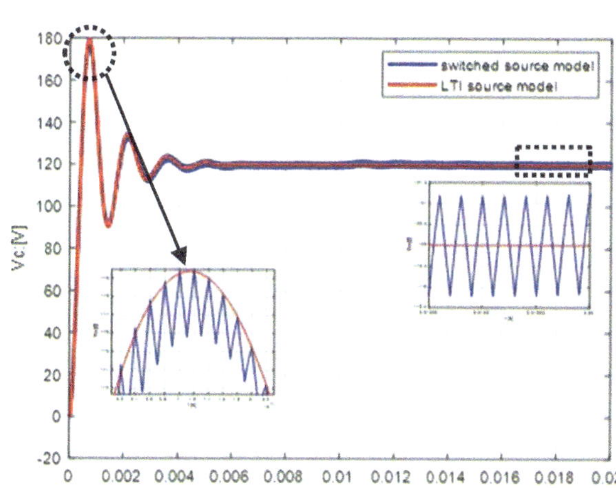

Fig. 9. The response of the Flyback Source under design conditions and that of its simplified linear model.

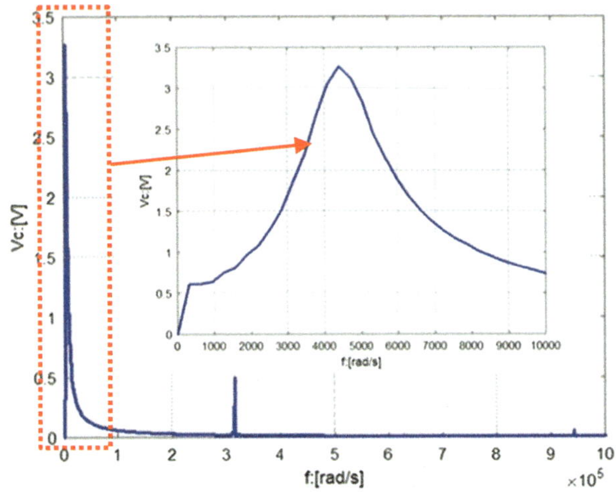

Fig. 10. The frequency response of the Flyback Source with the suppression of the continuous level.

Considering that the power supply is 12V and the graphic result shown in Fig. 9, for LTI source model, the damping constant is calculated with Eq. 19.

$$\xi = \frac{\left|ln\left(max(V_0) - \overline{V_0}\right)\right|}{\sqrt{\pi^2 - \left(ln\left(max(V_0) - \overline{V_0}\right)\right)^2}} = 0.21845489[-] \tag{19}$$

$$K = 10 \tag{20}$$

From the graphic result of the spectral study shown in Fig. 10, the natural frequency is calculated with Eq. 21.

$$\varpi_d = 4398.22971502 \left[\frac{\text{rad}}{\text{s}}\right] \tag{21}$$

Under the nominal design considerations of the source, the simplified model is expressed as shown in Eq. 22:

$$G(s) = \frac{K\varpi_n^2}{s^2 + 2\xi\varpi_n s + \varpi_n^2} \tag{22}$$

where K is the gain in the steady state, ξ the damping constant and ϖ_n the natural frequency of the LTI model of the switched source.

2.5 Development of the Entire PID Control for the Flyback Source

Only in cases when the entire PID is not able to satisfy the conditions of compensation, is when we proceed directly to design, starting from the considerations of the integral and/or fractal derivative. In order to achieve a complete PID satisfying the compensation conditions, filters are introduced, both for the measurement of the output variable as well as for the error signal. Additionally, a soft starting will be considered, establishing a controlled response based on the gradient. Figure 11 shows the blocks necessary to achieve the desired compensation effect.

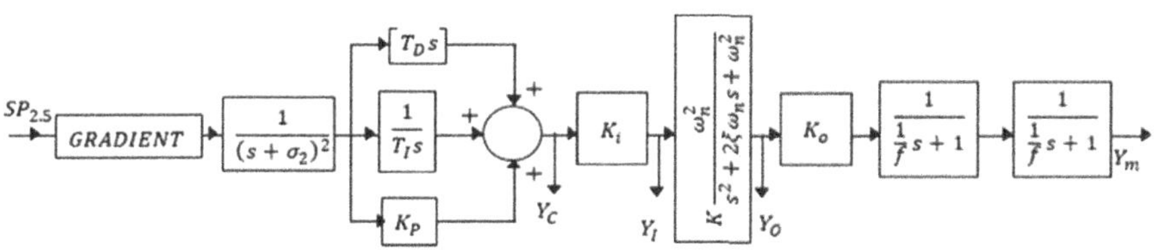

Fig. 11. Open control loop for controller design.

The system variables correspond to the control output Y_C, the physical input Y_I, the physical output Y_O, and the measurement output Y_m. In the design, it is considered the nominal voltage of the setpoint (in this case 2.5 [V]) to be able to plan the conditioning both for input gain Y_m and output gain K_O. Where K_I converts the output of the control, with values between 0 and 2.5 [V], in values seen by the source between 0 and 12 [V], and K_O converts the output voltage of the 120 [V] source to a proportional value of 2.5 [V] that they are measured from feedback. K_I and K_O are calculated by Eqs. 23 and 24, respectively.

$$K_I = \frac{12}{2.5}[-] \tag{23}$$

$$K_O = \frac{2.5}{120}[-] \tag{24}$$

An operation condition is established, which allows finding control parameters that the system can fulfill, for it is established a setup time of 0.01597063 s and an overshoot of 2%. Therefore, the parameters of the new pole S_n are defined by Eqs. 25 and 26. The new pole is defined in Eq. 27.

$$\xi_n = 0.779703267412072[-] \tag{25}$$

$$\omega_n' = 513.015677525179 \left[\frac{\text{rad}}{\text{s}}\right] \tag{26}$$

$$S_n = \omega_n' \angle \varphi_n \tag{27}$$

where ξ_n and ω_n' are parameters of the new pole S_n and φ_n is its phase.

The Flyback source is modeled as a transfer function corresponding to measurement and control signals ratio (see Fig. 11), as shown in Eq. 28.

$$G(s) = \frac{Y_m(s)}{Y_C(s)} \tag{28}$$

The parameters of the plant allows establishing Eqs. 29–35, to form a starting point to design the controller. The effect of the new pole S_n on the plant is represented in a phasor manner using the magnitude K_g and phase φ_g.

$$K_g = \left|\frac{G(S_n)}{S_n}\right| \tag{29}$$

$$\varphi_g = \angle\left(\frac{G(S_n)}{S_n}\right) \tag{30}$$

$$\sigma_1 = \omega_n' \frac{sin\left(\frac{4}{5}\varphi_n + \frac{\pi + \varphi_g}{4}\right)}{sin\left(\frac{1}{5}\varphi_n - \frac{\pi + \varphi_g}{4}\right)} \tag{31}$$

$$\sigma_2 = \omega_n' \frac{sin\left(\frac{4}{5}\varphi_n - \frac{\pi + \varphi_g}{4}\right)}{sin\left(\frac{1}{5}\varphi_n + \frac{\pi + \varphi_g}{4}\right)} \tag{32}$$

$$K_P = \frac{2\sigma_1\left(\left(\omega_n' cos(\varphi_n) + \sigma_2\right)^2 + sin(\varphi_n)^2\right)}{K_g\left(\left(\omega_n' cos(\varphi_n) + \sigma_1\right)^2 + sin(\varphi_n)^2\right)} \tag{33}$$

$$T_I = \frac{K_g\left(\left(\omega_n' cos(\varphi_n) + \sigma_1\right)^2 + sin(\varphi_n)^2\right)}{\sigma_1^2\left(\left(\omega_n' cos(\varphi_n) + \sigma_2\right)^2 + sin(\varphi_n)^2\right)} \tag{34}$$

$$T_D = \frac{\left(\left(\omega_n' cos(\varphi_n) + \sigma_2\right)^2 + sin(\varphi_n)^2\right)}{K_g\left(\left(\omega_n' cos(\varphi_n) + \sigma_1\right)^2 + sin(\varphi_n)^2\right)} \tag{35}$$

where σ_1 and σ_2 are parameters of the optimal location in the S plane, of the parameters of the controller and filter. Where K_P is the proportional gain, T_I the integral time and T_D the derivative time.

The verification of the control system in a closed loop, under the condition of integer integral, is implemented in a closed loop and is evaluated with the model proposed in Eq. 22, the results are observed in Fig. 12.

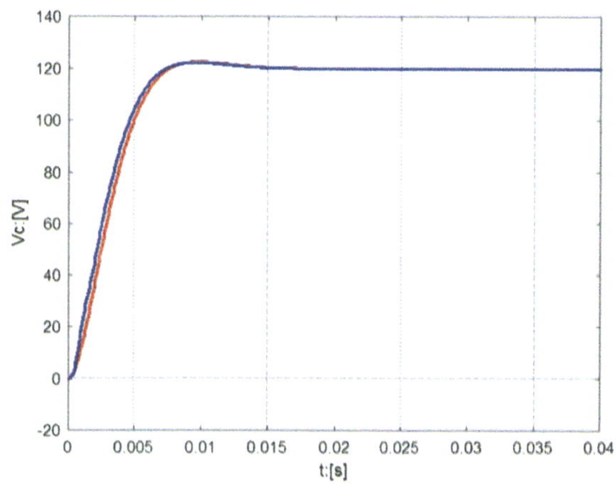

Fig. 12. System response: Ideal behavior (red curve), Flyback model with PID control (blue curve).

2.6 Development of the Fractal PID Control for the Flyback Source

Based on the development of the integer control, the implementation of the fractal control is made with the approximation of the Fractal Integral. It is discretized using Tustin, with the modification of adding a delay with the purpose of breaking algebraic loops, as shown in Eq. 36.

$$\frac{1}{s^{\alpha}} = \left(\frac{T}{2}\right)^{\alpha} \frac{\left(1 + \sum_{i=1}^{n} \frac{z^{-i}}{i!} \prod_{j=1}^{i} (\alpha - j + 1)\right) z^{-1}}{\left(1 + \sum_{i=1}^{n} \frac{(-1)^{i} z^{-i}}{i!} \prod_{j=1}^{i} (\alpha - j + 1)\right)} \tag{36}$$

where α is the fractal degree of the integral, T is the sampling time and n is the approximation order of fractal integral.

The representation in block diagrams for an approximation of the Fractal Integral [13, 14] of sixth order with an added delay, is as shown in the Fig. 13, being N_i the numerator coefficients and D_i the denominator coefficients ordered in descending order with respect to the degree of the polynomial term, where the literal u is the input of the fractal integral.

To achieve a better effect, the fractal integral is incorporated into the filter of the error signal and a multiple pole filtering for the retro feed. The output of the fractal control acts as a reference for comparison with a serrated tooth of amplitude 5, thus

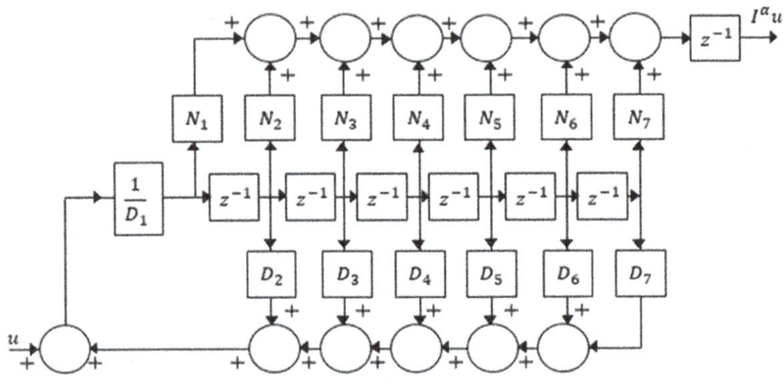

Fig. 13. Block diagram of a fractal integrator.

generating a PWM modulation. To avoid overvoltage and excessive integration, multiple anti-windup effects are incorporated, which considers both the control output and the voltage output of the source. The whole set operating in a closed loop is shown in Fig. 14, where Y_C is the control signal, Y_I is the PWM commutated signal, Y_m is the measurement signal, where Y_o is the output voltage of the switched source and the K_o the conversion factor of measure. It's clear that the parameters of the control PID are K_P, T_I, T_D.

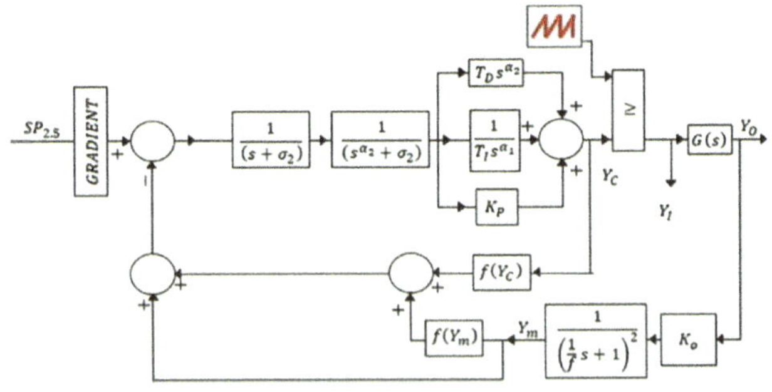

Fig. 14. Flyback source block diagram with control and fractal filter with anti-windup effect.

3 Results

Considering that the PID control starts with values of α equal to 1, the source the Flyback switched source takes charge by performing the output correction with respect to the setpoint, with the set point of 2.5 for a voltage output of 120 [V], this because the controller was initially parameterized to operate as an integer-order PID controller.

Once the starting condition has been evaluated, the values of each one of them are selected logarithmically until finding a better global performance based on the resistive values. The performance in open-loop observed in Fig. 7 is presented as improved when the loop is closed with the fractal controller and its respective integral and fractal filters, as shown in Fig. 15.

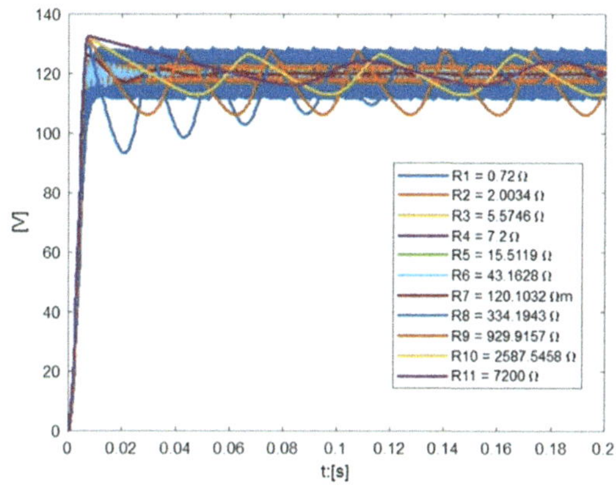

Fig. 15. The output voltage of the Flyback source for different values of R_O, for when it operates with the fractal PID controller.

Table 1 shows the resistive values used where 7.2 [Ω] is the nominal value to achieve the 2000 [W]. The square root of the mean square error, when the values of α start from the unit, and after adjusting, the comparison is made with respect to the ideal behavior taken as a reference of the response to a unit step of a second-order system with gain. These parameters being are defined in Eqs. 26 and 27 with a steady-state value of 120 [V].

Table 1. Results of the source performance before and after the adjustment of α_1, α_2.

Resistance [Ω]	$\sqrt{\frac{\varepsilon^2}{N}}$[V] $\alpha_1 = 1$ $\alpha_2 = 1$	$\sqrt{\frac{\varepsilon^2}{N}}$[V] $\alpha_1 = 0.988$ $\alpha_2 = 0,988$
0.7200	6.2189	6.0820
2.0034	3.1033	2.9678
5.5746	2.3149	2.2125
7.2000	2.2483	2.1422
15.5119	2.1328	2.0524
43.1628	2.1541	2.1108
120.1032	2.1968	2.1483
334.1943	9.4165	9.3769
929.9157	8.6189	8.4865
2587.5458	5.2037	5.2144
7200.0000	5.1431	4.8162

Due to the anti-windup effect, during the start of the Flyback source with different loads (72 Ω, 7.2 Ω, and 3.6 Ω), the reference value for the comparison with the sawtooth output is always the same with a value corresponding to the setpoint. This is why the trip command for the switching of the power element is the same for the three load cases as seen in Fig. 16.

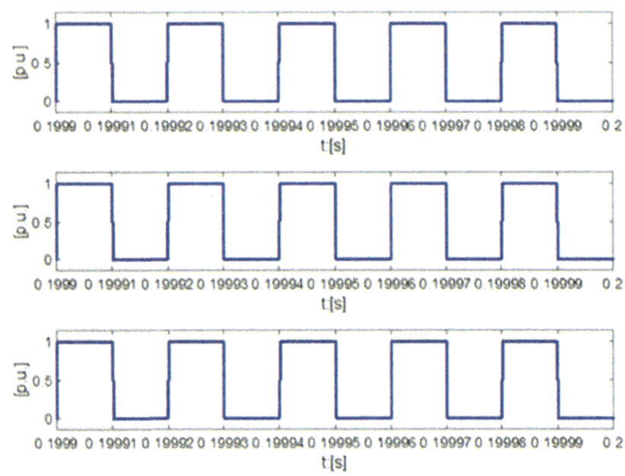

Fig. 16. During start-up, the source shows the same duty cycle pattern regardless of the load, due to the anti-windup effect (loads 72, 7.2 and 3.6 Ω).

Figure 17 is the result of the regulation with the output voltage of 120 [V]. In the voltage setting of the Flyback voltage source, the same firing pattern of the power switching element is observed, this behavior is shown in Fig. 18.

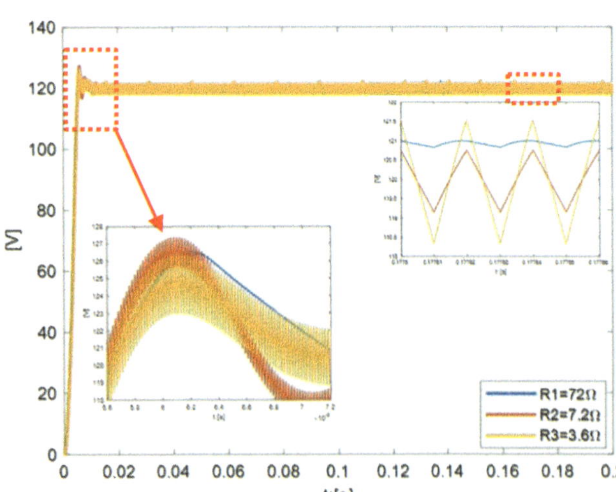

Fig. 17. Source voltage flyback with loads of 72, 7.2 y 3.6 Ω.

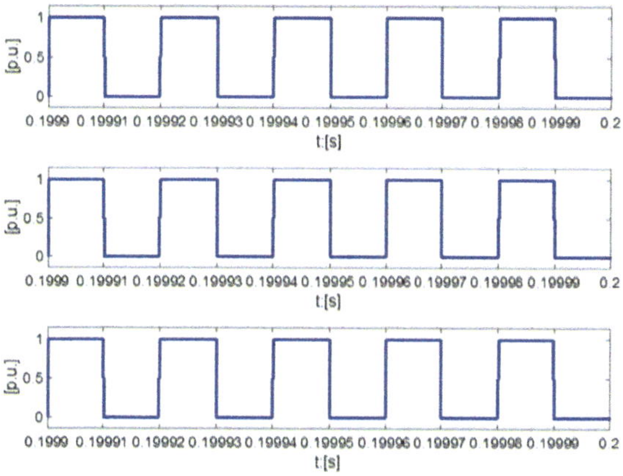

Fig. 18. At the steady state time, the source exhibits the same useful cycle pattern regardless of the load of 72, 7.2 y 3.6 Ω, for the switching of the power element.

The lower the power required to the source, the more complex is its regulation, being the double windup effect what manages to maintain the voltage in the environment of the set value. Figure 19 shows the behavior in the case of ohmic values of 7200 [Ω], 720 [Ω], and 72 [Ω]. Figure 20 shows the start pulses, which have a low duty cycle, and this results in a set point that rises slowly due to the gradient control.

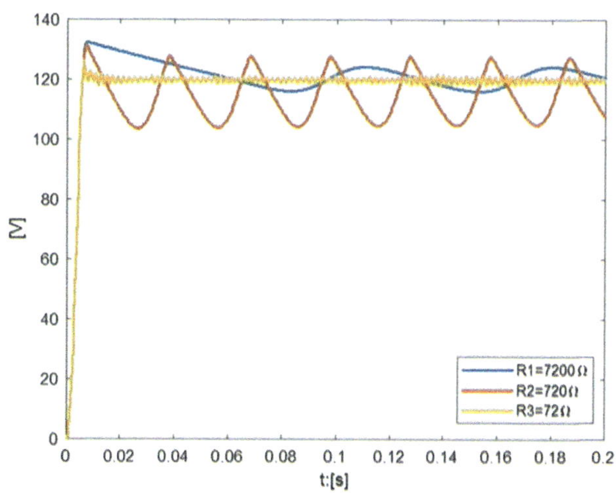

Fig. 19. Regulated voltage for loads of 1000, 100 and 10 times the nominal value, for each of the cases (R1, R2, R3).

However, once the 120 [V] is exceeded, the case of 7200 [Ω] and 720 [Ω] (1000 and 100 times the nominal ohmic value) the pulses disappear, as shown in Fig. 21. Finally, as can be observed in Fig. 22, during the steady-state, the load with ten times

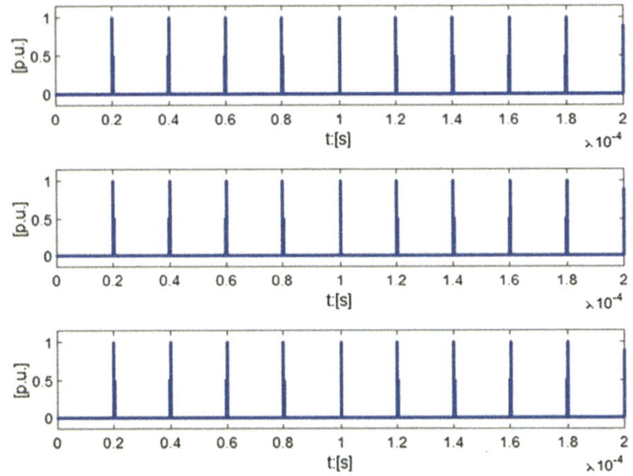

Fig. 20. Low duty cycle switching due to setpoint gradient control, for each of the cases (R1, R2, R3).

the nominal value presents a useful cycle in the environment of 50%. This value is much smaller for the ohmic value corresponding to 100 times the nominal value, and finally, not existing for value 1000 times the nominal ohmic value.

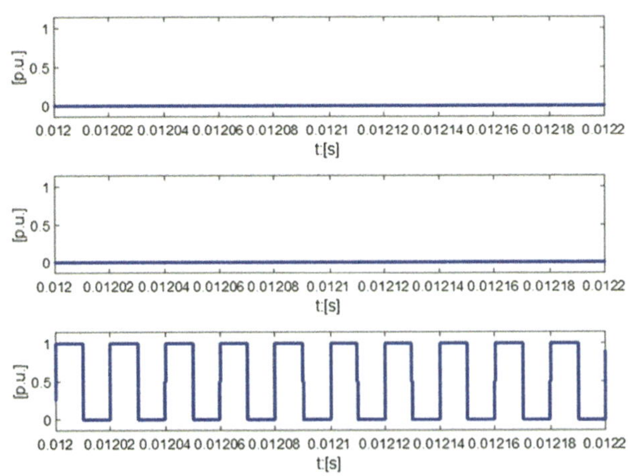

Fig. 21. Suppression of shots for loads of 1000 and 100 times the nominal value for when the output exceeds 120 [V] for the first time, for each of the cases (R1, R2, R3).

Regulation tests for a load variation between 72 [Ω] and 3.6 [Ω], with a frequency of 10 [Hz], can be observed in Fig. 23. They showed a deviation of 2.2243 [V] with respect to the second-order linear model.

Figure 24 depicts a more severe test performed with step changes, with load variation between 7.812 [Ω] and 6.588 [Ω] (variation of 17%), with a frequency of 10 [Hz]. This test showed a deviation of 2.2182 [V] respect to the linear model of second-order.

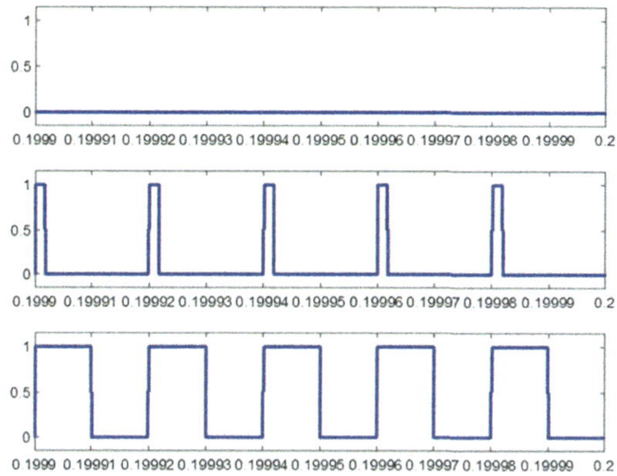

Fig. 22. Regulated commutation for the different loads, 1000, 100 and 10 times the nominal ohmic value, for each of the cases (R1, R2, R3).

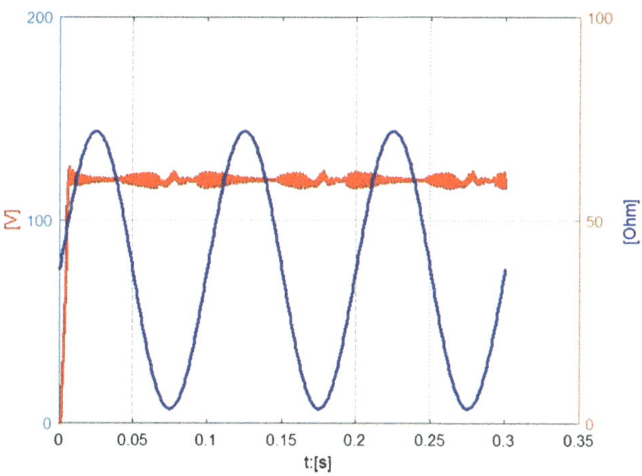

Fig. 23. Flyback source voltage, (red trace), load variation (blue trace).

4 Discussion

The findings of this study clearly show that a fractal control whose initial values are obtained from an integer-order PID controller with the same structure, present fractal values close to unity.

The Fractal Integral, in its definition, allows obtaining the fractal derivative. Considering that this is an approximate calculation, its discrete estimate is made using the Tustin approximation for the integer-order integral. However, in the present work, pure backwardness is incorporated, for any of the effects, both derivative and fractal integral, to avoid producing algebraic loops difficult to solve.

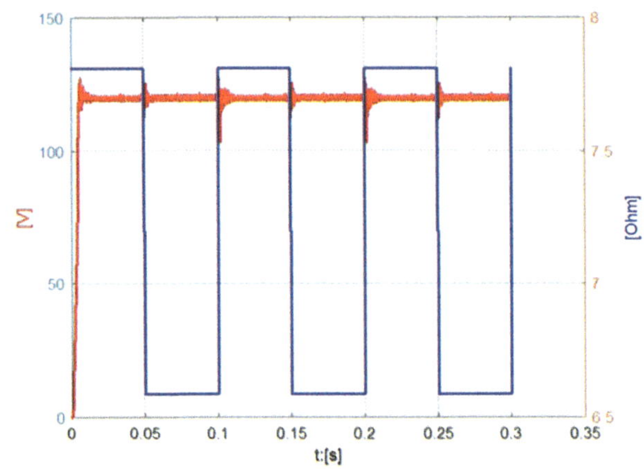

Fig. 24. Flyback source voltage, (red trace), load variation (blue trace).

The control action gets difficult when the power operation condition is distant from its nominal value, a situation that occurs when the load is reduced or when the ohmic value increases significantly.

When the source is working under load variation, the regulator is able to compensate with little deviation, even when the variation exceeds 1000 [%] of the design load resistive value, as long as the variation of the disturbance is smooth. However, this variation is significantly reduced to 17 [%] of the resistive design value when the variation is sudden, as shown in the step variation behavior.

The non-linear control that contemplates the anti-windup and setpoint with gradient control, allows maintaining the voltage value with little deviation during the ignition of the source.

The integration of filters both in the feedback as well as in the error signal allows reproducing a behavior of the controller with little disturbance to face changes in the switched source. Nevertheless, the non-integer integral order filters complicate to obtain the parameters of the controller, but they guarantee their operation and the capacity of the regulator to slow down the source response in conjunction with the gradient control of the setpoint change.

The reduction of noise in the feedback of the switched source is achieved by placing a filter with a response time equivalent to the inverse of the switching frequency, and whose multiplicity will depend on the quality of the filtering. In the present study, a multiplicity two filter was used in the feedback.

The design concept of the regulator must at all times consider that the control action is performed in low power or/and digitally. This way, the setpoint for the 120 [V] is 2.5, and when compared to a signal Sawtooth of amplitude 5 and minimum value 0, it generates a control signal for the switching of the power element. This concept must be present when determining the regulator parameters, as well as the voltage reduction at the source output to perform the feedback.

5 Conclusions

Although there are procedures to simplify the model of switched sources, these models do not reflect phenomena of interest that are sustained over time, of a non-linear nature, capable of destabilizing the control system. This is why non-linear modeling is used, considering each one of the operating states.

Is essential to obtain a linear model of the condition with greater overshoot than the one presented by the source in open loop. Considering the second-degree model obtained, and the one desired for its behavior, mathematical expressions were developed to obtain the parameters for a PID regulator of integer order and the filter of the error signal. The parameters seek to compensate the switched-source system in open loop; therefore, the switched source, the input and output gains, and the multiple-pole filtering must be considered.

The PID of integer order is a fractal PID with values for α equal to unity. The fractal implementation of the integrator and the derivative allowed a final adjustment in performance with respect to a standard second-degree behavior, for which the adjustments for each fractal effect were made separately.

Taking as reference the results presented in Table 1, starting from the values of an integer control and adjusting only the fractal effect, the integer control was improved by 2.40% by the fractal control.

The computational cost of the implementation of the fractal controller is relatively high with respect to the entire control; however, this is compensated with the goodness of fit that the system shows when modifying only the fractal indices of the control effects. Future work is focused on the implementation of the designed power flyback source on a test bench to experimentally evaluate the performance of the proposed fractal control. Besides, the ongoing work is focused on the use of fractal control in different power electronic applications such as hybrid systems.

References

1. Machado, J.A.T., Kiryakova, V.: The chronicles of fractional calculus. Fractional Calc. Appl. Anal. **20**(2), 307–336 (2017)
2. Chen, Y. Petras, I., Xue, D.: Fractional order control - a tutorial. In: American Control Conference, St. Louis, MO, USA, pp. 1397–1411 (2009)
3. Petráš, I., Vinagre, B.: Practical application of digital fractional-order controller to temperature control. Acta Montanist. Slovaca **7**(2), 131–137 (2002)
4. Matignon, D.: Stability results for fractional differential equations with applications to control processing. Comput. Eng. Syst. Appl. **2**, 963–968 (1996)
5. Dorcak, L., Petras, I., Kostial, I., Terpak, J.: State-space controller design for the fractional-order regulated system, vol. 1, pp. 15–20 (2002)
6. Hamamci, S.E.: Stabilization using fractional-order PI and PID controllers. Nonlinear Dyn. **51**(1–2), 329–343 (2008)
7. Beschi, M., Padula, F., Visioli, A.: The generalised isodamping approach for robust fractional PID controllers design. Int. J. Control **90**(6), 1157–1164 (2017)
8. Gutiérrez, R.E., Rosário, J.M., Tenreiro Machado, J.: Fractional order calculus: basic concepts and engineering applications. Math. Probl. Eng. **2010**, 19 (2010)

9. Rashid, M.H.: Power Electronics Handbook, 3rd edn. Elsevier, Burlington (2011)
10. Liang, T.J., Chen, S.M., Yang, L.S., Chen, J.F., Ioinovici, A.: Ultra-large gain step-up switched-capacitor DC-DC converter with coupled inductor for alternative sources of energy. IEEE Trans. Circ. Syst. I Fund. Theory Appl. **59**(4), 864–874 (2012)
11. Raj, A.S., Siddeshwar, A.M., Guruswamy, K.P., Maheshan, C.M., Vijay Sanekere, C.: Modelling of flyback converter using state space averaging technique. In: 2015 IEEE International Conference on Electronics, Computing and Communication Technologies, CONECCT, pp. 1–5 (2016)
12. Spruijt, H.J.N., O'Sullivan, D.M., Herman, J.N., Dermot, M.: PWM-switch modeling of DC-DC converters. IEEE Trans. Power Electron. **10**(6), 659–665 (1995)
13. Baleanu, D., Guvenc, Z.B., Machado, J.A.: New Trends in Nanotechnology and Fractional Calculus Applications. Springer, Netherlands (2010)
14. Tepljakov, A.: Fractional-Order Modeling and Control of Dynamic Systems. Springer, Estonia (2017)

Symptomatology of Musculoskeletal Pain Related to Repetitive Movements. Preliminary Study "Post-harvest in Floriculture Companies"

Luis Morales[1(✉)], Diana Silva[2], Víctor Moreno[2], and Santiago Collantes[3]

[1] Universidad Técnica de Ambato, Ambato, Tungurahua, Ecuador
luisamorales@uta.edu.ec
[2] Universidad Tecnológica Indoamérica, Ambato, Tungurahua, Ecuador
carito07012011@hotmail.com, victormoreno@uti.edu.ec
[3] Universidad de las Fuerzas Armadas ESPE, Latacunga, Cotopaxi, Ecuador
smcollantes@espe.edu.ec

Abstract. Musculoskeletal injuries affect the health of workers worldwide. This research relates the repetitive movements with the generation of musculoskeletal injuries in workers of the post-harvest area of a flower company. The evaluation methodology is the OCRA check list. The symptomatology of musculoskeletal pain was evaluated applying the Nordic questionnaire. The injuries were determined comparing morbidity rates and medical records of a sample of eighteen workers in Flores del Cotopaxi S.A. The results show that there is risk due to repetitive movements in activities like classification and bunching. Statistical tests established significant incidence ($p = 0.015$) for chi-square and ($OR = 15$), Odds Ratio. In both tests a confidence level of 95% was used. In the post-harvest area exist risks as a consequence of repetitive movements linked to the generation of musculoskeletal injury, mainly in the shoulders and wrists.

Keywords: Ergonomics · Repetitive movements · Musculoskeletal injury · OCRA check list

1 Introduction

Musculoskeletal disorders (MSD) represent a serious health problem in the working population [1] and the World Health Organization (WHO) has declared that MSDs constitute an important labor problem worldwide [2]. Epidemiological studies conducted in several countries show that this problem occurs in various human activities and in all economic sectors [3], and this situation also involves costs for employers and governments in those countries [4].

MSDs affect a large group of parts of the human body like nerves, tendons, muscles, and supporting structures of the locomotor system [5]; these affectations are associated with certain physical work factors present in many tasks such as:

J. Nummenmaa et al. (Eds.): CSEI 2019, AISC 1078, pp. 329–340, 2020.
https://doi.org/10.1007/978-3-030-33614-1_22

repetitiveness [6], force development [7], static and dynamic forced postures [8], exposure to vibrations [9], lifting loads [10], among others. This has an cumulative effect that can cause severe and debilitating symptoms such as pain, numbness, paresthesia and discomfort, in one or several body regions [11], as well as waste of time at work, temporary or permanent disability, difficulty to perform work duties and increase in compensation costs [12].

MSDs are associated to handwork [13]. Manual tasks are widely present in the industry which nature and sequence of tasks can vary greatly in the short and medium term [14]. This type of activity is characterized by being repetitive and complex, fundamentally because they are composed of subtasks that vary in strength, frequency and postural requirement [15]. The main affectations because of manual activities are injuries in the back and neck, carpal tunnel syndrome, rotator cuff, among others [13].

The repetitive movements and their influence in the MSDs is present in many labor activities of different industries such as: office work [16], fishing [17], assembly lines in manufacturing industries of any field [18], manual works in medicine as is the case of dentistry [19], agriculture [20], activities developed in greenhouses such as rose cultivation [21], among others.

Floriculture industry is characterized by manual tasks in many of its stages, especially in cultivation and post-harvest [22]; in this industry the reports of injuries are mainly from the upper extremities due to repetitive work with high energy demand [23]. In Ecuador, according to data from the Central Bank of Ecuador (BCE) to July 2017, the floriculture industry covers 8.3% of non-oil exports, besides, flower exports correspond to 9% of the world market share, behind Colombia (15%) and the Netherlands (52%) [24].

Tasks in the rose-growing industry especially in the post-harvest area require a lot of manual activity carried out by workers specifically of their upper extremities. In an attempt to address this problem, the objective of the research is to evaluate the ergonomic risks of this activity and its relationship with the generation of MSD.

2 Materials and Methods

2.1 Study Design

A cross-sectional correlational study was developed in the facilities of the company "Flores del Cotopaxi S.A" in Latacunga city in the province of Cotopaxi - Ecuador. Intentional or convenience sampling was used based on the evaluation of tasks related to the post-harvest area.

2.2 Participants

There were eighteen people, seven males and eleven females. Before starting the research, an informed consent form is signed by the workers to be evaluated based on the provisions of the Research Ethics Evaluation Committee (REC) of the World Health Organization [25]. The selection of these people for participation in the research was under the following eligibility criteria: working hours (8 h) and experience time (more than 1 year) based on that occupational diseases can appear after six months in the workspace [26]. Participants with a muscular or skeletal problem contracted outside of their workspace were excluded from this study in order to avoid confusion in the determination of pain symptoms due to other medical conditions.

2.3 Methodology

Below is shown, the techniques used in the evaluation of ergonomic risk caused by repetitive movements and musculoskeletal symptomatology.

OCRA Checklist Method. This method describes the risk of the workspace based on the value called OCRA Index. The result is the overall sum of factors (recovery factor, technical actions, intensity of effort, posture and additional factors), times the duration multiplier [27]. The method evaluates, in the first instance, the intrinsic risk of a workspace independently of the particular characteristics of the worker. Depending on the score obtained for this index, the method classifies as "optimal", "acceptable", "very slight", "slight", "medium" or "high". Finally, depending on the level of risk, the method suggests a series of basic actions, except in case of "optimal" or "acceptable" risk in which it is considered that actions on the position are not necessary. Finally, depending on the risk level, the method suggests a series of basic actions, except in cases as "optimal" or "acceptable" risk in which it is considered that actions on the workspace are not necessary. For the remaining cases, the method proposes actions such as a new analysis or improvement of the workspace ("very slight" risk), or the need for medical supervision and training for the worker in that workspace ("slight", "medium" risk) or "high") [28].

Kuorinka's Nordic Questionnaire. This questionnaire collects information about musculoskeletal symptoms such as: pain, fatigue or discomfort in different areas of the body; it studies nine anatomical regions of the body that are neck, shoulder, dorsal column, lumbar spine, hip, elbow, hand/wrist, knee, ankle/foot just for epidemiological purposes, not clinical; this method has of two sections, the first section applies a general questionnaire to identify areas of the body with discomfort in function of time (in the last 7 days, 12 months, etc.). In the second section questions are related to medical attention, rehabilitation, absenteeism and medication taken to control musculoskeletal conditions [29].

2.4 Data Collection and Analysis

The statistical data analysis was performed in SPSS version 21.0 (SPSS Inc., Chicago, IL, USA). Responses from the Kuorinka's Nordic questionnaire and OCRA checklist risk levels were tabulated. Frequency of ergonomic risks due to repetitive movements and pain symptomatology were established using statistical analysis. Chi-square variables were also tested with a significance of $p < 0.05$, and OR (Odds Ratio) with significance when the confidence interval does not contain the unit [30].

3 Results

The analysis of results involves both men and women equally since the research evidence shows that ergonomic risks caused by repetitive movements affect both genders in the same way as long as the risk factors are present in all work activities [31].

3.1 Socio-demographic Characteristics

The study was performed in the first semester of 2018. Table 1, summarizes the most relevant characteristics.

The analysis was applied to eighteen workers (seven men and eleven women) with more than one year working in the post-harvest area, most of them are married and have a stable working condition in the company. From the sample, fifteen people have pain in some part of their body associated with musculoskeletal symptoms (five men and ten women). This issue is found to a greater extent in bunching and classification workspaces, this is mainly due to the most activities in those workplaces are manual with forced standing positions and repetitive movements of upper extremities in 8-h days and up to six days a week.

Table 1 also reveals that the working population between one and five years of work experience presents more cases of musculoskeletal disease, both in men and women in the workplaces mentioned above.

Table 1. Socio-demographic characteristics.

		Gender											
		Female						Male					
		Work experience (years)						Work experience (years)					
		1–5		5–10		>10		1–5		5–10		>10	
		Presence of pain		Presence of pain		Presence of pain		Presence of pain		Presence of pain		Presence of pain	
		No	Yes	No	Yes	No	Yes	No	Yes	No	Yes	No	Yes
		f	f	f	f	f	f	f	f	f	f	f	f
Work position	Sprinkler	0	0	0	0	0	0	0	1	0	0	0	0
	Classifier	0	4	0	0	0	1	0	0	0	0	0	0
	Cutter	0	0	0	0	0	0	0	1	0	0	0	0
	Distributor	0	0	0	0	0	0	1	1	0	0	0	0
	Buncher	1	2	0	1	0	1	0	0	0	0	0	0
	Packager	0	0	0	0	0	0	0	1	0	1	0	0
	Labeler	0	0	0	1	0	0	0	0	0	0	0	0
	Dresser	0	0	0	0	0	0	1	0	0	0	0	0
	Total	**1**	**6**	**0**	**2**	**0**	**2**	**2**	**4**	**0**	**1**	**0**	**0**

3.2 Report of Musculoskeletal Pain Symptomatology

Table 2 shows the summary of the results obtained from the survey conducted using the Nordic test, which presents information about frequency (f) and percentage (%) of the musculoskeletal pain symptomatology in the surveyed workers. The response options are yes or no related to the presence of pain symptoms in the last twelve months and in the last seven days also if that symptoms have generated medical attention, besides include the intensity of the pain in the workers who have it.

The presence of symptoms of musculoskeletal pain in the last twelve months was reported in the neck, spine and joints of the shoulder, elbow and wrist in percentages of 17%, 6%, 39%, 17% and 22.2% respectively in the working population for this study. People who present this condition said that the symptoms are attributable to work activities with high physical demand in terms of movements in the performance of tasks in their workspace especially in activities such as flower stems classification and bouquet bunching. Pain manifestations can range from one week to several months, this is why workers have received medical and physiotherapy treatments. Consequence of this, 22.2% workers were medically treated for ailments in shoulder, elbow and wrist joints, while the 75%. On the other hand, pain intensity in people with symptoms goes from slight to very strong. In the last situation they usually go for medical assistance.

The indicators of musculoskeletal injuries in the post-harvest area, evidence the presence of two reports declared in the Ecuadorian Institute of Social Security (IESS).

Table 2. Symptomatology of musculoskeletal pain in female (F) and male (M) personnel

	Neck		Spin		Shoulders		Elbows		Wrists	
	f	%	f	%	f	%	f	%	f	%
Have you had symptoms in the last 12 months?										
Yes	2F, 1M	17	1F, 0M	6	6F, 1M	39	3F, 0M	17	4F, 0M	22.2
No	9F, 6M	83	10F, 7M	94	5F, 6M	61	8F, 7M	83	7F, 7M	77.8
Have you had symptoms in the last 7 days?										
Yes	3F, 1M	22	2F, 0M	12	6F, 0M	33	2F, 0M	11	4F, 1M	27.8
No	8F, 6M	78	9F, 7M	88	5F, 7M	67	9F, 7M	89	7F, 6M	72.2
Did you receive any treatment for the presence of pain in the last 12 months?										
Yes	0F, 0M	0	0F, 0M	0	1F, 1M	11	1F, 0M	6	0F, 1M	5.5
No	11F, 7M	100	11F, 7M	100	10F, 6M	89	10F, 7M	94	11F, 6M	94.5
What is the intensity of the perceived pain symptomatology?										
Slight	1F, 0M	5.5	0F, 0M	0	3F, 1M	22	0F, 0F	0	1F, 0M	5.5
Strong	0F, 1M	5.5	0F, 0M	0	2F, 0M	11	1F, 0M	5.5	1F, 1M	11
Very strong	2F, 0M	11	2F, 0M	12	1F, 0M	5.5	1F, 0M	5.5	2F, 0M	11
No pain	8F, 6M	78	9F, 7M	88	5F, 6M	61.5	9F, 7M	89	7F, 6M	72.5

3.3 Risk Assessment Report for Repetitive Movement OCRA Checklist

Table 3 shows the comparison of the results of the evaluation for repetitive movements, OCRA CHECKLIST, in the eight workspaces through the frequency of cases in the six risk categories for the right and left sides of the worker. Figures 1 and 2 illustrate the relationship between risk level and the presence of musculoskeletal pain in workers is evident.

Work positions with most frequency of risk levels correspond to classification and bunching with five cases each. The evaluation found risk on both sides of the body.

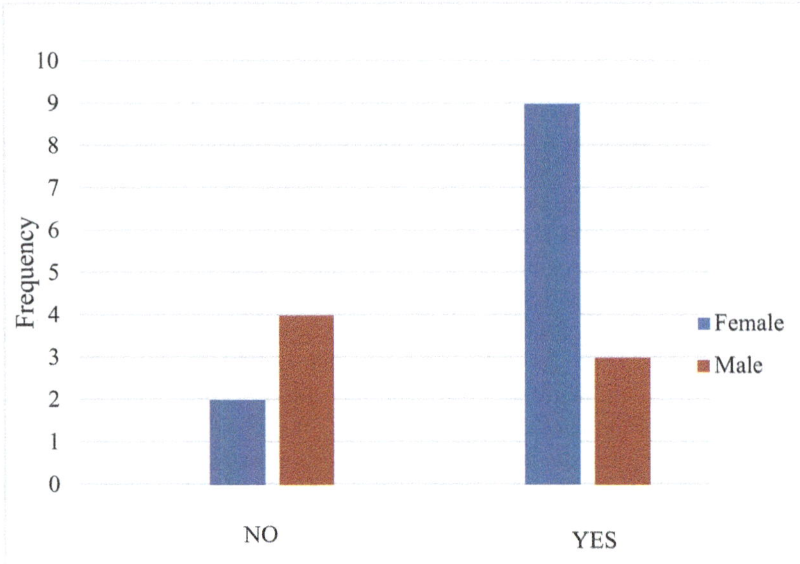

Fig. 1. Presence of pain in any part of the body.

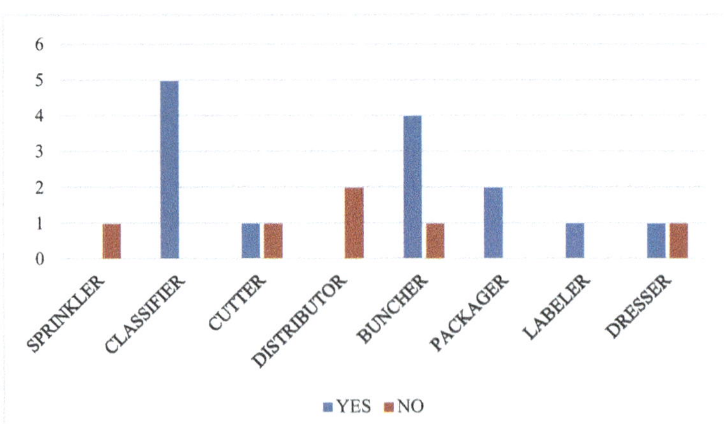

Fig. 2. Presence of musculoskeletal pain symptomatology in workers by job position.

Considering that from a very slight weighting the method recommends: an improvement of the workspace, medical supervision and training. On the other hand, in the classification and bunching positions only women work due to the fine motor skills needed for the activities in these tasks, hence workers who have presented musculoskeletal injuries correspond to these activities.

Table 3. OCRA risk category on the right and left sides of the body.

	OCRA result right side						OCRA result left side					
OCRA risk category	Optimal	Acceptable	Very slight	Slight	Medium	High	Optimal	Acceptable	Very slight	Slight	Medium	High
Work position	f	f	f	f	f	f	f	f	f	f	f	f
Sprinkler	1	0	0	0	0	0	1	0	0	0	0	0
Classifier	0	0	0	1	3	1	0	0	0	1	4	0
Cutter	1	0	0	0	0	0	1	0	0	0	0	0
Distributor	2	0	0	0	0	0	2	0	0	0	0	0
Buncher	0	0	2	0	3	0	0	0	1	2	2	0
Packager	2	0	0	0	0	0	2	0	0	0	0	0
Labeler	1	0	0	0	0	0	1	0	0	0	0	0
Dresser	1	0	0	0	0	0	1	0	0	0	0	0
Total	**8**	**0**	**2**	**1**	**6**	**1**	**8**	**0**	**1**	**3**	**6**	**0**

Table 4. Relationship of the OCRA risk category with musculoskeletal pain symptomatology

		Presence of pain in some part of the body		Total
		No	Yes	
OCRA final risk	Optimal, acceptable, very slight	5	3	8
	Slight, medium, high	1	9	10
Total		**6**	**12**	**18**

Table 5. Association of analyzed variable

Factor	Test type			
	Chi-square		Odd ratio	
	χ^2	$p < 0.05$	O.R	Confidence interval 95%
OCRA risk	5.51	0.019	15.0	1.215–185.198

3.4 Incidence of Risk Due to Repetitive Movement in Musculoskeletal Pain Symptomatology

Tables 4 and 5 present the association of variables for a confidence level of 95% and p-value < 0.05 for the chi-square and Odds Ratio statistical tests. The relationship establishes the OCRA's risk level which determines the intervention of a workspace and the presence of musculoskeletal pain symptoms in some part of the workers' body who perform repetitive movements. The results show that those with a slight, medium and high OCRA risk have a fifteen times higher incidence of pain symptoms in some part of their body than those with an optimal, acceptable and very slight evaluation. In addition, this association has a significance with a chi-square value of 5.51 and a p-value = 0.019.

4 Discussion

Results show the presence of musculoskeletal pain symptomatology in the post-harvest area in the floriculture company of study as it happens in researches in the same field in similar companies [32]. The highest frequency of these ailments occurs in work activities with demand of manual activity in long terms such as classification and bunching. In these positions people must use their upper extremities in a demanding regime in which they are exposed to risks of musculoskeletal injuries [33]. In the classification task, worker organizes the flower, in function of the size of the rose buttons, length of the stem and cut point. While in bunching, worker organizes the classified flowers into bunches of twenty-five, using plastic elements, cardboard, leagues, among others.

When performing a completely manual activity as in classification and bunching, in the post-harvest area, the staff very frequently performs pronosupination movements of the forearm and wrist, extensions and wrist flexions, besides they suffer radial or ulnar deviations, which were detected using the OCRA Checklist method [34]. These biomechanical effects bring as consequence presence of pain [2], so in other researches with similar characteristics, workers evidence musculoskeletal injuries such as: carpal tunnel syndrome [32], tendinitis and tenosynovitis [35], rotator cuff [36], epicondylitis [37], among others.

The presence of pain manifested throughout the year in evaluated workers in elbows, wrists and shoulders joints on the right and left side, evidenced a sustained exposure to ergonomic risks due to repetitive movements [14]. In these situations the joints, tendons, ligaments and nerves suffer a repetitive microtrauma [7], as a result of this the company presents a case of "bilateral carpal tunnel syndrome" and one of "painful arc syndrome in the right shoulder" approved in the Insurance of Occupational Hazards of the Ecuadorian Social Security Institute, the investigations of these two cases correspond to the situations mentioned above.

The solution of this issue can be focused in different ways. Nowadays, it is possible to use automated classification and bunching processes, in which, based on certain pre-established features and characteristics an image recognition of the flower can be implemented [38, 39]. By that, implementation of specific equipment called "bonchadoras" is possible. "Bonchadoras" requires less presence of people to operate and as result less exposure to repetitive movements in workers.

The fundamental factors in the generation of muscle and skeletal injuries due to repetitive movements correspond to the type of movement performed by the extremities [40], frequency and exposure times [41], in that sense a reduction of these factors minimizes the risk of injury. Therefore, controls are oriented to organizational safety measures and health monitoring such as: staff turnover [42], physical training of workers [43]. In the present, active pauses [44] and exercise at work [45] programs have been implemented. Monitoring the health of the staff is an obligation of the business owners and managers. For this reason, does not exist an efficient control without compromise and safety culture.

5 Conclusions

The presence of musculoskeletal pain symptoms, in post-harvest workers in this preliminary study, appear after the first year of work which is considered early. These symptoms are found particularly in women who, due to their fine motor skills, are required for manual activities.

Working days in the floricultural sector sometimes have more than eight hours of work during the whole week specially in worldwide festivities, such as Valentine's Day, Mother's Day, Day of the Death, among others. For this reason, ergonomic problems are considerable in this sector and it is recommended to extend the study considering other demographic variables and different physical and psychosocial working conditions.

References

1. Rojas, M., Gimeno, D., Vargas-Prada, S., Benavides, F.G.: Dolor musculoesquelético en trabajadores de América Central: resultados de la I Encuesta Centroamericana de Condiciones de Trabajo y Salud. Revista Panamericana de Salud Pública **38**, 120–128 (2015)
2. Thetkathuek, A., Meepradit, P., Sa-ngiamsak, T.: A cross-sectional study of musculoskeletal symptoms and risk factors in cambodian fruit farm workers in eastern region, Thailand. Saf. Health Work **9**, 192–202 (2018). https://doi.org/10.1016/j.shaw.2017.06.009
3. Mody, G.M., Brooks, P.M.: Improving musculoskeletal health: global issues. Best Pract. Res. Clin. Rheumatol. **26**, 237–249 (2012). https://doi.org/10.1016/j.berh.2012.03.002
4. Yazdani, A., Hilbrecht, M., Imbeau, D., Bigelow, P., Patrick Neumann, W., Pagell, M., Wells, R.: Integration of musculoskeletal disorders prevention into management systems: a qualitative study of key informants' perspectives. Saf. Sci. **104**, 110–118 (2018). https://doi.org/10.1016/j.ssci.2018.01.004
5. Gallagher, S., Schall Jr., M.C.: Musculoskeletal disorders as a fatigue failure process: evidence, implications and research needs. Ergonomics **60**, 1–45 (2017). https://doi.org/10.1080/00140139.2016.1208848
6. Shankar, S., Naveen Kumar, R., Mohankumar, P., Jayaraman, S.: Prevalence of work-related musculoskeletal injuries among South Indian hand screen-printing workers. Work **58**, 163–172 (2017). https://doi.org/10.3233/WOR-172612
7. Smith, T.G., Gallagher, S.: Impact of loading and rest intervals on muscle microtrauma. In: Proceedings of the Human Factors and Ergonomics Society 59th Annual Meeting - 2015, pp. 1217–1221 (2015). https://doi.org/10.1177/1541931215591191
8. Krishnakumar, V.B.R., Elavenil, P.: Reflected vision in surgical practice—a novel method to circumvent posture-related musculoskeletal disorders. J. Oral Maxillofac. Surg. **76**, 8–9 (2018). https://doi.org/10.1016/j.joms.2017.08.042
9. Xu, X.S., Dong, R.G., Welcome, D.E., Warren, C., McDowell, T.W., Wu, J.Z.: Vibrations transmitted from human hands to upper arm, shoulder, back, neck, and head. Int. J. Ind. Ergon. **62**, 1–12 (2017). https://doi.org/10.1016/j.ergon.2016.07.001
10. Tafazzol, A., Aref, S., Mardani, M., Haddad, O., Parnianpour, M.: Epidemiological and biomechanical evaluation of airline baggage handling. Int. J. Occup. Saf. Ergon. **22**, 218–227 (2016). https://doi.org/10.1080/10803548.2015.1126457
11. Kearney, G.D., Allen, D.L., Balanay, J.A.G., Barry, P.: A descriptive study of body pain and work-related musculoskeletal disorders among Latino farmworkers working on sweet potato farms in Eastern North Carolina. J. Agromedicine **21**, 234–243 (2016). https://doi.org/10.1080/1059924X.2016.1178613
12. Bellorín, M., Sirit, Y., Rincón, C., Amortegui, M.: Síntomas Músculo Esqueléticos en Trabajadores de una Empresa de Construcción Civil. Salud los Trabajadores **15**, 89–98 (2007)
13. Rahman, M.N.A., Zuhaidi, M.F.A.: Musculoskeletal symptoms and ergonomic hazards among material handlers in grocery retail industries. IOP Conf. Ser.: Mater. Sci. Eng. **226**, 1–14 (2017). https://doi.org/10.1088/1757-899X/226/1/012027
14. Boenzi, F., Digiesi, S., Facchini, F., Mummolo, G.: Ergonomic improvement through job rotations in repetitive manual tasks in case of limited specialization and differentiated ergonomic requirements. IFAC-PapersOnLine **49**, 1667–1672 (2016). https://doi.org/10.1016/j.ifacol.2016.07.820

15. Paulsen, R., Gallu, T., Gilkey, D., Reiser, R., Murgia, L., Rosecrance, J.: The inter-rater reliability of strain index and OCRA checklist task assessments in cheese processing. Appl. Ergon. **51**, 199–204 (2015). https://doi.org/10.1016/j.apergo.2015.04.019
16. Rahman, M.N.A., Masood, I., Awalludin, N.F., Hassan, M.F.: Ergonomic risk factors associated with musculoskeletal disorders in computer workstation. Int. J. Appl. Eng. Res. **12**(7), 1355–1359 (2017)
17. Silvetti, A., Munafò, E., Ranavolo, A., Iavicoli, S., Draicchio, F.: Ergonomic risk assessment of sea fishermen part II: upper limb repetitive movements. In: Goossens, R. (ed.) Advances in Social & Occupational Ergonomics, pp. 333–340. Springer, Cham (2017)
18. Botti, L., Mora, C., Regattieri, A.: Integrating ergonomics and lean manufacturing principles in a hybrid assembly line. Comput. Ind. Eng. **111**, 481–491 (2017). https://doi.org/10.1016/j.cie.2017.05.011
19. De Sio, S., Traversini, V., Rinaldo, F., Colasanti, V., Buomprisco, G., Perri, R., Mormone, F., La Torre, G., Guerra, F.: Ergonomic risk and preventive measures of musculoskeletal disorders in the dentistry environment: an umbrella review. PeerJ **6**(e4154), 1–16 (2018). https://doi.org/10.7717/peerj.4154
20. Neubert, M.S., Karukunchit, U., Puntumetakul, R.: Identification of influential demographic and work-related risk factors associated to lower extremity pain perception among rice farmers. Work **58**, 489–498 (2017). https://doi.org/10.3233/WOR-172649
21. Nasiri, S., Zamani, K., Ebrahimi, S., Ghazanfari, H.: Assessing body posture and skeleton - muscular disorders and related factors in workers of greenhouse in Khomeynishahr city. Int. J. Pharm. Technol. **7**(3), 9725–9734 (2015)
22. García-Cáceres, R.G., Felknor, S., Córdoba, J.E., Caballero, J.P., Barrero, L.H.: Hand anthropometry of the Colombian floriculture workers of the Bogota plateau. Int. J. Ind. Ergon. **42**(2), 183–198 (2012). https://doi.org/10.1016/j.ergon.2011.12.002
23. Barrero, L.H., Pulido, J.A., Berrio, S., Monroy, M., Quintana, L.A., Ceballos, C., Hoehne-Hueckstaedt, U., Ellegast, R.: Physical workloads of the upper-extremity among workers of the Colombian flower industry. Am. J. Ind. Med. **55**(10), 926–939 (2012). https://doi.org/10.1002/ajim.22102
24. Clúster Flor. http://flor.ebizor.com/como-van-las-exportaciones-de-flores-de-ecuador/
25. Council for International Organizations of Medical Sciences: International ethical guidelines for biomedical research involving human subjects. Bull. Med. Ethics **182**, 1–17 (2012)
26. SGRT IESS. http://sart.iess.gob.ec/DSGRT/norma_interactiva/IESS_Normativa.pdf. Accessed 20 Jun 2019
27. Soria, N., Basualdo, I., Ramoa, L., López de Silva, M.E.: Descripción de Tessaria dodeneifolia (Hook. & Arn.) Cabrera, (Asteraceae), “la planta dulce” como endulzante natural. Boletín Latinoamericano y del Caribe de Plantas Medicinales y Aromáticas **16**(2), 129–135 (2017)
28. Asensio, S., Basante, M., Diego, J.: Evaluación ergonómica de puestos de trabajo, 1st edn. Editorial Paraninfo, Madrid (2012)
29. Kuorinka, I., Jonsson, B., Kilbom, A., Vinterberg, H., Biering-Sørensen, F., Andersson, G., Jørgensen, K.: Standardised nordic questionnaires for the analysis of musculoskeletal symptoms. Appl. Ergon. **18**(3), 233–237 (1987). https://doi.org/10.1016/0003-6870(87)90010-X
30. Cerda, J., Vera, C., Rada, G.: Odds ratio: aspectos teóricos y prácticos. Revista Médica de Chile **141**(10), 1329–1335 (2013). https://doi.org/10.4067/S0034-98872013001000014
31. Lima, M., Coelho, D.A.: Ergonomic and psychosocial factors and musculoskeletal complaints in public sector administration–a joint monitoring approach with analysis of association. Int. J. Ind. Ergon. **66**, 85–94 (2018). https://doi.org/10.1016/j.ergon.2018.02.006

32. Piñeda, A.: El Túnel Carpiano: Riesgo ergonómico en trabajadoras de cultivo de flores. Revista de Ingeniería **1**, 11–20 (2013)
33. Wilder, H.D., Orjuela, R., Erley, M.: Factores laborales y extralaborales de floricultores con Síndrome del Túnel del Carpo. Medicina y Seguridad del trabajo **62**(244), 199–211 (2016)
34. Colombini, D., Occhipinti, E.: Scientific basis of the OCRA method for risk assessment of biomechanical overload of upper limb, as preferred method in ISO standards on biomechanical risk factors. Scand. J. Work Environ. Health **44**(4), 436–438 (2018). https://doi.org/10.5271/sjweh.3746
35. Micheli, G.J.L., Marzorati, L.M.: Beyond OCRA: predictive UL-WMSD risk assessment for safe assembly design. Int. J. Ind. Ergon. **65**, 74–83 (2017). https://doi.org/10.1016/j.ergon.2017.07.005
36. Castro Abril, H.A., Ramírez, A., Silva, L.: Modelo de elementos finitos del hombro: comparación de los esfuerzos mecánicos de un hombro sano y un hombro con síndrome del manguito rotador. Revista de Investigación **8**(1), 42–50 (2015). https://doi.org/10.29097/2011-639x.6
37. Valladares, Y.C., Fernández, I.Y.D., Eduardo, I.I., Garcí, L., Perdomo, I.V.C., Miguel, I.V., Martínez, V., Romero, K.M., Doris, I.V., Rodríguez, C.: Utilidad de las ondas de choque para la disminución del dolor en la epicondilitis. Revista Cubana de Medicina Física y Rehabilitación **9**(2), 1–13 (2017)
38. Xia, X., Xu, C., Nan, B.: Inception-v3 for flower classification. In: 2017 2nd International Conference on Image, Vision and Computing ICIVC 2017, Chengdu, China, pp. 783–787. IEEE (2017). https://doi.org/10.1109/icivc.2017.7984661
39. Krishnaveni, S., Pethalakshmi, A.: Toward automatic quality detection of Jasmenum flower. ICT Express **3**, 148–153 (2017). https://doi.org/10.1016/j.icte.2017.04.006
40. Singh, A., Singh, P., Ojha, P., Mishra, M.: Assessment of pain and discomfort among agricultural workers involved in floriculture. Asian J. Home Sci. **12**(1), 91–93 (2017). https://doi.org/10.15740/has/ajhs/12.1/91-93
41. Varmazyar, S., Choubdar, M.: Assessment of the risk of musculoskeletal disorders in the upper-limb in greenhouse workers by the OCRA and ACGIH-HAL methods. Int. J. Health Stud. **3**, 1–6 (2017). https://doi.org/10.22100/ijhs.v3i4.286
42. Ruíz Vargas, N.V.: Factores Asociados a La Ocurrencia De Accidentes De Trabajo En La Industria Manufacturera. Horizonte de Enfermería **29**(1), 41–54 (2018). https://doi.org/10.7764/horiz_enferm.29.1.41-54
43. Søgaard, K., Sjøgaard, G.: Physical activity as cause and cure of muscular pain: evidence of underlying mechanisms. Exerc. Sports Sci. Rev. **45**(3), 136–145 (2017). https://doi.org/10.1249/jes.0000000000000112
44. Cáceres-Muñoz, V.S., Magallanes-Meneses, A.A., Torres-Coronel, D., Copara-Moreno, P., Escobar-Galindo, M., Mayta-Tristán, P.: Efecto de un programa de pausa activa más folletos informativos en la disminución de molestias musculoesqueléticas en trabajadores administrativos. Revista Peruana de Medicina Experimental y Salud Pública **34**(4), 611–618 (2017). https://doi.org/10.17843/rpmesp.2017.344.2848
45. de Freitas-swerts, F.C.T., Dos Santos, S.V.M., Do Carmo, M.C.: Isostretching en la reducción del dolor, fatiga y aumento de la flexibilidad en los trabajadores. Salud de los Trabajadores **26**(2), 138–148 (2018)

Design of an Ergonomic Prototype for Physical Rehabilitation of People with Paraplegia

Franklin W. Salazar[1], Freddy Núñez[2], Jorge Buele[1,3](✉), Edisson P. Jordán[1], and Jeneffer Barberán[2]

[1] Universidad Técnica de Ambato, Ambato 180103, Ecuador
{fw.salazar, edissonpjordan}@uta.edu.ec, jorgebuele@uti.edu.ec
[2] Instituto Superior Tecnológico Tsa'chila, Santo Domingo 230109, Ecuador
{fnunez, jbarberan}@institutos.gob.ec
[3] SISAu Research Group, Universidad Tecnológica Indoamérica, Ambato, Ecuador

Abstract. Technology intervention within the field of physical rehabilitation allows to improve conventional processes and long-term results. Therefore, this research project proposes a design and implementation of an ergonomic device to exercise the upper and lower limbs of patients suffering from paraplegia, as part of their rehabilitation process. Despite attending conventional therapy sessions, the feeling of dissatisfaction is generalized by the need to complement these exercises. The general design of structure which includes the set of materials used, as well as the belt, power and control transmission systems. This design considers both weight of the structure exerted by gravity, as well as the exerted by the patient and the total dead load. In addition, the maximum stress analysis that supports the structure and the maximum deformation is carried out, taking into account that established safety factor is high. Experimental tests and surveys allow to corroborate the proper functioning and acceptance that this proposal has, given that its impact on rehabilitation process will be observed in long term.

Keywords: Biomechanics · Ergonomics · Physical rehabilitation · Paraplegia

1 Introduction

The term paraplegia comes from the Latin "paraplexia" [1] and as described in [2], it is used to name what a person suffers when the lower region of his body is paralyzed by some kind of damage [3]. This disorder can be caused by a hereditary or acquired disease, as a result of a spinal cord injury or, in most of the cases, an accident [4–6]. The spinal cord injury (SCI) is a pathology studied throughout the ages, described as damage in some part of the spinal cord or nerves of the end of the spinal canal [7–9].

The increasing advance of technology that fuses concepts with physical and sports rehabilitation has allowed physical exercises to be included as part of the treatment of diseases such as paraplegia [10, 11]. Despite being an idea that dates back to antiquity, studies have been conducted that have corroborated its positive influence for the

J. Nummenmaa et al. (Eds.): CSEI 2019, AISC 1078, pp. 341–353, 2020.
https://doi.org/10.1007/978-3-030-33614-1_23

improvement in life quality of the patient [12, 13]. In 2017 there were between 110 and 190 million people with some type of disability, it means, around 15% of the world population. These figures have been increasing due to the aging of the population and the increase in the suffering of chronic diseases [14].

This is why the exercise of upper and lower limbs is attractive in research field, due to its practical use for sports training and rehabilitation [15]. In [16] a study which describes technology impact on the process of therapeutic rehabilitation in paraplegic patients is carried out. In this study case, there are patients who have suffered a medullar disorder from a clinic in Santiago, Chile. In addition, several exercises and techniques that are part of multiple rehabilitation programs obtained from the documentary review are shown. For its part in [17] a feasibility study for a device to exercise the extremities of the human being is shown. In this context, this project proposes a novel system consisting of a portable manipulator guided by a low-cost cable.

Having the aim of improving the life quality of people with paraplegia, a technological proposal is presented that contributes to the physical rehabilitation process. This work is based on a survey carried out on patients, in which the need to complement the exercises with an additional device is determined. At design construction, all the previous study carried out is presented, to guarantee its resistance when used by patients. The simulations, experimental tests and surveys validate this proposal.

This article is organized as follows: the introduction in Sect. 1, in Sect. 2 the methodology is shown. In Sect. 3 the respective simulations are presented. In Sect. 4 the results are described and in Sect. 5 the conclusions.

2 Methodology

The focus of this research is mixed (quantitative-qualitative). It is qualitative because it is normative, explicit, realistic and the construction is made based on user needs [18]. In a parallel way it is quantitative, because design was made based on mathematical calculations and the respective mechanical analysis. Deduction and induction method are applied as Scientific methods in this work, through the bibliographic search to develop sustainable theory. In addition, the analysis and synthesis were applied to derive the necessary conclusions and obtain a design in accordance with the proposed objectives. Dedicated techniques were research and field work, which were reflected in the application of a survey to patients, gathering information on the frequency with which they perform rehabilitation exercises, their satisfaction, comfort, the need for a technological tool and executing complementary complex exercises, among others.

2.1 Structure

The prototype general design is described in Fig. 1. The structure is the main support of every single system and other physical components such as: motors, table, control panel, chair, foot supports, handle and rope. Weight calculation of the structure considering the gravity action as a function of the mass is carried out using (1).

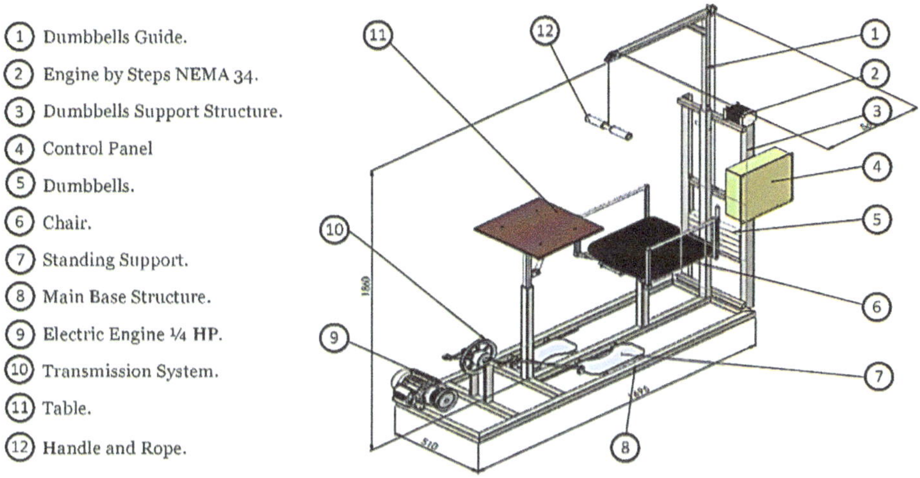

Fig. 1. General diagram of the prototype.

$$W_{me} = M_{me} * g \tag{1}$$

Where;
M_{me}: Structure Mass composed by several materials (kg).
g: Gravity acceleration (m/s^2).

In addition, dead load analysis is presented considering a critical factor of 20% in (2) and in (3) the calculation of the total dead load is shown.

$$W_S = 0.2 * W_{me} \tag{2}$$

$$C_{mt} = W_{me} + W_S \tag{3}$$

In (4) the approximate weight calculation of the person to be rehabilitated is presented. Then, the live load analysis considering a critical factor of 20% safety and the calculation of the total live load are described in (5) and (6) respectively.

$$W_P = M_P * g \tag{4}$$

Where;
M_P: Approximate person Mass (lb.).

$$W_{PS} = 0.2 * W_P \tag{5}$$

$$C_{Vt} = W_P + W_{SP} \tag{6}$$

To determine the shear stress of the pins, the calculation of the maximum over load that could be produced in the structure has been made in (7).

$$C_{\max} = C_{Vt} + C_{mt} \tag{7}$$

For profiles selection, the moment of inertia is made with respect to the section in (8) and the effort calculation in (9).

$$I = I_e - I_i \tag{8}$$

$$\delta = \frac{M * c}{I} \tag{9}$$

Where;
I_e: Exterior inertia.
I_i: Interior inertia.
M: Flexing moment.
c: Distance to the rotate center.

2.2 Belt Drive System

It consists on union of two or more pulleys, it means wheels held to a uniform circular rotation movement through a band, cable, etc. the same that embraces the pulleys exerting friction force by supplying power from the drive wheel, as shown in Fig. 2. For this design a first-gen lever has been used whose axis is fixed (fixed pulley), where the power is applied to one rope end and the resistance to the other. The pulley transmission ratio is shown in (10).

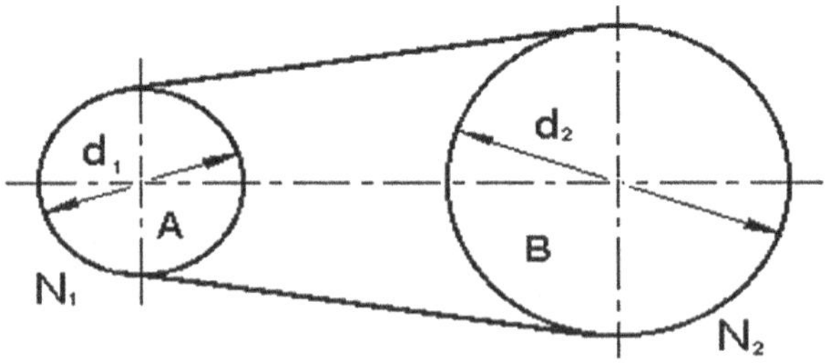

Fig. 2. Belt transmission system.

$$i = \frac{d_1}{d_2} \tag{10}$$

Where;
d_1: Pulley Diameter coupled to the motor (mm).
d_2: Pulley Diameter of the transmission system (mm).

2.3 Power Transmission System

It is based on a crank-crank system that transmits an engine effort by means of a shaft with pulley and two arms, it is subjected to a uniform circular movement (MCU). While a crank-crank system that transmits movement to the foot supports by means of two axes, which transforms the MCU into a uniform rectilinear motion (MRU), allowing movement of the lower extremities of the human body. The patient does not generate movement, therefore, the speed applied in the foot supports is a function of the angular position of the crank and the speed in it, as shown in Fig. 3.

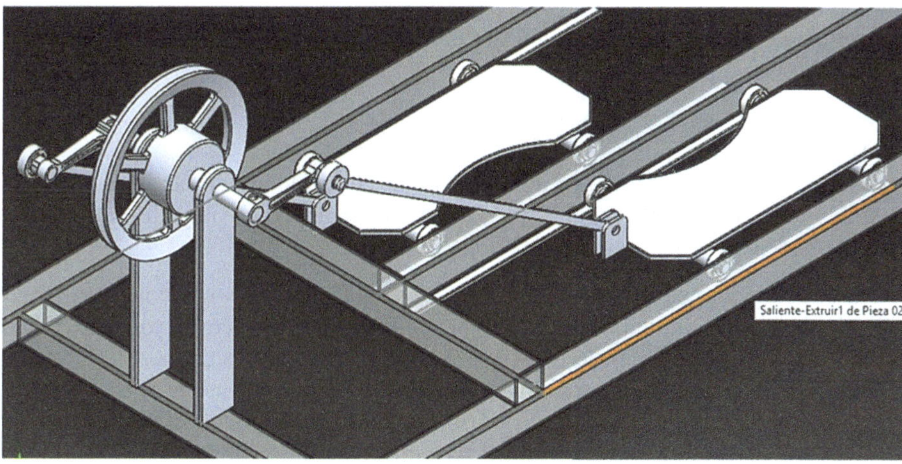

Fig. 3. Power transmission system.

The linear speed in the pedal, it is the same in the transmission system, that is, the same as in the 203 mm diameter pulley. To obtain this speed, the transmission ratio between the motor pulley (127 mm) and the pulley of the transmission system (203 mm) is calculated and calculated in (11). The value of the speed on the foot support varies depending on the angle, i.e. the position of the pedal, for example, the speed at a point of the foot support, was made using an angle of 250°.

$$v_p = i * v_m \tag{11}$$

Where;
i: Pulley length of the transmission system (203 mm).
v_m: Motor speed 1200 rpm.

2.4 Control System

The weights are approximately 20 kg, considering it is a rehabilitation machine, in this case, if it can support the 20 kg tension of mass and the structure can also support the same force, then it is concluded that the design is correct and design factor is oversized.

The rope tension is the same in all its trajectory and in (12) the necessary torque is determined to raise the force F, in this case a tension of 200 N.

$$T_0 = T * D \tag{12}$$

Where;

T: Rope tension.

D: Distance from the pulley center to the force point.

To avoid sudden movements, provide stability and have a relatively low speed, a stepper motor is used which maintains constant torque regardless of speed, unlike a common electric motor. The NEMA 34 motor has been selected since it supports the required load calculated in the design. Its main technical characteristics are described below:

- Up to 1710 oz-in. (8.9 lb.-ft) torsion.
- 1.8 grades of resolution.
- A single axis for open loop or double axis for closed loop with an encoder.
- Ability to operate with stepper drivers and NI movement controllers for improve the performance.

The structure was built in accordance with the planned design which has considered both the support of the appropriate weight, as well as the assembly of each of the pieces. The table has a dimension of 40 × 40 cm^2, which acts as support for the arms and allows to adjust its height through a pin to secure it. It also consists of an adjustable shaft that supports the inclination up to 80° and offers the possibility of using it in different ways. The design of the seat is ergonomically comfortable and safe, it consists of rotating arms and an adjustable shaft of 30 cm. The height is controlled with a variable pin, according to the height and comfort of the patient in the rehabilitation process. This device consists of 4 loads of 8 lbs. each, which contribute to the exercise of the lower limbs acting as a counterweight. To provide an aesthetic product, gray and blue paint has been used. In addition to using good resistance material, the ergonomic bases in force in the Ecuadorian regulations have been considered: UNE-EN 614-1: 2006 + A1: 2009, UNE-EN 547-2: 1997 + A1: 2009, UNE-EN ISO 12100-1: 2004 + A1: 2010 and ISO 26800: 2011.

3 Simulations

As can be seen in Fig. 4, the maximum effort that will be applied to this structure is 5,178 ksi, the tension that supports the material is 34 ksi and the maximum deformation is 0.87 mm, which demonstrates the robustness of this design. Figure 5 shows the simulation of the maximum deformation supported by the prototype.

The safety factor is high at all points due to the material strength and therefore the structure is oversized for the efforts it will support. As shown in Fig. 6, maximum axial and bending stress is equal to 11,541 ksi and material strength is 34 ksi, so the design is suitable to withstand stresses.

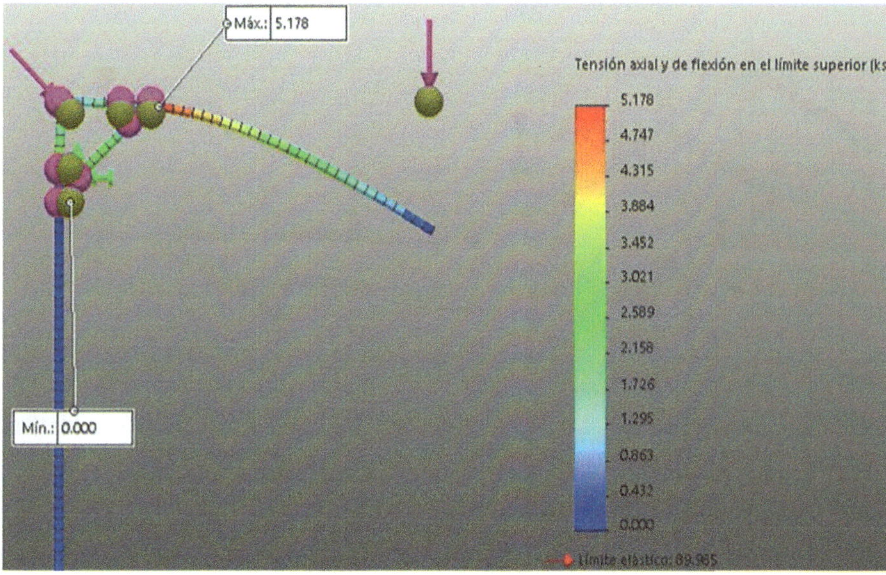

Fig. 4. Strength simulations: maximum effort.

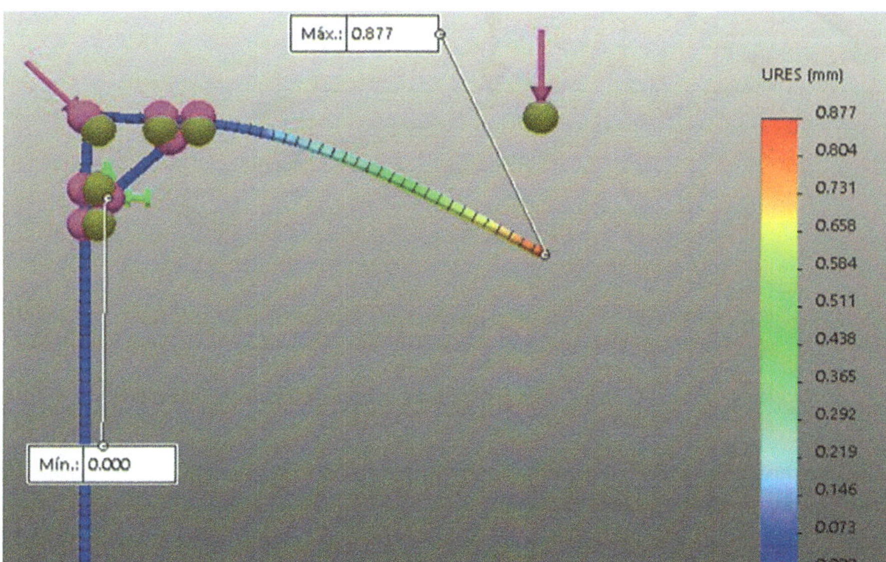

Fig. 5. Strength simulations: maximum deformation.

The maximum deformation, where the patient is located, is 1,348 mm, for the remaining points is less than 1 mm. It can be concluded that the design is oversized. In addition, the minimum structure safety factor that supports the greatest amount of effort

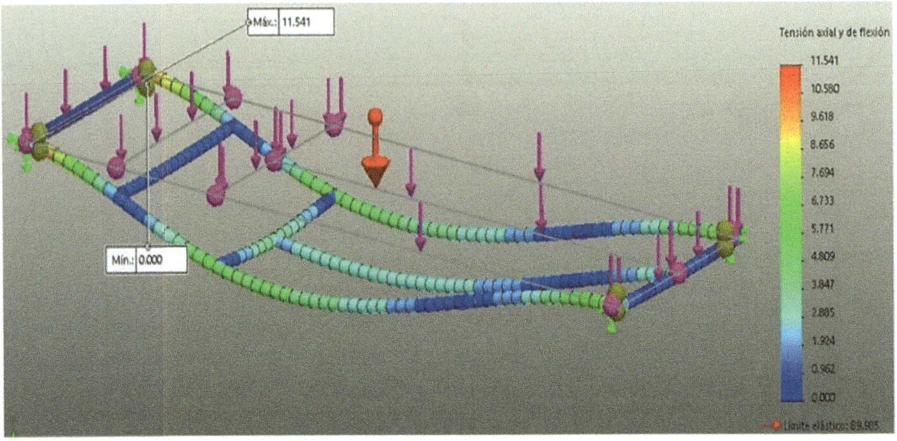

Fig. 6. Strength simulations: axial tension and maximum bending.

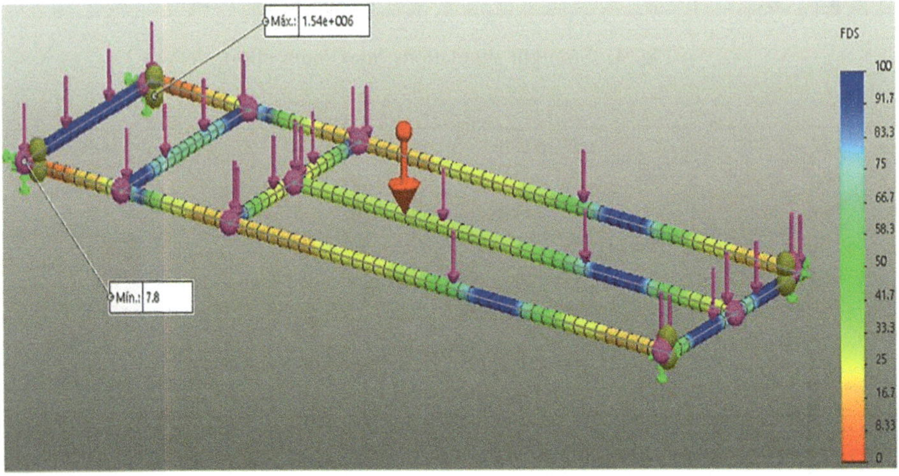

Fig. 7. Strength simulations: minimum safety factor of the structure.

is 7.8, as shown in Fig. 7. Therefore, the structures and elements that make up this prototype are oversized and support much greater efforts than those which were planned.

4 Experimental Results

4.1 A Subsection Sample

The functional tests were performed in a 1-h session, as shown in Fig. 8. The first 15 min were used to train the patient and remaining time was used for rehabilitation exercises. This process was performed 2 or 3 times a week (depending on the availability of patients), for 2 months' period. In the last session, timing was taken according to the series and the results are shown in Tables 1 and 2.

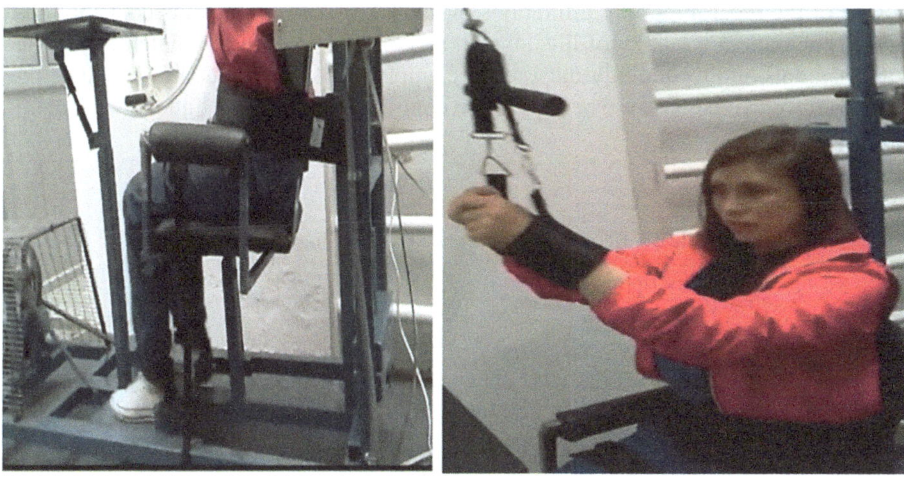

Fig. 8. Patient performing a physical rehabilitation routine.

Table 1. Lower extremities. Time entries in 1 series.

Item	Exercise	Time
1	Going	0.58 s
2	Return	0.51 s
Total	Series	1.09 s

Table 2. Superior limbs. Time entries in 1 series.

Item	Exercise	Time
1	Rise	1.20 s
2	Descent	1.08 s
3	Rest of 4 s between exercises	8 s
Total	Series	10.28 s

In order to evaluate the acceptance and limitations of this prototype, it was decided to submit to the criterion of experts (mechanical engineer (1), users (2), manager (1) and orthopedic specialist (1)) of a rehabilitation center, which is the aim of the study. Those who were directly linked to the design and construction process and who are in direct contact with patients and their families on a daily basis. These criteria are the basis to develop a general impact matrix, which is segmented into different fields such as: economic, social, environmental and technological. The rating is given by integer numbers in a range between −3 and 3, which symbolize a high negative and a high positive impact respectively. The information obtained by the respondents is tabulated for a better perception in Tables 3, 4, 5 and 6.

One of the most important limitations of this work is to evaluate its contribution to improve the patient's quality of life, since it is a preliminary study. Implementing state of art materials and technology must also be emphasized, due to its difficult since they come from a recycled base are being used to contribute to a low cost and environmental impact proposal.

Table 3. Economic impact obtained through a survey.

Item	Description	−3	−2	−1	0	1	2	3
1	Skilled labor						X	
2	Prototype easy access							X
3	Accessible rate							X
4	Access to low-income families							X
Total		11						

Table 4. Social impact obtained through a survey.

Item	Description	−3	−2	−1	0	1	2	3
1	Specialized care							X
2	Easy to use						X	
3	Social reintegration							X
4	Improvement in quality life						X	
Total		10						

Table 5. Environment impact obtained through a survey.

Item	Description	−3	−2	−1	0	1	2	3
1	4Rs application (Recycle, Reject, Reuse and Reduce)						X	
2	Wide work spaces						X	
3	Permanent cleaning							X
4	Smoke-free area							X
Total		10						

Table 6. Technological impact obtained through a survey.

Item	Description	−3	−2	−1	0	1	2	3
1	Innovative design							X
2	Reusable materials							X
3	Cutting-edge technology						X	
4	User comfort							X
Total		11						

From an economic point of view, it is determined that long-term items will decrease since the patient will be submitted to a more complete treatment which means a shorter period of time in the future. This will benefit users, their families, the Ministry of Health, regional governments and institutions providing health care (public and/or private). Socially it seeks to improve the life quality of the patient by contributing to the development of their therapies. In addition, this proposal will cause a positive environmental impact because all materials used to make the components are mostly recycled (4R phenomenon). All this at a low cost and with a novel and innovative design, which presents feasibility for its production, either in high-terms or in a personalized way.

A data average of the obtained results in surveys about the economic, social, environmental and technological impact has the following percentages of acceptance: 91.66%, 83.33%, 83.33% and 91.66% respectively.

5 Conclusions

This article deals with the design and manufacture of a device to exercise the upper and lower limbs. This prototype is developed using low cost materials and mostly recycled, as a solution with a reduced environmental impact. The bases of ergonomics are made taking the current Ecuadorian regulations: UNE-EN 614-1: 2006 + A1: 2009, UNE-EN 547-2: 1997 + A1: 2009, UNE-EN ISO 12100-1: 2004 + A1: 2010 and ISO 26800: 2011. A survey made to users and staff working in the rehabilitation center where the experimental tests carried out, it shows in average an 87.495% acceptance of this proposal.

Within the design, a maximum effort applied to the structure of 5,178 ksi was obtained and the tension that the material supports is 34 ksi. The maximum deformation is 0.87 mm, being less than 1 mm so it can be concluded that the structure will not suffer significant deformations. The minimum safety factor is 7.8 (where the reference value is 2.5), the maximum axial and bending stress is equal to 11,541 ksi and material strength is 34 ksi. Therefore, structures and elements that make up this prototype are oversized and support much greater efforts for which it was designed.

This proposal will be evaluated in long term to determine, from medical point of view, the impact it has had on rehabilitation process of patients. Through the feedback obtained, it is considered as future work, to establish a comparison between this proposal and a conventional device in this type of treatment. In addition, it is proposed to conduct a feasibility study to determine the possibility of developing this product on an industrial scale.

Acknowledgments. This work was financed in part by Universidad Técnica de Ambato (UTA) and Dirección de Investigación y Desarrollo (DIDE) under project PFISEI 26. Special thanks to the support provided by SISAu Research Group of the Universidad Tecnológica Indoamérica and Instituto Superior Tecnológico Tsa'chila.

References

1. Giesser, B.S.: The history of multiple sclerosis: from the age of description to the age of therapy. Primer Multiple Sclerosis **2**, 3–10 (2016)
2. Yáñez, V.B., Batista, Y.E.B., Montero, M.G.: Caso clínico: paciente con implante de electrodo medular con paraplejia posquirúrgica y fístula de líquido cefalorraquídeo, importancia de la atención de enfermería. Revista Científica de la Sociedad Española de Enfermería Neurológica **49**(C), 23–27 (2019)
3. Gill, M.L., Grahn, P.J., Calvert, J.S., Linde, M.B., Lavrov, I.A., Strommen, J.A., et al.: Neuromodulation of lumbosacral spinal networks enables independent stepping after complete paraplegia. Nat. Med. **24**(11), 1677–1682 (2018)
4. Martino, G., Ivanenko, Y., Serrao, M., Ranavolo, A., Draicchio, F., Casali, C., Lacquaniti, F.: Locomotor coordination in patients with Hereditary Spastic Paraplegia. J. Electromyogr. Kinesiol. **45**, 61–69 (2019)
5. Manole, A., Männikkö, R., Hanna, M.G., Kullmann, D.M., Houlden, H.: De novo KCNA2 mutations cause hereditary spastic paraplegia. Ann. Neurol. **81**(2), 326–328 (2017)
6. de Souza, P.V.S., de Rezende Pinto, W.B.V., de Rezende Batistella, G.N., Bortholin, T., Oliveira, A.S.B.: Hereditary spastic paraplegia: clinical and genetic hallmarks. Cerebellum **16**(2), 525–551 (2017)
7. Hurlbert, R.J., Hadley, M.N., Walters, B.C., Aarabi, B., Dhall, S.S., Gelb, D.E., et al.: Pharmacological therapy for acute spinal cord injury. Neurosurgery **76**(1), 71–83 (2015)
8. Mehta, S., McIntyre, A., Janzen, S., Loh, E., Teasell, R., Team, S.C.I.R.E.: Systematic review of pharmacologic treatments of pain after spinal cord injury: an update. Arch. Phys. Med. Rehabil. **97**(8), 1381–1391 (2016)
9. Assinck, P., Duncan, G.J., Hilton, B.J., Plemel, J.R., Tetzlaff, W.: Cell transplantation therapy for spinal cord injury. Nat. Neurosci. **20**(5), 637 (2017)
10. da Fonseca, L.O., Ferreira, B.M., Paredes, M.E., Freire, J.P., Sanches, P., Bó, A.P.L.: Towards indoor rowing assisted by electrical stimulation for persons with paraplegia. In: XXVI Brazilian Congress on Biomedical Engineering, pp. 391–395. Springer, Singapore (2019)
11. Zbogar, D., Eng, J.J., Miller, W.C., Krassioukov, A.V., Verrier, M.C.: Movement repetitions in physical and occupational therapy during spinal cord injury rehabilitation. Spinal Cord **55** (2), 172–179 (2017)
12. Pilatásig, M., Tigse, J., Chuquitarco, A., Pilatásig, P., Pruna, E., Acurio, A., et al.: Interactive system for hands and wrist rehabilitation. In: International Conference on Information Theoretic Security, pp. 593–601. Springer, Cham (2018)

13. Varela-Aldás, J., Palacios-Navarro, G., García-Magariño, I., Fuentes, E.M.: Effects of immersive virtual reality on the heart rate of athletes warm-up. In: International Conference on Augmented Reality, Virtual Reality and Computer Graphics, pp. 175–185. Springer, Cham (2019)
14. Selfslagh, A., Shokur, S., Campos, D., Donati, A., Almeida, S., Yamauti, S.Y., et al.: Non-invasive, brain-controlled functional electrical stimulation for locomotion rehabilitation in individuals with paraplegia. Sci. Rep. **9**(1), 9–16 (2019)
15. Sánchez, Z.A., Alvarez, T.S., Segura, F.R., Núñez, C.T., Urrutia-Urrutia, P., Salazar, L.F., et al.: Virtual rehabilitation system using electromyographic sensors for strengthening upper extremities. In: Smart Innovation, Systems and Technologies, pp. 231–241. Springer, Singapore (2019)
16. Martínez, Á.H.: La rehabilitación terapéutica a pacientes parapléjicos: impacto desde las tecnologías. PODIUM: Revista de Ciencia y Tecnología en la Cultura Física **12**(1), 21–30 (2017)
17. Carbone, G., Cavero, C.A., Ceccarelli, M., Altuzarra, O.: A study of feasibility for a limb exercising device. Mech. Mach. Sci. **47**, 11–21 (2017)
18. Glaser, B.G., Strauss, A.L.: Discovery of Grounded Theory: Strategies for Qualitative Research. Routledge (2017)

Author Index

A
Aldas, Darwin, 260
Alvarez, Kevin, 260
Angulo, Karen, 3

B
Barberán, Jeneffer, 341
Barona-Oñate, Ruth Viviana, 83
Benalcazar, Freddy, 184
Benalcázar, Marco, 226
Benalcázar, Marco E., 155
Borge-Diez, David, 287
Buele, Jorge, 129, 341
Buenaño, Hernando, 99

C
Caiza, Gustavo, 143
Castelo, Homer, 260
Castillo Sequera, José Luis, 168
Castillo, Franklin, 71
Collantes, Santiago, 329
Comas Rodríguez, Raúl, 36
Cruz, Patricio J., 226

D
De Giusti, Armando, 51
de la Iglesia Campos, Manuel, 22
Duque Méndez, Néstor Darío, 111

E
Enríquez, Víctor, 129
Espitia, Helbert, 3

G
García Lorenzo, María Matilde, 36
Garcia, Carlos A., 199
Garcia, Marcelo V., 199
García-Santillán, Iván D., 51
Gil, Danilo, 3
Gómez Soto, Jacobo Andrés, 111
Gordón, Carlos, 184
Granizo, César, 184
Guevara, Cesar, 71
Guilcapi, Jaime, 184

H
Herrera, Víctor Isaac, 308
Herrera-Granda, Erick P., 51
Hojas-Mazo, Wenny, 22

I
Icaza, Daniel, 287
Ilvis, David I., 199

J
Jordán, Edisson P., 341

L
Leon, Carlos S., 199
Llivisaca, Juan, 247
López López, Jimmy P., 83
López-Pérez, Sonia de los Angeles, 83
Loya, Hugo, 129

M
Mantilla-Brito, Juan, 214
Manzano, Santiago, 184

J. Nummenmaa et al. (Eds.): CSEI 2019, AISC 1078, pp. 355–356, 2020.
https://doi.org/10.1007/978-3-030-33614-1

Mariño, Christian, 274
Merino, Christian, 308
Miranda-Quintana, Oswaldo, 71
Mocha-Bonilla, Julio A., 83
Molina, Cristian, 143
Morales, Luis, 214, 329
Moreno, Víctor, 329
Morocho-Caiza, Andrés, 308

N
Núñez, Freddy, 341

O
Oña, Alex F., 226

P
Palacios-Navarro, Guillermo, 71
Paredes Larroca, Fabricio, 168
Paucar, Wilman, 143
Peñafiel, Víctor, 99
Pozo-Espín, David, 214
Pusdá-Chulde, Marco R., 51

Q
Quezada, Jonnathan, 247
Quino Favero, Javier, 168
Quintero Lorza, Diana Patricia, 111

R
Remache, Edison G., 199
Reyes, John, 260
Rodríguez, Nancy, 260
Rodríguez-Flores, Jesús, 308
Rosales Huamaní, Jimmy, 168
Ruíz-Carrera, Juan Carlos, 22

S
Saeteros, Morelva, 143
Saettone Olschewski, Erich, 168
Salazar, Franklin W., 129, 341
Salazar-Fierro, Fausto A., 51
Sánchez Fleitas, Nayi, 36
Sánchez, Carlos, 129
Sandoval-Pillajo, Lucía, 51
Siguenza-Guzman, Lorena, 247
Silva, Diana, 329
Simón-Cuevas, Alfredo, 22, 36
Solórzano, Santiago, 214
Suarez, Cristian, 260

U
Urrutia, Fernando, 129

V
Varela-Aldás, José, 71
Vargas, Javier, 274
Velasteguí, Homero, 184
Velasteguí, Rommel, 260
Villalba, Williams R., 199
Vimos, Victor H., 226

Z
Zea, Jonathan A., 155

The manufacturer's authorised representative in the EU is Springer Nature Customer Service Centre GmbH, Europaplatz 3, 69115 Heidelberg, Germany. If you have any concerns regarding our products, please contact ProductSafety@springernature.com

Printed and bound by CPI Group (UK) Ltd, Croydon, CR0 4YY
30/09/2026
02225238-0009